The Red-cockaded Woodpecker

NUMBER FORTY-NINE
The Corrie Herring Hooks Series

THE
Red-cockaded Woodpecker

SURVIVING IN A FIRE-MAINTAINED ECOSYSTEM

RICHARD N. CONNER,
D. CRAIG RUDOLPH,
AND JEFFREY R. WALTERS

FOREWORD BY FRANCES C. JAMES

UNITED OF TEXAS PRESS
AUSTIN

Requests for permission to reproduce material from this work should be sent to Permissions, University of Texas Press, P.O. Box 7819, Austin, TX 78713-7819.

♾ The paper used in this book meets the minimum requirements of ANSI/NISO Z39.48-1992 (R1997) (Permanence of Paper).

Library of Congress Cataloging-in-Publication Data

Conner, Richard N.
  The red-cockaded woodpecker : surviving in a fire-maintained ecosystem / Richard N. Conner, D. Craig Rudolph, and Jeffrey R. Walters ; foreword by Frances C. James.
      p.   cm. — (The Corrie Herring Hooks series ; no. 49)
Includes bibliographical references (p.   ).
  ISBN 0-292-71234-0 (hardcover : alk. paper)
    1. Red-cockaded woodpecker.   2. Endangered species—Southern States.
3. Southern pines—Ecology.   4. Birds, Protection of—Southern States.
I. Rudolph, D. Craig.   II. Walters, Jeffrey Ray, 1952–   III. Title.   IV. Series.
  QL696.P56   C66   2001
  598.7′2—dc21
  00-012823

We dedicate this book to

J. DAVID LIGON

*one of the first and
one of the best to study
Red-cockaded Woodpeckers,
a pioneer in studies of
cooperative breeding.
His motivation has always
been pure: conservation of
the bird rather than personal
ambition, and the joy of
biological discovery rather
than political agendas.*

Contents

Illustrations

Tables

Foreword

Here is a true story about the adaptations of a small, inconspicuous wood-pecker to its environment—about how scientists have progressed in their understanding of its biology and how this understanding is gradually being incorporated into policies for its management. Research on this species in the last 15 years has made it possible for the three authors of this work to de-velop an insightful new management strategy that, if implemented, could lead to recovery of the Red-cockaded Woodpecker. Even so, the final chap-ter of this story cannot yet be written. Its outcome will depend largely on whether policy makers and managers use the information we now have. I think it will also depend on the outcome of some of the other research that is now in progress.

The authors of this book are the major researchers among a large cadre of people working on this endangered species. Dick Conner and Craig Rudolph work for the Southern Research Station of the U.S. Forest Ser-vice in eastern Texas. With their collaborators, they have produced many papers on the dependence of the Red-cockaded Woodpecker on specific as-pects of old, living pine trees. Unless a tree has a high resin flow and suf-ficient heartwood, often softened by fungus, this species will not be able to excavate a roosting or nesting cavity. The birds regularly wound old rel-ict pine trees, inducing a resin flow that serves as a barrier to predators. Highly resinous old pine trees were common enough in loblolly, shortleaf, and longleaf pine forests in presettlement times, but they are rare today. In addition, there is increasing evidence that Red-cockaded Woodpeckers prefer larger and older trees for foraging; the males often excavate insects from dead branches, and the females hitch themselves up the trunk, flaking the loose bark for ants. These special adaptations to old-growth conditions mean that most current populations of Red-cockaded Woodpeckers are un-likely ever to have an optimal habitat. Isolated populations remain from

Virginia to Texas on public and private land, but before 1990 only two of these at most were not in decline.

A major breakthrough in understanding of the biology of the Red-cockaded Woodpecker came in the early 1990s with the publication of several papers by Jeff Walters and his collaborators, then at North Carolina State University. Jeff's research on the cooperative breeding system of the bird led him to test experimentally his idea that cavity limitation was the factor keeping birds from establishing new social groups in otherwise suitable habitat. He found that artificially drilled cavities could reduce this constraint. This result came just in time, because in September 1989 Hurricane Hugo destroyed 80% of the cavity trees in the large population of woodpeckers in the Francis Marion National Forest in South Carolina. Managers there and elsewhere added hundreds of artificial cavities to Red-cockaded Woodpecker habitat in the 1990s. Jeff's theoretical analysis was tested by experiment, just the way scientific progress is supposed to work. Now at Virginia Tech University, Jeff is continuing his collaborative research in the Carolina Sandhills and on Camp Lejeune in North Carolina and at Eglin Air Force Base in northwestern Florida.

Thus three very active biologists working on Red-cockaded Woodpeckers are the authors of this book. They have put together a remarkably complete case history, beginning with prehistoric evidence that the longleaf-pine ecosystem, with its diverse ground cover of grasses and forbs, attained its full distribution across the uplands of the southern United States 5,000 years ago. That forest was cut over in the early part of the twentieth century, and much of the land was cleared for agriculture and other purposes. Then, beginning about 1940, vast areas of both public and private land were converted to short-rotation forestry. Most areas that have Red-cockaded Woodpeckers today are also being managed with the goal of producing saw timber. They have a few relict trees that were spared in the initial harvest, and some of those trees are the current cavity trees.

The management strategy recommended here as a package includes a combination of tactics. Emphasis is on the provision of high-quality cavities, the protection of relict trees, the control of hardwood midstory (especially with growing-season burning), and the translocation of birds either to fill breeding vacancies or to establish new social groups adjacent to current ones. Where this strategy has been implemented, woodpecker populations have increased dramatically. Five separate cases are described in which formerly declining populations are now increasing. Nevertheless, the last chapter of the book is titled "An Uncertain Future." Although recent guidelines for the management of the Red-cockaded Woodpecker in national forests and on military bases have incorporated most elements of the new strategy, their implementation is often left to the discretion of local land managers. On private lands, the only program that is designed to encourage increases in woodpecker populations is the "Safe Harbor" program promoted by the Environmental Defense Fund. It seems that the scientists have done their part, but now the managers must develop policies that accommodate conflicting interests.

In addition, I think that new research concerning fire ecology and the long-term restoration of the full diversity of southern pine ecosystems must be added to the management strategy proposed here. Our work in the longleaf pine forests of the Apalachicola National Forest in northern Florida and the research by Jeff Hardesty at Eglin Air Force Base shows that Red-cockaded Woodpecker social-group size and productivity are negatively related to the density of pine trees in current forests. To achieve the long-term goal of an all-aged, self-regulating forest will require not just eliminating the hardwood midstory but also establishing an herbaceous ground cover that will carry fire and will support natural pine regeneration. Current forests, even those that have never been plowed, are overstocked, choked with the even-aged closed-canopy stands of trees that regenerated after the initial cut more than 70 years ago. New research is needed to test alternative methods of silviculture and alternative burning regimes, but the objective is clear: attainment of that all-important open structure of the original ecosystem.

To me, the story of the endangered Red-cockaded Woodpecker is more interesting than any mystery novel. The reader will be astonished at the amount of material covered and the depth of ecological understanding evident in the following pages. I have the greatest admiration for this work, presented as it is in such a readable style and with such a convincing argument. I think the book sets a new standard for works of this type.

Frances C. James

Acknowledgments

Discussions through the years with Jerry Jackson, Bob Hooper, Dave Ligon, Jay Carter, Phil Doerr, Fran James, Dan Lay, Ron Escaño, Dennis Krusac, Joe Dabney, Todd Engstrom, Roy DeLotelle, Rick Schaefer, Dan Saenz, Bob McFarlane, Ron Billings, Dave Kulhavy, Bill Ross, Bob Coulson, Forrest Oliveria, and Pete Lorio have helped formulate many of the ideas presented in this book. The lab group at Virginia Tech (Jennifer Allen, Caren Cooper, Sue Daniels, Sergio Harding, Memuna Khan, and Jim Lyons) provided helpful comments on Chapters 6, 9, 11, and 12. We thank Ronald Billings and Forrest Oliveria for their scientific review of Chapter 8. We thank Jennifer Matos for her review of Chapter 2; Clifford Shackelford for his review of the manuscript and preparation of the initial sonograms; Toni Trees for reviewing, proofing, and editing all parts of the manuscript; Ron Thill and Daniel Saenz for reviewing the manuscript; and Chris Collins and Shirley Burgdorf for preparation of some of the graphs and diagrams. Jeffrey Conner provided expertise and help with computer image scanning and computer graphics. Howard Williamson and Dan Saenz helped with photo lab work. We thank Ron Thill, Stan Barras, and Tom Ellis for their approval and support of this endeavor when we were invited to write the book 10 years ago. We thank Shannon Davies for her editing help on early versions of the book as our initial sponsoring editor, Bill Bishel and Carolyn Wylie at the University of Texas Press for patiently guiding us through the steps preparing the book for publication, and Letitia Blalock for her outstanding work as copyeditor of the book. Finally, we especially thank Todd Engstrom and Fran James for their critical review of an early draft of the complete book and many helpful suggestions.

The Red-cockaded Woodpecker

CHAPTER ONE

An Introduction

When you see a Red-cockaded Woodpecker (*Picoides borealis*) for the first time, it is not obvious why this bird has received so much attention. It is a small, rather nondescript species. It does not have brilliant plumage or engage in spectacular displays. It utters no bizarre calls. Only when you begin to understand the complexities of its ecology and social behavior do you come to appreciate what a gift of nature the Red-cockaded Woodpecker represents.

Its social system is as complex as that of any North American animal, more like that of a primate than of a bird. Its ecological uniqueness has led to its endangerment, and its endangerment has earned it the title "Spotted Owl of the Southeast," because, like the Spotted Owl (*Strix occidentalis*), its conservation affects land use across millions of hectares in the region.

A Brief History

To understand the history of the Red-cockaded Woodpecker is to understand the history of the Southeast itself. In this book we attempt to tell the story of the Red-cockaded Woodpecker, and thereby the story of the great pine forests of the Southeast, of the evolution of complex social behavior, and of conservation in the United States. This first chapter presents a brief overview of the story told in detail in the rest of the book.

Ancestors of the Red-cockaded Woodpecker most likely lived in what is now central and southern Florida. Over geological time they had arrived in this southern forest refuge as a result of a series of Pleistocene glaciers that periodically dominated the North American continent. Exactly when the Red-cockaded Woodpecker's ancestors began to use the open pine woodlands characteristic of Florida at this time, rather than the hardwood forests more typically inhabited by woodpeckers, is not precisely known. The open pine landscape had been produced and maintained by frequent growing-season fires that burned for days and covered vast areas. As the woodpeckers adapted to forests relatively devoid of hardwood trees and dead trees (snags), which were the domain of other species of woodpeckers, behavioral changes occurred that would make this one of the world's most unique bird species. It had already acquired a wedge-shaped bill with a chisel-like tip, a tool that it used to excavate cavities in trees for nesting and roosting, and to flake off the bark of trees to extract insect prey. The species also had the typical woodpecker tail, equipped with special stiff feathers that the bird used as a prop against the tree bole when it excavated a cavity, pecked to expose prey, or climbed up a trunk or limb.

While adapting to the open pine habitat, ancestors of the present-day Red-cockaded Woodpecker may have developed a behavioral aversion to forests with hardwoods, a trait that appears to be rigidly fixed within the species today. Because fire in southern pine ecosystems had removed most of the dead trees that were typically used for cavities in hardwood forests, natural selection apparently favored a woodpecker species that used live pines for nest and roost trees. The firmness of the undecayed sapwood of living pines, as well as the pine gum that flowed copiously when the birds pecked on the trees, posed problems for individuals attempting to excavate a cavity in a live pine. Pine gum stuck to their bills and feathers, and many woodpeckers likely were trapped and perished over the centuries while an adaptation to handle resin developed. But the pine gum eventually became an ally. Some woodpeckers began to peck small wounds around their cavity entrances, probably to prevent the tree from growing over the cavity opening, and streams of pine gum flowed down the boles of pines from these excavations. The pine gum that flowed from these small wounds, however, proved to be an effective barrier against predatory raids by the bird's greatest enemy, rat snakes (*Elaphe* spp.). Thus natural selection favored woodpeckers that pecked these "resin wells," and their construction became a characteristic and unique adaptation of the species.

A major problem faced by this evolving woodpecker was the long time required to excavate a cavity in the old-growth pines of this early forest. But as pines aged, the woodpeckers gained yet another ally. Red heart fungus (*Phellinus pini*) helped the woodpecker by decaying the central heartwood of older pines, making it easier for the woodpeckers to chisel out the softened wood tissue while making their cavity chamber. The birds still had to excavate an entrance tunnel through the outer sapwood, which was undecayed and sticky. But once into the heartwood, the presence of decay reduced the time needed to make a cavity, likely by several years in most cases. Spores of the fungus gained access to the heartwood of the pine through dead branch stubs. After a decade or so of growth within the heartwood of the pine, the fungus produced a cinnamon-colored conk (spore-producing fruiting body) on the bole of the pine where the dead branch stub had been. During warmer months, millions of spores produced by conks were dispersed by the winds and bathed the boles and branches of pines throughout the forest, some finding conditions suitable for germination in the heartwood of a large broken branch. Older pines with well-developed heartwood decay were well marked by the presence of fungal conks on their boles, possibly identifying themselves as suitable to woodpeckers searching for potential cavity trees.

Eventually, the species became dependent on a sufficient supply of old pines for cavity trees. Longleaf pines (*Pinus palustris*) were the best, because they were very resistant to diseases and produced large quantities of the pine gum that helped deter predatory rat snakes, which were agile climbers on the boles of pines. The pines had to be old; old pines had greater diameters of heartwood in which cavities could be excavated and a higher probability of containing the helpful red heart fungus. As natural selection forged the new woodpecker species, it became, through adaptation, closely linked to the frequently burned pine landscape.

In the early upland forests of the South, old pines were common but cavity trees were in limited supply. Even with decayed heartwood, cavity construction in living pines was a major life commitment, typically taking a fourth to a half of the bird's natural life span. A young woodpecker was unable to disperse from its natal cluster of cavity trees and excavate a nest cavity in an old, decayed snag in a week or two, like woodpeckers of other species. This constraint led to major changes in the woodpecker's social system. Competition was focused on existing cavities, which could be used for two decades or more, rather than on open space in which new territories, with new cavities, could be carved out. Under these conditions, natural selection favored young birds, especially males, that stayed at home with their parents and waited for a breeding vacancy to arise in an existing territory in the vicinity. While waiting, it was to their genetic advantage to serve as "helpers," helping to raise their brothers and sisters. A male that bided his time might take over his father's cluster of cavity trees when his father passed into the food chain, or a neighbor's cluster. As a result of this behavioral adaptation, members of the new woodpecker species began to live together in family groups. They would gather as a group each morning after they emerged from their roost trees and move together to forage for arthropods on pines in the open forest.

The Southern Pine Ecosystems

Following the Pleistocene glaciations (8,000–12,000 years ago), the climate of North America moderated and warming trends prevailed. The lands to the north and west of Florida became drier as the southern forests warmed. Fires ignited by Native Americans and lightning repeatedly consumed the leaves, resinous pine needles, and grasses of the ground cover. As hardwood trees became less abundant, fire-maintained pine ecosystems spread to replace them. Between 4,000 and 12,000 years ago, pines from refugia in Florida, and possibly southern Texas, spread across the entire southeastern United States, dominating most of the drier habitats. The frequent fires forged a path for the woodpecker by consuming most hardwoods that got in the way or those that tried to reclaim the land. With fire habitually sweeping through the uplands, hardwoods thrived only on the sites that were too wet to burn with any regularity. Intense fires that occurred predominantly during the spring and summer exposed the ground to the sun and allowed autumn-dropped pine seeds direct access to the warm soil. Seeds and seedlings were not immediately destroyed because winter fires were rare. (The relatively cool winter fires of the modern managed pine forests are not natural but are set by humans during the wet, cool months—after seed germination—to reduce fuel loads and the potential for wildfire.)

Longleaf pines dominated the uplands in the new forest landscape that emerged. Loblolly pines (*Pinus taeda*) typically grew in the dynamic transition zones, which burned less frequently and were located between the longleaf pines and the hardwoods of the bottomlands and riparian zones. Loblolly seedlings were consumed by the frequent fires of the uplands, surviving only where time between successive fires was sufficiently long to enable the plants to grow to the somewhat fire-resistant sapling stage.

As southern pine ecosystems matured and spread, so did the Red-cockaded Woodpecker. This social woodpecker reached the northern limits of its expansion in southern New Jersey and Delaware and its western limits in Oklahoma and Texas. The woodpecker flourished and became one of the more abundant avian species in the southern pine forests.

Change across the Landscape

Ten thousand years after the Red-cockaded Woodpecker spread from Florida, another species began an additional wave of its dispersal. Three hundred years ago the European version of *Homo sapiens* was becoming well established in the New World's sparsely populated lands. Native Americans quickly fell against the expanding tide of this new society that forged metal and conquered the lands with plowshares, gunpowder, and steam-driven locomotives. As the European surge pushed across the new continent, the face of the landscape changed. The southern pine forests were not spared from these forces.

Timber was needed to fuel the modern machines and to build dwellings and ships. Wood fiber was needed to make paper to spread the news of human events and conquests. The torrent of human expansion initiated a wave of timber harvesting that reached eastern Texas by the mid-1800s. Harvesting of the South's forests reached its peak around 1900 and lasted two more decades. The longleaf pines of the South's pristine forests were ideal to harvest. The natural fires of the past had cleared the lands of the tangles of restraining vegetation, and unlike many loblolly pines, longleaf pines were straight and tall with ideally spaced growth rings, characteristics needed for high-quality timber (Figure 1.1). Only the grass that covered the open forest floor lay between the pines and the teeth of the long crosscut saws. People needed wood, and at the turn of the century the forests seemed endless. By 1930, the lands had yielded their bounty to hand-powered saws, and oxen, mules, and steam-powered tram locomotives had conveyed the fallen timber to mills that had sprouted in the region. The bonanza era of harvesting was over (Figure 1.2, Figure 1.3).

Woodpecker Populations Plummet

As the timber vanished, populations of Red-cockaded Woodpeckers plummeted. Most likely, single birds and family groups traveled widely in attempts to find habitat to replace lost trees. Time was against them because they required such a long time to make new cavities to replace felled cavity trees. The woodpeckers were spared a few trees in some locations, because resin-coated cavity trees and other stunted pines with fungal conks on them were not worth the effort required to fell them with hand-powered saws. Past experience had taught the sawyers that wood within these "peckerwood" cavity trees was also likely riddled with decay, which greatly reduced their value. The woodpeckers likely were able to survive, albeit in reduced numbers, on landscapes with greatly diminished pine density as long as their cavity trees were not all lost. In some areas only the larger pines were taken, and those passed over had a head start toward re-creating a forest. Such lands were extremely valuable to Red-cockaded Woodpeckers and probably served as centers for ancestors of some of the larger populations existing today.

FIGURE 1.1 At the beginning of the past century, rail spur lines carried log trains hauling tall, straight longleaf pines out of the open, fire-maintained virgin forest. Courtesy of the East Texas Research Center, Steen Library, Forest History Collections, Thompson Family Lumber Enterprises Collection, P90T:202, Stephen F. Austin State Univ., Nacogdoches, Texas.

The mid-1900s saw a continued decline of woodpecker populations as they conformed to the new carrying capacity of each region. Some populations probably began to grow as the pines that had not been cut became sufficiently old for woodpeckers to excavate cavities in them. These pines had been released from competition by the cutting of the more dominant pines, and they swelled in diameter and expanded their crowns.

By the 1950s, a more subtle threat than the harvesting of old-growth pines crept through the South. The expanding network of roads, lands cleared for agriculture, and human-dammed lakes served as firebreaks. This, and the intentional suppression of forest fires, eliminated fire as a dominant ecological feature and permitted the insidious encroachment of hardwood vegetation into the once predominantly grass-covered understory of fire-maintained pine ecosystems. The prescribed fires that were intentionally set by humans were now typically started during winter, a time when the ground was wet and the air was cold. Cool winter fires consumed the pine needles and grass fuels, but the invading hardwoods were minimally impacted. The face of the forest landscape was again changing, and as before to the detriment of the Red-cockaded Woodpecker. The aversion to hardwood vegetation that had helped it survive in the past now excluded the woodpecker from many of the second-growth forests.

The ever-expanding agricultural openings and lakes created problems beyond serving as simple barriers to the spread of fire. Woodpecker populations were wedged apart. Distances between populations increased as forest loss and fragmentation spread like a cancer throughout the land-

FIGURE 1.2 Around 1908 in Trinity County, Texas, oxen were used to haul logs to rail spur lines. Courtesy of the East Texas Research Center, Steen Library, Forest History Collections, Sawdust Empire Collection P90s:29, Stephen F. Austin State Univ., Nacogdoches, Texas.

FIGURE I.3 Steam-powered locomotives hauled logs cut from the pristine forests to the sawmills in eastern Texas around 1908. Courtesy of the East Texas Research Center, Steen Library, Forest History Collections, Stephen F. Austin State Univ., Nacogdoches, Texas.

scape. Openings within forest tracts further aggravated the isolation of woodpecker groups from each other within populations. Helpers had fewer neighboring territories to monitor for breeding vacancies, and dispersers had more difficulty locating territories. The process of breeder replacement became less efficient, so that several unproductive years often passed in perfectly good territories with good cavity trees because of the absence of a breeder of one sex or the other. When distances between groups became too great, demographic collapse was catalyzed by the rarity of old pines for cavity excavation and the invasion of hardwoods. Isolated groups and populations disappeared.

During the 1960s yet another threat emerged. Efforts to "sanitize" the southern forests commenced. Pines with signs of red heart fungus were actively sought for removal to eliminate disease. The fungus decayed the pine's heartwood, which was the best part of the tree for quality lumber; timber volumes and profits were at stake. Cavity trees and other old pines that had initially evaded the saw were fairly old now. The fungi that softened the heartwood and helped the woodpecker during cavity excavation had come to fruition and grown their spore-disseminating conks on the boles of the old trees. Now these conks became signposts for removal. Sawyers assaulted the forests throughout the South to eliminate the spore-producing sources of the red heart fungus. Cavity trees were felled by the thousands.

During the 1960s progress changed the way pines were harvested. Efficient gasoline-powered chain saws became common, while the teeth of two-person crosscut saws collected rust and cobwebs rather than wood chips. Harvesting of single or small groups of pines in the second-growth forests gave way to the economically efficient clear-cutting of larger blocks of trees on public and private lands. Removal and fragmentation of the maturing second-growth forest increased, and fewer and fewer remnant tracts of old-growth forest remained. Distances between populations of the woodpecker continued to increase, and the demographic viability of more populations declined as they were further fragmented into increasingly smaller subsets, separated by increasing distances. As woodpecker populations became smaller, the specter of inadequate genetic variation appeared in some of the smaller populations that were able to cling to existence. In such small woodpecker populations, chance mortality can eliminate beneficial genes (alleles) and reduce genetic viability.

The Tide Begins to Turn

During the late 1960s and early 1970s, political pressure mounted to halt the loss of our wildlife heritage through species extinction. The primary concern was the increasing rate of human-caused extinctions as industrialized societies overexploited their own lands and resources as well as those of distant third-world countries. The sacrifice of entire species for economic expediency came under attack. The Red-cockaded Woodpecker was listed by the United States federal government as an endangered species in 1968. With the passage of the Endangered Species Act in 1973, the Red-cockaded Woodpecker received legal protection. The act directed that recovery plans be drafted for all threatened and endangered species, plans that would bring the species back from the edge of extinction. The first Red-cockaded Woodpecker symposium was held in 1971 and revealed great voids in our knowledge of the ecology of the species. Not until 1979 was the first recovery plan drafted and approved by the federal government. By that time, research on the woodpecker had provided some new information, but research takes time. Many aspects of the ecology and behavior of the bird were still not completely understood.

Subsequently, guidelines were drafted by different agencies to manage the woodpecker on various types of public and private lands. Unfortunately, a lack of full implementation of guidelines and inadequacies in our knowledge of the woodpecker resulted in ineffective management, and population declines continued. By 1985, a second recovery plan was approved by the U.S. Fish and Wildlife Service. But the same basic problems that haunted the first recovery plan were again present: implementation was incomplete and our knowledge of the biology of the species was inadequate. Although a few of the largest woodpecker populations showed some signs of stability or increase, the vast majority continued to dwindle in numbers. The smaller populations suffered the most. But even large populations experienced decreases in abundance on the fringes of their range and internally where family groups had become isolated from one another. This trend continued throughout the 1980s. During September 1989 Hurricane Hugo cut a path through the Frances Marion National Forest on the Atlantic Coastal Plain of South Carolina and devastated the second-largest population of Red-cockaded Woodpeckers. About 87% of the woodpeckers' cavity trees were destroyed in a single day, and estimates indicated that more than one-half of that woodpecker population was lost during the storm.

In the early 1990s the bad news continued. Fran James, a professor with Florida State University, discovered that even portions of the largest Red-cockaded Woodpecker population, on the Apalachicola National Forest in northern Florida, were declining. New management plans were obviously needed, and this time correct implementation would be essential if the woodpecker was to survive.

In the late 1980s Jeff Walters and coworkers at North Carolina State University had managed to solve the most significant problem associated with understanding Red-cockaded Woodpecker demography. They described in detail the dispersal of young and adult woodpeckers and determined the crucial role of completed cavities in the formation of new woodpecker groups and persistence of old ones. Additional research by Dick Conner and Craig Rudolph at the U.S. Forest Service's Southern Forest Experiment Station (now the Southern Research Station) in Nacogdoches, Texas, indicated how isolation and habitat fragmentation affected successful woodpecker dispersal for mate replacement; their results were consistent with the emerging new view of population dynamics. Carole Copeyon, with Walters and others, also developed and tested a major tool for woodpecker management: the artificial creation of Red-cockaded Woodpecker cavities in live pines through a drilling technique. David Allen, at the U.S. Forest Service's Southeastern Forest Experiment Station in South Carolina, developed and refined an additional technique for creating artificial cavities, which involved inserting a nest box within a live pine. Both the insert and drilling techniques received intensive testing as a result of Hurricane Hugo. Necessity in this case truly became the mother of invention as hundreds of artificial cavities were installed in a little over a year, providing roosting and nesting sites for hundreds of Red-cockaded Woodpeckers that might have died without them.

An Unknown Future

Today the stage is set for the next act of the Red-cockaded Woodpecker story. The new research information and management techniques have given us the ability to bring this unique species back from near extinction. Populations that were stable or declining are now increasing wherever the new management strategy is being employed. But it is not being employed everywhere. How we choose to act remains to be written in the script. Back in 1982, Jerry Jackson, who is now the Whitaker Eminent Scholar at Florida Gulf Coast University and who has studied the woodpecker for decades, appeared in a film advocating support for the recovery of endangered species. The film, produced by Marty Stouffer, was entitled *At the Crossroads: The Story of America's Endangered Species*. We remain at the crossroads. We still have a chance to recover the Red-cockaded Woodpecker through active, positive management for the bird, although at some economic sacrifice. New management guidelines have recently been developed by several agencies, notably the U.S. Forest Service and Department of Defense. These guidelines, unlike previous ones, have the potential to recover the species. Yet in the 26 years since the Red-cockaded Woodpecker received legal protection under the Endangered Species Act, we have yet to successfully implement management guidelines to save this species. Populations have dwindled to their lowest levels (Figure 1.4). Will it be any different this time?

The Red-cockaded Woodpecker is a symbol. It represents the old-growth, fire-maintained pine forests of the southeastern United States, the devastation of which has been much greater than that of habitats generally perceived as most subject to destruction, such as wetlands. The Red-cockaded Woodpecker has suffered more than the other species inhabiting these forests because it is so highly adapted to this particular ecosystem and is especially sensitive to the changes wrought on the remaining habitat—loss of old growth and fire suppression. It is also symbolic of the difficult issues of conservation in United States society because its territories are large and therefore its requirements for space are great. No other endangered species in the Southeast affects so much land, both public and

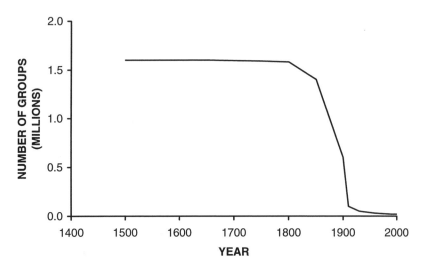

FIGURE 1.4 Estimated Red-cockaded Woodpecker population over the past 400 years, showing the dramatic decline during the past century.

private. We know how to save the Red-cockaded Woodpecker and the economic costs. The actions we take as a society over the coming decades with respect to this species will reveal to the world our true colors where conservation is concerned and our true values.

It is the collective intent of the authors of this book that the information here will contribute to a fuller understanding of the biological and ecological needs of this endangered woodpecker. Through education we can meet the challenges of the future. It is our hope that the more detailed information that follows in subsequent chapters will help us as a society make an informed and correct choice of path at the crossroads we now face.

CHAPTER TWO

Fire-Maintained Pine Ecosystems

The Red-cockaded Woodpecker is exquisitely adapted to the fire-maintained pine ecosystems of the southeastern United States. Many aspects of Red-cockaded Woodpecker biology are a consequence of evolution in these forests. In order to appreciate the biology of the Red-cockaded Woodpecker and understand the reasons for the decline of the species, one must examine the biological and physical interactions that have shaped these forests.

History of the Southeastern Forests

PLEISTOCENE FORESTS

The ecosystems of the southeastern United States, like all ecosystems, are dynamic. Massive changes occur in the plant communities on time scales of thousands to hundreds of thousands of years. Such changes have been especially dramatic during the Pleistocene glaciations and the relatively brief interval since the most recent Wisconsinan glaciation. These changes have dramatically affected the extent of pine-dominated habitat suitable for Red-cockaded Woodpeckers.

Recent research indicates that regular cyclic changes in Earth's orbit about the Sun result in changes in insolation that regulate the buildup of ice sheets in high latitudes. Changes in ice volume directly affect sea level and result in major alterations of the arrangement of sea and land. These factors interact in complex and poorly understood ways to alter climate and ultimately to determine the composition of plant communities.

Our understanding of the history of Pleistocene plant communities throughout the southeastern United States comes primarily from the study of plant fossils (especially pollen) preserved in the sediments of natural ponds and swamps. Data from many sites throughout eastern North America have allowed a reasonably detailed analysis of vegetation changes during the last 18,000 years. Prior to 18,000 years ago, changes of a similar magnitude undoubtedly occurred during the waxing and waning of several Pleistocene glaciations.

The details of these floristic changes are determined by compiling the data from pollen records of many sites. One such site is Goshen Springs, located near the Conecuh River in southeastern Alabama. This site was investigated by Paul Delcourt of the University of Minnesota and provides a remarkable history of changes in the local vegetation. The site is a natural pond created by the collapse of cavernous limestone beneath the surface. The resulting depression, fed by surrounding seepage springs, has been collecting and preserving pollen from the local plant community for the last 33,000 years.

In the Goshen Springs area, longleaf pine/wiregrass (*Aristida* spp.) communities dominated the uplands when Europeans arrived. On the slopes, forests of southern magnolia (*Magnolia grandiflora*) and American beech (*Fagus grandifolia*) transitioned into mixed-hardwood forests in the poorly drained bottomlands. Fires, judging by analogy with better-known areas, were undoubtedly frequent in the longleaf uplands. The resulting longleaf pine/wiregrass community would have provided excellent habitat for Red-cockaded Woodpeckers.

Analysis of pollen preserved in the sediments of the pond at Goshen Springs, obtained from drilling cores, provides a record of past changes in the surrounding plant communities. Comparison of this record with radiocarbon dates obtained at various levels in the sediment cores provides a dated sequence of the extensive changes in the vegetation. Because of increasing aridity, oaks (*Quercus* spp.) and hickories (*Carya* spp.) replaced the mesic-temperate forests present prior to 28,000 years ago. This vegetation persisted with minor changes throughout the glacial maximum. Approximately 5,000 years ago, pine increased dramatically and the local flora achieved its modern composition.

In the time interval from 18,000 years ago to the present, the major changes in plant community distributions in North America have been a direct result of the retreat and disappearance of the Laurentide ice sheet of the Wisconsinan glaciation. The Laurentide ice sheet, near its maximum development 18,000 years ago, is difficult to envision from our current interglacial perspective. This immense ice sheet reached an estimated maximum height of over 3 km in central Canada and extended south to 41°N latitude in the vicinity of what is today southern Missouri. The impact of this ice sheet and the accompanying climatic changes shifted the range of plant communities far to the south. During the glacial maximum, plant communities resembling the present boreal forests dominated by spruce (*Picea* spp.) and jack pine (*Pinus banksiana*) occurred as far south as Tennessee and northern Georgia. At the present time these species are found hundreds of kilometers to the north, in the vicinity of the Great Lakes, and, in the case of spruce, also at higher elevations in the Appalachians.

Deciduous forests of oaks, hickories, and many other species extended to the Gulf of Mexico. At the height of glaciation 18,000 years ago, the forests of southern pines—and by implication the habitat of Red-cockaded Woodpeckers—most likely were restricted to peninsular Florida and the coastal regions from northern Florida to the Carolinas, and extended west perhaps no further than portions of southern Alabama.

The changes in distributions of major plant communities from the full glacial condition 18,000 years ago to the present have been primarily northward shifts. Boreal forests have shifted to the north of the Great Lakes, and the deciduous forests that occurred south to the Gulf of Mexico now extend south only as far as the confluence of the Mississippi and Missouri Rivers. At the time of the arrival of Europeans, a vast area dominated by fire-maintained pine forests extended from southern Virginia south through Florida and in a broad band west to eastern Texas. This area was not, however, one vast sea of pines when Europeans arrived. Pines were dominant in the drier, better-drained and extensive uplands, whereas deciduous forests dominated the bottomlands of the major rivers and riparian zones adjacent to the many thousands of lesser streams. The valley of the Mississippi River, in particular, was a deciduous forest belt separating pine habitat to the east and west by 50–100 km. In addition, significant areas of swamp forest and prairies occurred throughout the region.

The pollen record at Goshen Springs thus reflects the passage of these plant communities as they shifted northward. However, regionally the changes from 18,000 years ago to the time of arrival of Europeans were not as direct and simple as they might seem from the pollen record at Goshen Springs. Two major complicating factors were the fluctuating climate and the influence of human populations. The climates from the glacial maximum (18,000 years ago) to about 8,000 years ago were progressively warmer but still mesic enough to preclude extensive pine forests on much of the southeastern Gulf Coastal Plain. Deciduous forests continued to dominate, and pine became even more restricted in distribution. Approximately 10,000 years ago southern pines were probably restricted to a

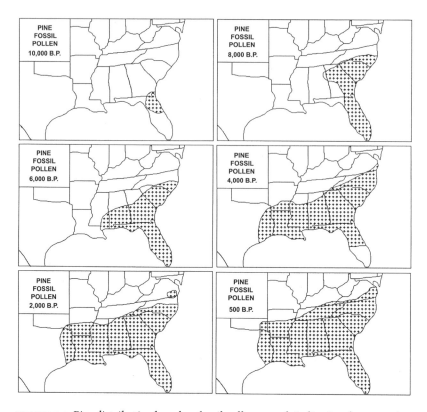

FIGURE 2.1 Pine distribution based on fossil pollen records indicating the expansion of pines from a glacial refugium in Florida and Georgia during the last 10,000 years. Adapted from Webb 1986.

limited area near the coast in northern Florida and southeastern Georgia (Figure 2.1). The range of Red-cockaded Woodpeckers was presumably similarly restricted at this time. A warmer, drier interval from approximately 8,000 years ago to 4,000 years ago, known as the Hypsithermal Interval, was coincident with a major spreading of pine habitats out of the Florida-Georgia region. By 5,000 years ago pine-dominated communities had attained essentially their modern distribution, extending from the Carolinas throughout Florida and west to Texas. Red-cockaded Woodpeckers presumably expanded their range coincidentally with the greatly expanded pine habitat.

Recent research by two U.S. Forest Service geneticists has suggested the possibility of a quite different Pleistocene scenario for the history of pines in the southeastern United States. R. C. Schmidtling and V. Hipkins examined genetic diversity of allozymes in longleaf pine and found the highest diversity in the western portion of the range. These data suggest that the Pleistocene refugium for longleaf pine may have been in the west—either southern Texas or northeastern Mexico—rather than in Florida. O. O. Wells, G. L. Switzer, and R. C. Schmidtling had previously proposed Pleistocene refugia for loblolly pine in both southern Florida and southern Texas–northern Mexico. Obviously, if a southern Florida–northern Mexico refugium for these pine species is accepted, it would substantially expand the possibilities for the late Pleistocene biogeography of Red-cockaded Woodpeckers and the pine ecosystems of the southeastern United States.

Maximum warmth and aridity occurred about 5,000 years ago as the southern pine forests attained near maximal expansion. Since then, the climate of the region has been somewhat cooler and more mesic. These changes have not, however, resulted in major adjustments in the range of southern pine forests, due in part to the influence of humans during this period. The impact of fire on southern pine forests and the adaptations of individual species, especially longleaf pine, to frequent fire will be discussed shortly. Frequent natural fires were undoubtedly involved in the expansion and maintenance of pine forests throughout the region. However, a complicating factor is the occurrence of fires of human origin during much, if not all, of the period since the last glacial maximum.

THE HUMAN INFLUENCE

When Europeans began their colonization of North America following the voyages of Columbus, the southeastern portion of what is today the United States was heavily forested. Pines of several species were an important component of these forests (Figure 2.2). Exclusive of the extensive river bottoms, pines dominated many of the forests from Virginia and the Carolinas south to Florida and west to Texas and Oklahoma. Important outliers also occurred in New Jersey and Delaware. The arrival of Europeans precipitated major changes in these forests. In little more than 400 years, essentially all of the forests throughout this extensive region were cut, ending with the logging of the pine forests of eastern Texas during the first decades of the 1900s. Large portions of these lands, following the cutting, were converted to nonforest uses. Agricultural fields, improved pastures, and urban areas fragmented the remaining forest ecosystems. Areas that remained in forest were also extensively altered. Tree age, species composition, and forest structure all changed as a result of conscious management and un-

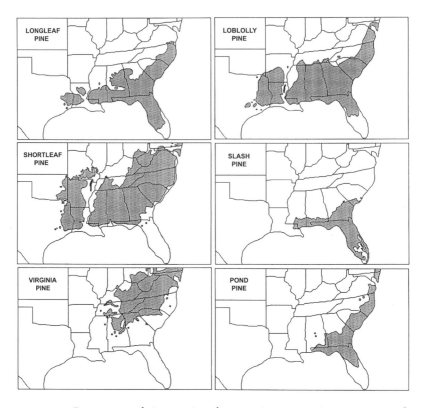

FIGURE 2.2 Range maps of pine species of greatest importance in supporting Red-cockaded Woodpecker populations. Adapted from E. L. Little Jr. 1971.

planned alteration of the ecological relationships that had prevailed prior to the landing of Columbus in the Antilles.

The landing of Columbus in 1492 was not the first human dispersal event to impact the pine forests that Red-cockaded Woodpeckers occupied. Humans had crossed the Bering Land Bridge from Asia to North America during the Wisconsinan glaciation and established populations in the Arctic. The conventional hypothesis, and the one that best fits subsequent events, is that between 12,000 and 13,000 years ago an ice-free corridor opened between the occupied areas north of the Wisconsinan ice sheet and the areas south of the ice. Human populations moved south between the ice-bound Rocky Mountains to the west and glacial Lake Agassiz to the east. These populations rapidly expanded throughout both North and South America. An alternative hypothesis suggests that this dispersal event occurred earlier (24,000–40,000 years ago) and possibly by sea

along the west coast of the continent. However, archaeological sites supposedly documenting the earlier arrival dates are subject to considerable
controversy.

Regardless of the exact timing of the arrival of humans to North America south of the Pleistocene ice sheets, by 10,000–12,000 years ago human
populations based on big game hunting were widespread. Archaeological
sites of the Clovis culture are abundant in North America and clearly document the hunting of Pleistocene large mammals. By 5,000 years ago cul-
tural change resulted in a settled population based on the cultivation of
corn, squash, and other crops in what is now the southeastern United
States.

Contrary to popular myth, the Native Americans did not exist in ecological harmony with the biological communities they colonized. The initial spread of the Clovis big game hunters constituted an additional stress
for the existing prey populations. Paul Martin, at the University of Arizona,
a research scientist closely associated with developments in this field, and
Klein hypothesize that hunting pressures by the Paleo-Indian populations
were responsible for the extinction of numerous species of large vertebrates
at the conclusion of the last glaciation. This Pleistocene overkill hypothesis is supported by (1) the timing of the extinction increase that corresponds
with the arrival of the big game hunters, and (2) the ecological position of
the affected species, primarily large herbivores, and the large carnivores
and scavengers dependent on them.

During a relatively short time interval a diverse array of large herbivores
suffered extinction, leaving few ecological replacements. The large herbivores that became extinct just within the historical range of the
Red-cockaded Woodpecker include a mastodon (*Mammut americanum*),
horses (*Equus* spp.), giant ground sloths (*Megalonyx jeffersonii, Eremothe-
rium rusconii*), glyptodonts (*Glyptotherium floridanum*), camelids (*Hemi-
auchenia macrocephala, Palaeolama mirifica*), and giant tortoises (*Geo-
chelone* spp.). The extinction of the sabertooth cat (*Smilodon fatalis*), dire
wolf (*Canis dirus*), and other large predators in this same region was presumably a result of the collapse of their large vertebrate prey base. The extinction of several species of vultures, including Merriam's Teratorn (*Tera-
tornis merriami*) with its wingspan of nearly 4 m, and the range restriction
of the California Condor (*Gymnogyps californianus*) to the western portion
of North America were probably further consequences of the extinction of
large herbivores.

One large herbivore, the American bison (*Bison bison*), has had a complex recent history in this region. When Europeans first entered what is

now the southeastern United States, bison were apparently absent from the coastal plain from the Carolinas through Florida and west to Louisiana. Expeditions led by Hernando de Soto (1540s), Tristán de Luna (1550s), and Juan Pardo (1560s) failed to observe bison during their extensive travels in the region. After the departure of Pardo in 1567, there are no other accounts of the region until the 1670s. From this time forward, there is abundant evidence, both archaeological data and written accounts, of bison throughout the region and deep into the Florida peninsula. Between the 1560s and the 1670s, the primary ecological change in the region was the disruption of the Native American population precipitated by the colonial expansion of Europeans. Archaeologists and anthropologists have concluded that between the mid-1500s and 1700 Native American populations declined by up to 80%, and extensive portions of the Southeast were essentially depopulated. It has been hypothesized by E. Rostlund that the resulting decrease in hunting pressure allowed the rapid expansion of bison into the region from the north and west. The fire-maintained pine forests would have provided abundant forage for a large grazing mammal such as the bison.

Bison were abundant during the early 1700s, as numerous reports attest. Mark Catesby, author of *The Natural History of Carolina, Florida, and the Bahama Islands,* wrote of bison in South Carolina in 1722: "They range in droves, feeding in the open savannas morning and evening, and in the sultry time of the day they retire to shady rivulets and streams." The European advance continued, and increasing settlement eliminated the bison once again from the region in the ensuing decades. William Bartram foresaw this event when he wrote of Georgia as it was in 1774: "The buffalo, once so very numerous, is not at this day to be seen in this part of the country."

The extinction of a major portion of the large herbivore fauna must have had a large influence on the plant communities of North America. Recent findings on the role of grazing pressure in maintaining grasslands and the impact of African elephants (*Loxodonta africana*) on woodlands provide a graphic illustration of the shaping of plant communities by large herbivores.

Herbivore extinction resulting from the Pleistocene overkill was not the only impact Native Americans had on plant communities. Of equal or greater significance was the widespread use of fire as a management tool. Fire opens up dense vegetation to facilitate travel, and in the process provides improved foraging conditions for many herbivores. Consequently, Native Americans used fire extensively and the plant communities of North America were drastically altered as a result. Natural fire-maintained communities expanded, and fire-tolerant and fire-dependent species increased. These changes were nowhere as dramatic as in the pine and hardwood forests of the southeastern United States.

The timing of fire in these pine communities has significant ecological consequences. Natural fires were a result of lightning strikes. The timing and frequency varied somewhat across the region but everywhere exhibited a strong peak in the late spring, summer, and early fall months. Many plant species, including pines, in these communities are apparently adapted to a regime of growing-season fires. Most woody species, however, are vulnerable to killing by growing-season fire. Consequently, alteration in the timing as well as frequency of fire has consequences for the structure of the forest. Pine-dominated forests with a primarily herbaceous understory are more apt to be maintained if growing-season fires are frequent than if fire is infrequent or occurs mostly in the nongrowing season. Thus the use of fire by Native Americans, as well as changes in climate, presumably promoted the postglacial spread of pine forests in the southeastern United States and their maintenance over the last 5,000 years.

When Europeans arrived in what is now the southeastern United States, pines dominated vast areas. Seven species occurred on the coastal plains and Piedmont of the Southeast, and additional species occurred in the Appalachian Mountains and plateaus. Most of these species provided habitat for Red-cockaded Woodpeckers. These southeastern pine species are ecologically diverse, but the continued existence of most, if not all, of these species is dependent on fire and to a lesser extent on other types of disturbance.

The Longleaf Pine Forest

The most important species of pine relative to the biology and distribution of Red-cockaded Woodpeckers is longleaf pine. This is a result, among other factors, of the extensive range of longleaf pine and its tendency to occur in relatively pure, open parklike stands (Figure 2.3). Prior to the extensive logging activities initiated by European colonists, longleaf pine occurred abundantly on the coastal plains and Piedmont from Virginia to central Florida and west to eastern Texas (Figure 2.2). Significant extensions were found in more mountainous regions, especially on ridges where the influence of fire was more extreme. It has been estimated that more than 25 million hectares were occupied by longleaf pine–dominated communities, making this forest type one of the most extensive plant communities of relatively homogeneous composition on the North American continent.

Early accounts of the longleaf pine forests typically describe them as essentially single-species forests with a minimum of understory (Figure

FIGURE 2.3 Photograph of a longleaf pine forest in eastern Texas (ca. 1908) illustrating the open aspect of the understory and large tree sizes in the original upland pine forests of the South that were maintained by frequent fires. Courtesy of the East Texas Research Center, Steen Library, Forest History Collections, Sawdust Empire Collection P90s:6, Stephen F. Austin State Univ., Nacogdoches, Texas.

2.3). William Bartram, writing of his travels in the 1770s, describes the vast forests he encountered in the following terms:

> We now rise a bank of considerable height, which runs nearly parallel to the coast, through Carolina and Georgia: the ascent is gradual by several flights or steps for eight or ten miles, the perpendicular height whereof, above the level of the ocean, may be two or three hundred feet (and these are called the sand-hills), when we find ourselves on the entrance of a vast plain, generally level, which extends west sixty or seventy miles, rising gently as the former, but more perceptibly. This plain is mostly a forest of the great long-leaved pine (*P. palustris* Linn.), the earth covered with grass, interspersed with an infinite variety of herbaceous plants, and embellished with extensive savannas.

Also prominent in the early accounts were references to fire in longleaf pine habitats. Presettlement fire frequencies have been estimated at several per decade, and early accounts are consistent with this estimate.

Occasional early accounts hint at the dependence of the longleaf pine ecosystem on frequent fire. Sir Charles Lyell, a proponent of geological uniformitarianism and supporter of Charles Darwin, made the connection in 1845. Concerning the vicinity of Tuscaloosa, Alabama, he wrote as follows:

> These hills were covered with longleaved pines, and the large proportion they bear to hardwoods is said to have been increased by the Indian practice of burning the grass; the bark of the oaks and other kinds of hardwoods being more combustible, and more easily injured by fire, than that of the fir tribe. Everywhere the seedlings of the longleaf pine were coming up in such numbers that one might have supposed the ground to have been sown with them; and I was reminded how rarely we see similar self sown firs in English plantations.

FIRE AND LONGLEAF PINE

Much has been learned about the biology of longleaf pine and its associated ecosystem since Lyell wrote of Alabama longleaf forests. The impressive open forests with a dense herbaceous stratum of grasses and forbs—Bartram's "open airy grove"—were a result of the biological attributes of longleaf pine, especially its adaptations to fire. Pines as a group exhibit several strategies that adapt them to survive fire, and longleaf pine practices an extreme version of one of these strategies.

Adult longleaf pines are large; heights of 37 m and diameters at breast height (DBH) of 90 cm were not unusual in the unlogged forests (Figure 2.3, Plate 1). The presence of relatively thick bark confers considerable resistance to fire damage. In addition, the terminal twigs are very stout (approximately 2 cm in diameter) and able to withstand considerable heat stress. Consequently healthy longleaf pines can sustain total defoliation by fire with minimal mortality. In frequently burned habitats, the open structure of the original longleaf pine forests would readily permit low-intensity ground fires, but more destructive crown fires were rare (Plate 2). Adult longleaf pines in such situations, resistant to fire mortality and relatively resistant to disease and insect attack, frequently reached advanced ages. Individual trees have the potential to survive for several centuries, and ages in excess of 450 years were not uncommon in the original forests.

A remarkable physiological adaptation of longleaf pine, abundant resin production, is a significant factor in the ability of longleaf pine to occur in extensive stands that approach a natural monoculture and to survive for more than four centuries. In the warm and mesic climate of the southeastern United States, tree growth is rapid, though longevity is not extreme. Fungal decay and insect attack are prevalent, and few species reach the advanced ages characteristic of many tree species in western states. Bald-cypress (*Taxodium distichum*) is an exception, reaching ages in excess of 1,700 years, due primarily to extensive production of cypresene and other chemicals that provide resistance to fungal and insect damage.

Similarly, the exceptional longevity of longleaf pine is in large part a function of the greater production of pine resins and various terpenes that provide a defense against fungal and insect attack. These terpenes were also responsible for the extensive naval stores industry, discussed in Chapter 4, that flourished in the Southeast and depended primarily on longleaf pine.

In the southeastern United States the insect species most often respon-sible for widespread pine mortality is the southern pine beetle (*Dendroc-tonus frontalis*). As detailed in Chapter 8, epidemics are apparently cycli-cal and have the potential to kill large areas of pine forest. Older forests and those more heavily dominated by pines are more susceptible. South-ern pine beetle outbreaks are currently, and presumably were historically, a major factor in limiting the age and extent of pine-dominated forests in the southeastern United States. Longleaf pine transports abundant resin to the sites where southern pine beetles are attempting to chew through the bark to reach the cambial region. This resin provides an effective physical barrier and reduces beetle success to such an extent that sustained beetle outbreaks in longleaf pine are rare. It is this prolific resin production that Red-cockaded Woodpeckers exploit to protect their cavities from predatory rat snakes (see Chapter 5).

It is in the reproductive strategy of longleaf pine that the most signifi-cant adaptations to frequent fires are evident. Longleaf pines release their seeds in the fall. Individual seeds are larger than those of other southeast-ern pines and have a substantial wing that facilitates dispersal by wind cur-rents. Successful germination and subsequent survival are enhanced if the seeds are dispersed to bare mineral soil. The primary provider of bare min-eral soil is recent fire. In the absence of recent fire, the longleaf pine seeds frequently lodge in the thick herbaceous vegetation and do not reach the soil surface. If they overcome this obstacle and manage to reach the soil surface and germinate, subsequent survival is poor due to shading and com-petition from other plants.

Following successful germination, the young longleaf pines enter a period of unusual vegetative growth termed the grass stage (Figure 2.4). During the first several years of growth, height increase is suppressed and the apical meristem, the actual growing tip, remains close to the soil sur-

FIGURE 2.4 Responses of pine seedlings to fire. *(A)* Longleaf pine grass stage prior to fire. *(B)* Longleaf pine grass stage after fire, showing survival of the terminal bud and rapid initiation of growth. *(C)* Shortleaf pine after fire, showing vigorous sprouting. *(D)* Loblolly pine after fire, showing lack of survival mechanisms. Photos by D. C. Rudolph.

face. A dense cluster of needles at ground level, resembling a clump of grass, eventually results. The photosynthetic products are stored in a rapidly developing root system. During the grass stage, the young pine is remarkably resistant to the damaging effects of fire. The large diameter terminal stems characteristic of longleaf pine are present in the grass stage and provide considerable protection from the effects of heat. In addition, the dense cluster of green pine needles serves to protect the growing tip. As a consequence, most grass-stage individuals survive a fast-moving ground fire with minimal damage. Even though the needles may be burned, leaving a short blackened stem protruding from the soil, the apical meristem usually survives, and within a few weeks a new complement of needles is produced from the nutrients stored in the massive root system.

After several years (sometimes as many as 10–25 years) spent in the grass stage, rapid height growth is initiated. Within 3–4 years of initiating height growth, the young pine reaches a height of 2 m or more using the large supply of nutrients sequestered in the roots. This strategy allows the plant to pass rapidly through the intermediate heights where it is most vulnerable to fire.

This remarkable growth history allows longleaf pine to reproduce abundantly in an environment where fires may occur several times per decade. No other tree species in the region can match this ability. The ultimate result is a mosaic of impressive forests composed almost exclusively of longleaf pine with an herbaceous understory. These pines are capable of replacing themselves as openings in the canopy occur with no intervening successional stages, producing a remarkable fire-maintained plant community.

In old-growth longleaf pine forests, regeneration is suppressed in the immediate vicinity of mature trees, and thus is concentrated in gaps where trees have fallen or died. Hurricanes may occasionally produce large gaps, but most gaps are small, resulting in a landscape of unevenly spaced trees of mixed ages. Many individual trees exist in a suppressed state, inhibited by a larger neighbor, waiting to grow and take their place among the large, dominant individuals when the neighbor dies. Of course, few longleaf tracts fit this description today. Instead, the trees are younger, denser, and more evenly spaced, and regeneration is suppressed by litter accumulation in the absence of fire.

Longleaf pine forests are best developed on dry, sandy soils throughout their extensive range. However, the occurrence of fire has a strong controlling influence on the species, and if fires are frequent enough, longleaf pine can dominate in other soil and moisture conditions. Thus longleaf

A B

FIGURE 2.5 *(A)* Longleaf pine/bluestem habitat in eastern Texas and *(B)* longleaf pine/wiregrass habitat in northern Florida. Photo *A* by R. N. Conner; photo *B* by D. C. Rudolph.

pine forests occur in wet sites, generally as open savanna communities and along ridgelines in areas generally dominated by other species of pines or hardwoods.

THE PLANT COMMUNITY

The plant and animal communities associated with longleaf pine forests are of considerable interest. Woody species other than longleaf pines are generally uncommon, except in areas protected from the influence of fires. The well-developed herbaceous stratum is heavily dominated by grasses and contains a very diverse mixture of monocots and forbs. In much of the eastern portion of the range of longleaf pine, wiregrasses (*Aristida stricta* and *A. beyrichiana*) are, or were, the dominant grass species (Figure 2.5). Adapted to periodic drought and fire, wiregrass covered thousands of hectares in the southeastern United States. West of the range of wiregrass, from eastern Mississippi to eastern Texas, species of bluestem grass (*Andropo-*

gon and *Schizachyrium*) dominate the longleaf pine understory (Figure 2.5). Under the influence of frequent fire that suppresses most woody vege- tation, longleaf pine forests are basically tallgrass prairies with an over- story of pines, more properly termed savannas. The diversity of herba- ceous species is truly remarkable in these savannas. Over 150 herbaceous plant species have been found in areas of a few hectares. Grasses, sedges, legumes, and composites predominate, but a wide variety of taxa are repre- sented. On wetter sites, such as flatwoods and hillside seepage bogs, orchids and insectivorous plants are often abundant (Plate 3). Included in this large array of species are many rare or localized species.

A reduction in natural or anthropogenic fire frequency initiates an in- vasion by a large number of woody species. These species are present, al- though generally uncommon, in frequently burned savannas. They also exist on sites less frequently burned because of topography, isolation, or chance. Consequently, invasion of the grassy savannas by woody species is rapid when fires are suppressed. Ten to 15 years without fire is suffi- cient to initiate successional changes and drastically alter the open aspect of longleaf pine savannas. With an increasing invasion of woody species, the diverse herbaceous component of the flora is increasingly suppressed and many species become locally extinct. Within a relatively short period of time, the longleaf pine savanna changes via succession to a mixed-pine and hardwood forest with the loss of many of the distinctive species that formerly occurred.

The plant species occurring in the longleaf pine savannas are them- selves adapted to frequent fire. Wiregrass, the dominant grass in many longleaf communities, flowers and produces abundant viable seed only fol- lowing growing-season fires. Herbaceous annuals persist in this commu- nity by maintaining a seed bank in the soil that survives the fires and maintains populations. The more abundant perennial species generally re- sprout from underground roots, tubers, and rhizomes. Many species of forbs that contribute heavily to the high species diversity in these habi- tats tend to increase with frequent fire. William Platt and coworkers at Tall Timbers Research Station and Louisiana State University are conduct- ing a long-term study of the effects of fire season and frequency on the herbaceous community in a longleaf savanna in the Florida panhandle. Under a growing-season fire regime corresponding to the natural peak in fire frequency, they have found increased synchronization of flowering and increased abundance of fall flowering forbs compared to a nongrowing- season fire regime. Increased seed production and stimulation of vegeta- tive reproduction through the killing of growing apical meristems by fire, thereby releasing dormant buds, increases the abundance of fall flowering species in the community and contributes significantly to the high plant species diversity.

Other Southern Pines

Longleaf pine, due to its biology and unique adaptations allowing it to survive frequent fires, historically provided the most extensive habitat for Red-cockaded Woodpeckers and probably the optimum habitat for the species. However, several other species of pines provide significant habitat for Red-cockaded Woodpeckers. Each of these pine species has unique biological attributes that influence the forest communities in which it occurs and the Red-cockaded Woodpecker populations that occupy them.

LOBLOLLY PINE

Loblolly pine ranges from New Jersey to central Florida and west into eastern Texas. It occurs essentially throughout the range of longleaf pine, with significant extensions to the north and west of the range of longleaf pine. In most parts of its extensive range, loblolly pine originally was characteristic of wetter sites adjacent to streams, even bottomland situations where flooding was not excessive (Figure 2.6). Occurrence in these habitats was responsible for the common epithet "loblolly," a topographic term also applied to moist depressions or baygalls in the Mid-Atlantic states. In these habitats loblolly pine typically occurs with a variety of hardwood species, often as a minor component of the community. In many areas, especially in the western portion of its range, loblolly pine also occurs more extensively in upland situations, either in pure stands or mixed with other pines—primarily shortleaf pine (*Pinus echinata*)—and hardwoods. In general, with numerous exceptions, loblolly pine tends to occur on sites that are more mesic, with tighter soils (less sand, more clay), higher fertility, and less frequent fires than those supporting longleaf pine.

In marked contrast to longleaf pine, loblolly pine is unlikely to persist as a fire-maintained community. Lacking the fire-resistant grass stage, loblolly pine has difficulty regenerating under a frequent fire regime (Figure 2.4), although the mature trees are moderately resistant to fire. With less frequent fires, additional species, especially more shade-tolerant seedlings of hardwoods, begin to establish in the understory. The accumulation of hardwood leaf litter, which is less flammable than pine needles, and the reduction in herbaceous vegetation, especially highly flammable grasses,

FIGURE 2.6 Loblolly pine habitat in eastern Texas. Loblolly pine typically grows in more mesic sites that burn less often than sites with longleaf pine. Consequently, hardwood vegetation is more abundant. Photo by D. C. Rudolph.

reduce the susceptibility of the resulting plant community to the low-intensity ground fires so important to the perpetuation of longleaf pine communities.

Consequently, loblolly forests and forests with a loblolly pine component are generally experiencing a successional transition. In the absence of disturbance, loblolly pine would decline in abundance through time. However, loblolly pine is the most shade-tolerant of the widespread southern pines, and even minor disturbances (treefalls, minor storm damage) are probably sufficient to maintain loblolly pine as at least a minor component of these forests indefinitely. Catastrophic disturbances (beetle outbreaks, tornadoes, hurricanes, large crown fires) are sufficient to allow loblolly pine to regain importance and even dominance in the regenerating forest. Across a landscape, these interactions were presumably sufficient to produce the pattern of loblolly pine distribution and abundance present when Europeans colonized North America.

The combination of medium-weight seeds with medium-size wings provides loblolly pine with significant dispersal abilities. This attribute was important in allowing loblolly pine to persist in a forest matrix by colonizing disturbances both large and small. It was also an important factor in the remarkable spread of loblolly pine into habitats altered by European colonists. Abandoned croplands are characteristically colonized by loblolly pine in many areas of the Southeast, which is the basis for its alternate name, "old-field pine." Loblolly pine is also an aggressive colonizer of long-leaf sites, if the fire frequency is reduced. As a result of these processes and its preferred status in forestry plantings (see below), loblolly pine has increased greatly in the southeastern United States since the arrival of Europeans, largely at the expense of longleaf pine.

Loblolly pine reaches moderate ages (250+ years); however, due to relatively rapid growth, often on productive sites, it achieves by far the largest size of the southern pine species. Individuals 1.5 m DBH and 45 m in height were not uncommon in the original forests. Although less common, individuals 2 m DBH and over 70 m tall have also been recorded. Unfortunately, extremely few trees approaching these sizes survive at the present time.

In the existing loblolly pine forests, and presumably in the original forests as well, bark beetle attack is a very significant factor, especially attack by southern pine beetles (see Chapter 8). Loblolly pine is one of the most susceptible of the southern pines to bark beetle attack, due in large part to the lower resin production capabilities of the species. Southern pine beetle epidemics and lower intensity chronic infestations presumably helped prevent loblolly pine from achieving dominance over larger expanses of the original forests. In a sense, loblolly pine acted as a fugitive species in both space and time in its interaction with southern pine beetles. In the currently existing forests, loblolly pine is very often favored by timber management due to its rapid growth. Consequently, loblolly pine is often found as a virtual monoculture over large areas. That devastating southern pine beetle outbreaks occur in these altered situations is not unexpected.

In spite of its tendency to occur in mixed stands of pine and hardwood, its vulnerability to hardwood succession, and its reduced resin production, loblolly pine still provides abundant habitat for Red-cockaded Woodpeckers. In significant portions of the range of Red-cockaded Woodpeckers, especially large portions of the Piedmont, woodpecker populations are dependent on loblolly pine. However, for several reasons—discussed in Chapter 5—loblolly pine has probably always represented poorer habitat for Red-cockaded Woodpeckers than has longleaf pine.

SHORTLEAF PINE

A third species of southern pine, similar to loblolly pine in many respects, is shortleaf pine. Shortleaf pine has the most extensive range of the southern pines. Except for peninsular Florida, its range encompasses the known historical range of the Red-cockaded Woodpecker. Like loblolly pine, short-leaf pine tends to occur in mixed forests of hardwoods—primarily oaks—and loblolly pine. The predominance of shortleaf pine in the western and northern portions of the historical range of the Red-cockaded Woodpecker resulted in its particular significance in these areas. In the Interior Highlands, which includes the Ouachita and Ozark Mountains, shortleaf pine was the only pine species occurring naturally, and Red-cockaded Woodpeckers in this extensive portion of their historical range were dependent on this species.

Shortleaf pines are similar in size to longleaf pines (1 m DBH and 40 m in height) and intermediate between longleaf and loblolly in age potential (350+ years). Resin flow and other attributes suggest that shortleaf pine may constitute better habitat for Red-cockaded Woodpeckers than loblolly pine, although not as good as longleaf pine. Shortleaf pine tends to occur on less mesic sites than loblolly pine and is tolerant of nutritionally poorer soils. In heavier, more mesic sites, it is susceptible to fungal infections of the root system that can result in little-leaf disease. In the Interior Highlands shortleaf pine was originally dominant on the southern slopes of the mountains, where periodic fires were an important ecological factor in reducing competition from oaks and other hardwoods.

Shortleaf pine is very similar to loblolly pine in its susceptibility to bark beetles and fire. Consequently it acts somewhat like loblolly pine in the forest landscape. However, several aspects of its biology result in significant differences. The restriction of shortleaf pine to drier sites, partly as a result of little-leaf disease and its greater tolerance of dry sites, results in subtle but consistent differences in its local distribution in relation to loblolly pine in spite of the frequent co-occurrence of the two species. In the drier sites, hardwood competition is reduced and fires have more long-lasting consequences, providing additional opportunities for shortleaf pine to maintain importance or dominance in the forest community. In addition, shortleaf pine seedlings develop a J-shaped crook at the ground surface by 2–3 months of age. Dormant buds in this region are capable of initiating growth if the developing stem is killed by fire or other causes (Figure 2.4). This important adaptation provides juvenile shortleaf pines with a mechanism to regenerate in a fire-influenced community somewhat analogous to the grass stage of longleaf pine, although not as effective.

Shortleaf pine is also more resistant than loblolly or longleaf pines to the detrimental effects of ice storms. Because of its shorter needles, ice buildup during these events is less massive and causes less breakage of limbs. This is a significant factor in the wider and more northerly distribution of the shortleaf pine (Figure 2.2).

In portions of its range, shortleaf pine has occurred in fire-maintained forests in nearly pure stands. In the Ouachita Mountains of Oklahoma and Arkansas, for example, south-facing slopes were originally dominated by shortleaf pine in a fire-maintained community with a little bluestem (*Schizachyrium scoparium*) understory.

A globally significant tract of the Ouachita Mountains containing such habitat has been preserved as a wilderness area (Plate 4). The 5,700-ha McCurtain County Wilderness Area in southeastern Oklahoma has never been subjected to significant timber harvest. This is one of the largest such tracts in eastern North America and preserves the full range of habitats present in the region. In addition, a remnant population of Red-cockaded Woodpeckers still occurs on the site. Unfortunately, management has included fire suppression, and the magnificent virgin shortleaf pine forests are experiencing severe hardwood succession; pine reproduction is minimal. As a consequence, the best surviving example of old-growth shortleaf pine forest is being lost, along with its remnant Red-cockaded Woodpecker population.

Fortunately, the managing agency (Oklahoma Department of Wildlife Conservation) is demonstrating a willingness to confront the political and practical difficulties involved in returning fire to the ecosystem. We wish them well in their efforts to restore the shortleaf pine ecosystem and the Red-cockaded Woodpecker population.

OTHER PINE SPECIES

The three pine species discussed above (longleaf, loblolly, and shortleaf) provide the majority of Red-cockaded Woodpecker habitat. An additional four species provide moderate amounts of habitat and, in addition, are important components of southern pine ecosystems.

Slash pine (*Pinus elliottii*) ranges from South Carolina through Florida and west to the Mississippi River bottomlands. Slash pine resembles longleaf in some respects, and often occurs in mixed stands with longleaf pine. Two significant differences are its greater tolerance for extremely wet conditions and the much greater susceptibility of the seedlings, which typically lack a grass stage, to fire. Consequently, slash pine occurs on the margins of flatwoods ponds and other wet sites in pure stands or mixed with a variety of hardwoods. It also occurs mixed with longleaf on transitional sites, and is an aggressive invader of longleaf habitat if the fire frequency is reduced.

In the southern half of the Florida peninsula and the Florida Keys, a form of slash pine (*P. elliottii* var. *densa*) occurs that develops a grass stage and thick taproot similar to longleaf pine, and is strikingly convergent with longleaf pine in its general appearance. The result is a slash pine population with a greatly increased ability to reproduce in habitats subject to frequent fire. In the wet prairies of southern Florida, this variety replaces longleaf pine.

Slash pine provides significant habitat for Red-cockaded Woodpeckers in many wetter sites marginal to the extensive stands of longleaf pine. In addition, the southern Florida variety, *densa*, is the only pine species naturally occurring in the southern one-third of the Florida peninsula and provides the only habitat for Red-cockaded Woodpeckers in this portion of their range. Due to the rapid growth of slash pine, the species has also been widely planted by the timber industry beyond its natural limits, typically in former longleaf pine habitat.

Pond pine (*Pinus serotina*) occurs from New Jersey south, on the coastal plain, to northern Florida. Pond pine is typical of wetter sites—stream margins, baygalls, swamps, and flatwoods ponds—and occurs as scattered trees or mixed with a variety of hardwoods. Pond pine is also characteristic of the extensive pocosins of the Carolinas. Pocosins are upland shrub bogs, often characterized by the accumulation of peat, which are maintained by recurrent catastrophic fires. Pond pine has two characteristics that are responsible for its success under a catastrophic fire regime. First, as the specific name implies, the cones are serotinous, tending to remain unopened on the tree for several years. Following exposure to the heat of fires, the cones open and release several years' accumulation of seeds to germinate in the mineral soil exposed by the fire. Second, mature trees also have considerable ability to sprout from the trunk and major branches following scorching and defoliation by fire.

Pond pine is relatively uncommon throughout much of its range and provides limited habitat for Red-cockaded Woodpeckers. However, in the pocosins of the Carolinas, where other pines cannot grow, significant habitat is provided by pond pine. Such areas are characterized by low densities of both pines and Red-cockaded Woodpeckers compared to other habitat types.

Virginia pine (*Pinus virginiana*) is a characteristic species of the Piedmont and lower elevations of the Appalachian Mountains from Pennsylvania to northern Alabama and Georgia. Virginia pine attains up to 32 cm

DBH and 34 m in height but is generally much smaller. It is relatively short-lived, and few trees exceed 100 years in age. The species typically occurs in relatively pure stands and is a common colonizer following disturbance, especially by catastrophic fire. As a result of its small stature and limited age potential, Virginia pine provides only marginal habitat for Red-cockaded Woodpeckers.

Pitch pine (*Pinus rigida*) is a widespread species of the Appalachian Mountains at moderate elevations. The species is resistant to fire because it resprouts vigorously and has a tendency to develop serotinous cones. In favorable situations, pitch pine achieves moderate size (1 m DBH and 30 m in height). Because pitch pine has a peripheral range, compared to that of the Red-cockaded Woodpecker, and a fragmented distribution, the species provides very limited and far from optimal habitat for Red-cockaded Woodpeckers. Virginia pine and pitch pine are far less significant to the woodpeckers than the other species of pines discussed.

The Animal Community

A considerable number of vertebrate species, in addition to Red-cockaded Woodpeckers, are intimately associated with the open fire-maintained pine forests of the southeastern United States. As one might predict, a number of these species are declining and even endangered due to habitat changes and related causes.

Amphibians are not uncommon in southeastern pine forests, but most species—even if endemic to the region—are not restricted to pine habitats. The flatwoods salamander (*Ambystoma cingulatum*) is closely associated with pine-and-wiregrass habitats and breeds in flatwoods ponds. Today it is critically endangered and breeds in very few of the many flatwoods ponds within its former range from South Carolina to northern Florida. The pine woods treefrog (*Hyla femoralis*) is also characteristic of these habitats from Virginia to eastern Louisiana. The pine barrens treefrog (*Hyla andersonii*) has a disjunct distribution in New Jersey, the Carolinas, and western Florida and southern Georgia. This rare species inhabits pine barrens, pocosins, and similar habitats in southern pine ecosystems (Plate 5).

Among reptiles, a number of species were once characteristic of open fire-maintained pine habitats; abundant sunlight reaches the ground in such habitats, often an important consideration for large ectothermic vertebrates. In particular, gopher tortoises (*Gopherus polyphemus*) (Figure 2.7), pine snakes (*Pituophis ruthveni* and *P. melanoleucus*) (Plate 6), eastern indigo snakes (*Drymarchon couperi*), and eastern diamondback rattlesnakes

FIGURE 2.7 The gopher tortoise is a characteristic, and once abundant, species associated with fire-maintained pine ecosystems east of the Mississippi River. Photo
by D. C. Rudolph.

(*Crotalus adamanteus*) were once common and widespread inhabitants of
fire-maintained pine forests. All of these species have their closest relatives in areas to the west and south. The occurrence of these species in a
mesic warm temperate region is dependent on relatively open habitats that
are most abundantly provided by fire-maintained pine forests. All of these
species have declined dramatically due to changes in the fire regime, and
their continued existence is in question.

The avian fauna of southeastern United States pine forests, especially of
the more open longleaf pine savannas, has traditionally been considered de-
pauperate compared to that of the deciduous and mixed forests. However,
Todd Engstrom, a biologist with the Tall Timbers Research Station, examined a unique old-growth longleaf pine savanna habitat (the Wade Tract)
in southern Georgia (Plate 1). Engstrom found avian diversity that compared favorably with that of other forest types of similar latitudes in North
America. Cavity nesting species, including Red-cockaded Woodpeckers,
and ground nesting species were particularly well represented. The mature
structure of the forest, including the presence of many large dead trees, was
a prime factor in maintaining this diversity. However, the abundance of
such snags in the pre-Columbian forests may have been lower. Hot head-
fires (burning with the wind) may have consumed more snags than the controlled backfires that are currently typical of the Wade Tract (see Chapter 5).

Among the diversity of avian species inhabiting southern pine ecosystems, several deserve particular mention. The endemic Red-cockaded Woodpecker and the nearly endemic Brown-headed Nuthatch (*Sitta pusilla*) are, or were, characteristic throughout the range of southern pines, and are completely dependent on pine ecosystems. The Pine Warbler (*Dendroica pinus*) breeds in pine habitats from Florida and Texas to the Great Lakes region. It is a migratory species and in winter is confined to southern pine ecosystems. Each of these species is arboreal and primarily dependent on relatively mature pines for foraging substrate. The Brown-headed Nuthatch and Pine Warbler are not as dependent on specific attributes of forest structure as the Red-cockaded Woodpecker, and consequently remain fairly common to abundant.

Several terrestrial species are also characteristic of the open fire-maintained pine forests of the southeastern United States. Bachman's Sparrow (*Aimophila aestivalis*), often called the pine-woods sparrow, is intimately associated with fire-maintained pine forests. This ground-nesting species, associated with grassland vegetation, is generally abundant in open fire-maintained pine forest. Bachman's Sparrows are rapidly eliminated by encroaching woody vegetation following alteration of the fire regime. A grass-dominated understory is a critical habitat requirement of Bachman's Sparrows, and temporary populations can colonize early successional habitats—for example, recent clear-cuts. Typically these populations persist for only 2–4 years. Growing-season fire suppression has resulted in population declines of this species throughout its range. Henslow's Sparrow (*Ammodramus henslowii*) is a species similar to Bachman's Sparrow in its dependence on grass habitats. It breeds in prairie openings and fields of the northeastern United States. Henslow's Sparrow is migratory and winters exclusively in the southeastern United States, originally primarily in the open fire-maintained pine forests. Loss of suitable habitat, primarily in the breeding range, has resulted in precipitous population declines.

A third terrestrial species, the Northern Bobwhite (*Colinus virginianus*) is also characteristic of open pine forest. A grassy understory with abundant legumes provides excellent habitat, and fire-maintained pine forests throughout the region once supported abundant populations. Northern Bobwhites were avidly hunted for decades throughout the southeastern United States, but currently, huntable populations are nonexistent in many areas due to habitat loss and reduction of fire. The hunting of "partridges" with well-trained dogs is a declining tradition throughout the South. Currently, hunting is primarily restricted to local areas where management is intensive, and includes the use of prescribed fire and captive-raised birds. The link between declines in quail populations and alteration of pine communities due to fire suppression is not well recognized in the sporting community. Indeed, many hunters continue to advocate the policy that led to the original population declines—fire suppression and restriction of prescribed burning—as a means to increase populations.

Another avian species of note is the American Kestrel (*Falco sparverius*). This small cavity-nesting falcon is widespread in open habitats throughout nearly all of North and South America. Winter resident kestrels are common in the southeastern United States in open habitats of various types. There is, however, a resident subspecies, *F. s. paulus*, that is, or was, primarily restricted to the open pine forests of the Gulf Coastal Plain from Georgia to Texas. Red-cockaded Woodpecker cavities enlarged by Pileated Woodpeckers (*Dryocopus pileatus*) and other species provide, or provided, an important source of nest cavities for the southeastern subspecies of American Kestrels in many areas. Frequent fires provided the open habitat with herbaceous vegetation of low stature necessary for hunting. Populations of the subspecies declined due to lack of nest cavities, and presumably to changes in forest structure resulting from changes in the fire regime.

Other notable bird species especially common in, but not restricted to, some fire-maintained pine communities include Chuck-will's-widow (*Caprimulgus carolinensis*), Common Nighthawk (*Chordeiles minor*), Eastern Wood-Pewee (*Contopus virens*), Great Crested Flycatcher (*Myiarchus crinitus*), Summer Tanager (*Piranga rubra*), and Eastern Bluebird (*Sialia sialis*). For these and other species, pine savannas may well represent primary habitat, even though these species are more often thought of as associated with hardwood habitat types. Other species such as White-eyed Vireo (*Vireo griseus*), Common Yellowthroat (*Geothlypis trichas*), and Eastern Towhee (*Pipilo erythrophthalmus*) may be very common in dense vegetation along drains and streams, although not in the adjacent pine uplands. The tremendous research effort currently under way to document population dynamics of Neotropical migrants will soon provide the data needed to evaluate the value of fire-maintained pine ecosystems in relation to other habitat types for the many habitat generalists that occur in pine forests.

The pine specialists such as Brown-headed Nuthatches, Pine Warblers, and Bachman's Sparrows obviously are highly dependent on fire-maintained pine forests. But none have as many unusual adaptations to these pine ecosystems as the Red-cockaded Woodpecker. Adaptations for foraging in pines are interesting but not remarkable. The remarkable adaptations are those related to cavity excavation in living pines, which are discussed in detail in Chapter 5. We can only speculate about the selective pressures that caused this bird to evolve such behavior.

The mammal fauna characteristic of southern pine ecosystems is composed primarily of widespread species not particularly dependent on southern pines. The major exceptions are certain subspecies of the wide-ranging fox squirrel (*Sciurus niger*) (Plate 7). Ten subspecies of the fox squirrel are currently recognized and collectively range throughout most of the eastern United States. Throughout most of its range, the fox squirrel is relatively abundant. Most subspecies are generally reddish in color and weigh 600–900 g. Four subspecies (*cinereus, niger, shermani, bachmani*) occupy southern pine ecosystems from the Delmarva Peninsula in New Jersey south through much of Florida and west to the bottomlands of the Mississippi River. These subspecies are much more variable in color; are often strikingly patterned in gray, white, and black; and are generally larger, up to 1,200 g. These subspecies are characteristic of pine forests, especially longleaf pine. Their larger size may be an adaptation to allow efficient handling of the large cones of longleaf pine that smaller squirrels have difficulty carrying.

Studies by Peter Weigl of Wake Forest University and others have shown that the combination of open forests, seasonal food shortages, and large body size in these subspecies results in low population densities and large home ranges, compared to those of other fox squirrels. The ecological changes of recent decades, as discussed previously, have precipitated severe declines in range and population size of these beautiful and distinctive subspecies and their widespread replacement by gray squirrels (*Sciurus carolinensis*). This significant change has gone largely unnoticed by the general population and by many biologists.

A Summary Perspective

Southern pine ecosystems, formed by several species of pines with distinct biological characteristics, were very diverse. Coastal plain habitats were dominated by longleaf pine, and to a lesser extent by slash pine and other species. Upland pines were often supplemented by loblolly and pond pines on associated slopes and drainages. These forests provided the most extensive pine-dominated habitat in the historical range of the Red-cockaded Woodpecker. On the coastal plains, only extensive bottomland and transitional areas were dominated by other nonpine species.

Inland of the coastal plain, pines (primarily loblolly and shortleaf) were widespread. In the original Piedmont forests, however, various hardwood species were dominant over vast areas, and mixed-pine and hardwood forests were abundant. For example, the original forests of the Georgia Piedmont were estimated to consist of over 50% mixed forest dominated by oaks, and only 15% was thought to have been predominantly pine. Further inland in the Appalachian Ridge and Valley Region and the Interior Highlands, pine-dominated areas were primarily confined to ridges and south-facing slopes.

Overall, these forests provided abundant and relatively continuous habitat for Red-cockaded Woodpeckers on the Gulf and Atlantic Coastal Plains and increasingly fragmented habitat inland on the Piedmont and more mountainous regions. This original pattern has been greatly altered by events since the arrival of Europeans. Essentially all of the original forests have been logged at least once, so that very little old-growth forest remains. Much of the logged area has been converted to other uses and no longer supports forest.

The existing forest habitats of the region are also very different from the original forests. Age is one important difference, with very little existing forest dominated by trees more than 90 years of age. Many thousands of hectares are harvested on very short rotations (20–50 years) for pulpwood used in paper production. The ecological impact of fire has also been drastically reduced, with profound impacts on the ecosystem. Species conversion and hardwood succession due to reduced fire impact are widespread. Conversely, in many areas of mixed-pine and hardwood forest, hardwoods have been greatly reduced by girdling, herbicide injection, and cutting, so that many areas are now more heavily dominated by pine, especially loblolly and shortleaf, than previously.

Timber management practices have altered forest ecosystems in many ways. One such practice is the planting of certain species in place of others, often far beyond their natural range and ecological amplitude. Loblolly pine and slash pine, in particular, are extensively planted, often in areas that were formerly longleaf pine. Conversion to nonforest use and replacement by other species of pine have reduced the extent of longleaf communities to less than 5% of their original range. The surviving longleaf communities are often highly altered due to changes in the fire regime and no longer resemble the fire-maintained ecosystem once present. Regionally, the alterations in extent and condition of fire-maintained pine ecosystems since the arrival of Europeans dwarf those in other habitats, even wetlands, which receive much more popular attention. It therefore is not at all surprising that Red-cockaded Woodpeckers and a plethora of other plant and animal species have declined dramatically. If anything, the surprise is that these species have been able to persist, given the extensive loss and degradation of their ecosystems.

PLATE 1. Original-growth longleaf pine on the Wade Tract in Thomas County, Georgia. This is one of the very few remaining examples of old-growth longleaf pine forest. A frequent fire regime has been maintained on this site resulting in a continuation of the fire-maintained ecosystem. This site is currently managed by the Tall Timbers Research Station. Photo by R. N. Conner.

PLATE 2. Frequent, low-intensity ground fires are critical in the maintenance of longleaf pine ecosystems. The fire maintains the open understory dominated by herbaceous vegetation in the short term, and the dominance of longleaf pine in the overstory in the long term. Photo by D. C. Rudolph.

PLATE 3. Pitcher plant (*Sarracenia flava*) bog in a Texas longleaf pine savanna. Bogs and other included plant associations occur frequently in fire-maintained pine communities and support many plant species of conservation concern. Photo by D. C. Rudolph.

PLATE 4. McCurtain County Wilderness Area in southeastern Oklahoma, the best remaining example of shortleaf pine forest. The area has suffered from fire exclusion since shortly after its acquisition by the state in 1918, resulting in massive increases in hardwood density. However, increased emphasis on Red-cockaded Woodpecker management has resulted in the implementation of a prescribed burning program that has the potential to restore and maintain this unique shortleaf pine community. Photo by D. C. Rudolph.

PLATE 5. Pine barrens treefrog (*Hyla andersonii*), a species distributed in isolated habitats in New Jersey, the Carolinas, and the Florida panhandle and a small area of adjacent Alabama. The pine barrens treefrog is one of several vertebrate species limited to southeastern pine forests, and is typically found in pine barrens, pocosins, and shrub bogs. Photo by C. Harrison.

PLATE 6. The Louisiana pine snake (*Pituophis ruthveni*) is one of two species of pine snakes characteristic of pine forests of the southeastern United States. The Louisiana pine snake, due to its limited range in eastern Texas and western Louisiana, as well as the alteration of its habitat resulting in large part from changes in the fire regime, is one of the rarest vertebrate species in the United States. Photo by D. C. Rudolph.

PLATE 7. A fox squirrel (*Sciurus niger niger*) from the Carolina Sandhills. Coastal plain populations of the widespread fox squirrel are adapted to pine forests, especially longleaf pine, and are extremely polymorphic in color and pattern. Courtesy of North Carolina Wildlife Resources Commission.

PLATE 8. A breeding male Red-cockaded Woodpecker with red "cockade" clearly visible. The breeding male is the socially dominant member of a woodpecker group, typically selecting the best cavity tree available in the cavity tree cluster. Photo by D. C. Rudolph and R. N. Conner.

PLATE 9. (*right*) A nestling male Red-cockaded Woodpecker peering out of its nest cavity. Note the red crown patch, which is clearly visible on the head of the young woodpecker. Photo by Derrick Hamrick.

PLATE 10. A red heart fungus (*Phellinus pini*) fruiting body, or conk, on the bole of a pine tree indicates that the heartwood of the pine is decayed. Red-cockaded Woodpeckers preferentially select pines with decayed heartwood for their nest and roost trees, because cavities are more easily excavated in pines with decayed heartwood. Photo by D. C. Rudolph.

PLATE 11. Rat snakes (*Elaphe obsoleta*) are excellent tree climbers, using furrows between bark plates and bark ridges as support points while climbing. Fresh pine resin that flows from regularly pecked resin wells coats the bole of Red-cockaded Woodpecker cavity trees with a sticky barrier that can deter rat snake predation on eggs and woodpeckers in cavities. Photo by R. N. Conner.

PLATE 12. Pileated Woodpeckers (*Dryocopus pileatus*) account for substantial annual losses of Red-cockaded Woodpecker cavities. Pileated Woodpeckers regularly excavate and enlarge Red-cockaded Woodpecker cavity entrances, rendering the cavities unsuitable for the endangered woodpecker. Such enlarged cavities, however, do provide cavity sites for the larger secondary cavity users. Photo by R. N. Conner.

PLATE 13. The southern flying squirrel (*Glaucomys volans*) is a frequent occupant of Red-cockaded Woodpecker cavities. Similar to Red-cockaded Woodpeckers, flying squirrels prefer unenlarged cavities and often acquire active Red-cockaded Woodpecker cavities. Anecdotal reports indicate that the squirrels prey on eggs and nestling woodpeckers in some instances. Photo by D. C. Rudolph.

PLATE 14. Red-cockaded Woodpeckers typically select the older pines within a stand for their cavity trees. If enough older pines are available within a stand, the cavity trees are often clumped together in a cluster. The accumulation of hardened pine resin from resin wells over the years makes the boles of cavity tees look like white candles. Photo by R. N. Conner.

PLATE 15. An aerial view of a large, expanding southern pine beetle infestation (spot) reveals dead pines in the center surrounded by infested dying pines, attacked red-topped pines, and green uninfested trees. Southern pine beetles spread from one pine to another by following a chemical pheromone scent trail through the air to mass-attack new pines. Photo by R. F. Billings, Texas Forest Service.

Threats to the Fire-Maintained Ecosystem

Human activities pose numerous potential threats to fire-maintained pine ecosystems of the southeastern United States. These threats might more properly be termed alterations, since an ecosystem will still be present. In the present context, we are referring to changes in species populations, community structure, and ecosystem processes compared to the pre-European condition of the ecosystem.

SITE CONVERSION

Many hectares of fire-maintained pine forest have been eliminated, with the lands now used for agriculture, grazing, urbanization, and numerous other purposes. Historically this has been the major impact of humans on this ecosystem. The process began with the Native Americans and accelerated with the arrival of Europeans. The process of deforestation began on the eastern seaboard with the first colonists and progressed westward, with the last large-scale elimination of forest occurring in the early decades of the 1900s in Texas. Due to widespread depletion of soils under cultivation, substantial areas subsequently have reverted to forest across the region. Currently, approximately 60% of the original 100 million hectares of fire-maintained pine forests of the region remain in some type of forest.

Removal of forest followed by conversion to other land uses has had the greatest impact on the ecosystem. Relatively few species of plants or animals characteristic of southeastern forests occur in nonforested habitats consisting of agricultural fields, improved pastures, and urban areas. Native pastures, suburban areas, and many areas with some surviving trees support many more species. In all cases, however, species composition and basic ecosystem processes are highly altered.

ALTERATION OF FIRE REGIME

With the exception of deforestation, alteration of the fire regime has resulted in the most dramatic change in fire-maintained pine ecosystems. Fire suppression and concentration of prescribed fire in the nongrowing season have greatly reduced the frequency of fire and changed its ecological effects. The primary result has been a massive increase in hardwood abundance, generally restricted to a dense midstory that eliminates the once diverse herbaceous understory. Ironically, mature canopy hardwoods are at the same time much less abundant in the highly altered forests of the present than they were in pre-Columbian pine forests of the South. Throughout the region, fire-maintained ecosystems have lost the single most important ecological process that sustained their evolution and existence. As a result, the general condition of most remaining pine forest in the region is rapid succession to some type of hardwood forest. Pines, however, remain a dominant feature of the landscape primarily due to silvicultural manipulations for timber production.

Complete loss of pines from the ecosystem obviously impacts a large number of species, beginning with the pines and species directly dependent on pines. A greater number of species are impacted as effects cascade through the ecosystem and basic ecosystem processes change.

Even if pines remain a dominant part of the ecosystem because of silvicultural manipulations, extensive ecological changes occur. Tens of thousands of hectares of pine forest in the region suffer heavy encroachment by hardwoods, primarily in the midstory, and the once extremely diverse herbaceous flora is heavily suppressed. Numerous species of plants and animals adapted to fire-maintained ecosystems have declined precipitously and been replaced with other species. Longleaf pine has experienced tremendous declines as a result of these processes because of its extreme adaptation to fire. Numerous plant species, especially herbaceous perennials, have declined precipitously due to successional changes and the resulting suppression of the herbaceous flora. Such species form a major portion of lists of threatened and endangered species throughout the region. In addition to effects on abundance, fire performs subtle functions in the biology of many plant species adapted to these habitats. Two examples are lower incidence of brown spot fungus (*Scirrhia acicola*) on longleaf pine in the grass stage resulting from the reduction of the duff and grass layer, and the stimulation of flowering in wiregrass by growing-season fires.

Important vertebrate species in these fire-maintained systems also decline or disappear with alteration of the fire regime. Included in this list are gopher tortoises, pine snakes, Bachman's Sparrows, and pocket gophers (*Geomys* spp.), as well as Red-cockaded Woodpeckers. Most of these species seem to decline because of changes in forest structure related to increases in midstory vegetation and suppression of the herbaceous understory. Invertebrate communities are presumably highly altered as well, although little is known. Certainly the conspicuous butterfly populations characteristic of the open, well-burned savannas with abundant nectar-producing flowers change in species composition and diversity with the increase in woody understory and midstory vegetation.

SILVICULTURAL IMPACTS

Most silvicultural activities alter fire-maintained ecosystems, the nature of the changes depending on the particular harvest method used. There are a number of methods, ranging from even-age methods (clear-cut, seed tree, shelterwood) to uneven-age methods (group selection, single-tree selection), that are discussed more fully in Chapter 11. The degree to which the harvest method deviates from natural regeneration processes and interferes with natural ecosystem processes determines the impact on fire-maintained pine ecosystems. Species of pines adapted to catastrophic fires that remove all or most above-ground vegetation are generally less impacted by an even-age harvest method such as clear-cutting. Of the species within the range of Red-cockaded Woodpeckers, this most applies to sand pine (*Pinus clausa*) but also to Virginia pine and pond pine. These species are adapted to intense fires that remove adult trees. Natural regeneration is accomplished by massive reseeding from serotinous cones or by vigorous stump sprouting. With these species, clear-cutting best mimics the natural regime and perpetuates natural ecosystem processes. In the sand pine/scrub communities of central Florida, many endemic species would be negatively impacted if uneven-age harvest methods altered the cycle between open scrub and closed-canopy forest originally driven by catastrophic fire.

Longleaf pine exemplifies the other extreme. This species is adapted to very frequent low-intensity ground fires. Both adult trees and young seedlings in the grass stage are very resistant to these low-intensity fires. Consequently, some form of uneven-age harvesting best mimics natural fire-driven processes. Numerous other species in this community also thrive under such a system.

Several important species are intermediate in regard to adaptations to fire, notably loblolly and shortleaf pine. Both species resemble longleaf in that the adult trees are resistant to low-intensity ground fires. However, they lack adaptations to resist fire in the seedling stages, although shortleaf resprouts when young. These species were most successful in situations where fires were less frequent than in longleaf communities and less catastrophic than in sand pine communities. These species present the greatest challenge in attempting to use harvest methods that mimic natural processes. Clear-cutting severely alters the ecosystem to the detriment of many species that, unlike those in sand pine communities, are not adapted to widespread total forest removal. Single-tree selection with frequent regeneration presents additional difficulties. It is often difficult to obtain adequate regeneration due to the intolerance of pines to shading by residual canopy trees, and nearly impossible to maintain fire in the ecosystem without destroying the continually present young regeneration. Generally, hybrid harvesting techniques such as group selection or shelterwood methods allow the closest approach to natural ecological processes. Failure to adjust harvesting techniques to mimic natural processes in the various pine-dominated communities can significantly alter the structure of the forest, with major impacts analogous to those discussed above concerning alteration of the fire regime.

A second aspect of harvesting techniques relates to the way in which pines that are dying due to lightning, beetles, disease, and other factors are managed. In the original forests, pine mortality produced the snags that many cavity-dependent species required. The extent to which snags are protected in harvesting operations, and especially the vigor with which they are removed in salvage operations, has a major impact on many species. In many management situations, periodic searches detect even single dead or dying trees, which are then salvaged. Reduction in numbers of snags impacts numerous species, especially those requiring larger cavities (Pileated Woodpeckers, American Kestrels, Wood Ducks [*Aix sponsa*], and tree squirrels).

Following harvest, the methods used to prepare the site for pine regeneration have different impacts on the ecosystem. Obviously, fire is the method that most closely resembles natural processes. If fire was essential to maintain a vigorous ecosystem prior to harvest, it is generally sufficient to prepare an adequate seedbed or planting surface after harvest. If other methods are chosen, the disturbance of the ecosystem is more extreme. Mechanical methods (windrowing of logging debris, chopping of logging debris) or use of herbicides result in major impacts, especially to the important herbaceous layer. Wiregrass, once a dominant species in longleaf pine habitats east of the Mississippi River, is now much less common because of replacement by other species after extensive mechanical site preparation. The potential for erosion is also greater if mechanical disturbance of the soil occurs.

Following harvest and site preparation, the various methods used to regenerate the pines have very different impacts on the ecosystem. The original ecosystem is best regenerated by encouraging natural regeneration. This obviously requires that a seed source be retained, and therefore harvest methods other than clear-cutting are generally required. Artificial regeneration, generally by the planting of seedlings, results in much greater impacts. Site preparation to facilitate planting activities is often quite severe. However, the most significant impact is the potential to drastically alter the genetic composition of the regenerating pines, or even replace the species of pine. Millions of hectares of longleaf pine have been converted to either slash pine or loblolly pine by this process. Slash pine, in particular, has been extensively planted beyond its natural range. Very little attention has been given to the origin of seed sources used to regenerate large areas of the region, and selected genetic stocks have been used extensively. As a result, the natural genetic makeup of pine populations across the region has also been severely disrupted, and the original genetic composition of the forests has been greatly altered. The impacts of this alteration have received little attention.

Biomass removal, implicit in harvesting, occurs in a variety of forms in the fire-maintained ecosystems of the southeastern United States. Obviously wood products for lumber or pulp are the primary form of biomass removal. The contribution of this biomass removal to observed declines in site productivity are the focus of ongoing research. Other forms of biomass and nutrient removal occur, and their contribution can be substantial. Two in particular deserve special mention. Increased erosion following silvicultural treatments is often substantial, leading to nutrient loss and productivity declines in the regenerating forests. The harvesting of fallen pine needles (pine straw) of longleaf pines for horticultural uses (decorative mulch) is also increasing. There is growing concern that in addition to the contribution to biomass removal, removal of the highly flammable pine straw lowers the pyrogenicity of the site and ultimately alters the fire regime. The method of raking could also impact the ecosystem through soil disturbance and compaction.

The removal of longleaf pine stumps that survive from the original harvests decades ago also supports a small industry. The recovered stumps are used as a source of resins and other products. The widespread removal of these stumps results in the loss of habitat for numerous species that find shelter in rotting stumps. This may be significant, especially since the present short-rotation practices do not produce a supply of large long-lasting stumps to replace those removed.

Fragmentation of the forest landscape also has a major impact on these fire-maintained ecosystems. Many factors contribute to this fragmentation, among which conversion to nonforest is the most significant. Activities within managed forests, as well as natural processes, contribute additional fragmentation, which is generally temporary. These contributions can be significant, especially given the highly fragmented landscape already present due to conversion to nonforest. Harvesting, especially by clear-cutting, increases fragmentation of forest habitats. The cumulative impact can be substantial and is highly dependent on rotation age. For example, if the rotation is 50 years, 20% is harvested each decade, whereas only 10% is harvested each decade if the rotation is 100 years. Natural contributions to fragmentation, primarily windstorms, tornadoes, and hurricanes, are minor contributors to forest fragmentation except locally, where the effects, especially of hurricanes, can be severe.

In summary, fire-maintained pine ecosystems have been greatly reduced in the southeastern United States since Europeans arrived, and much of what remains has been tremendously altered, most importantly by changes in the fire regime, past agricultural use, and short-rotation silviculture. Many species, including the Red-cockaded Woodpecker, have declined precipitously due to these changes. It is possible, however, to manage the remaining habitat so that the basic ecosystem processes continue to function. If this goal is realized, most declining species, including the Red-cockaded Woodpecker, will maintain viable populations in many areas.

CHAPTER THREE

Evolution, Taxonomy, and Morphology of the Red-cockaded Woodpecker

Evolution

Lester Short, with the American Museum of Natural History, indicated that the greatest number of woodpecker species is found in the New World tropics. Woodpeckers occur on every continent except Australia and Antarctica, but for unknown reasons there are relatively few species in Africa. The distribution of woodpeckers suggests that ancestral woodpecker species had evolved prior to the breakup of Gondwanaland and the drifting of the present Neotropics from Africa but after the Australian continent had separated, making the woodpecker lineage perhaps 65 million years old.

Other evidence points to a more recent origin. Fossil records for the woodpecker family (Picidae) are relatively rare, so a detailed history of woodpecker evolution cannot be accurately reconstructed. Earliest records reported by Storrs Olson at the Smithsonian Institution indicate that woodpeckers were present in the New World during the middle Miocene, about 25 million years ago. In Arizona, a recently discovered fossil tree with cavity and entrance hole, apparently constructed by a woodpecker, was dated to be from the Eocene, about 40–50 million years ago. Charles Sibley, with Yale University, and Jon Ahlquist, with Ohio University, used the molecular technique of DNA-DNA hybridization to examine the evolutionary history of all avian groups. They suggested that woodpeckers may have branched from their closest living relatives more than 50 million years ago. Thus, the limited evidence available is consistent with the hypothesis that woodpeckers have been present as a distinct lineage for approximately 50 million years.

The evolution of Red-cockaded Woodpeckers as a distinct species is likely more recent. Most evidence suggests that Red-cockaded Woodpeckers evolved sometime during the Pleistocene geological epoch, which spanned the period from 7,000 to about 2 million years ago. The Pleistocene was marked by a series of glaciations, the last of which, the Wisconsinan, receded in North America about 8,000–10,000 years ago.

At some time during the Pleistocene, a Red-cockaded Woodpecker died and fell into a stream in the vicinity of what is now Orange County, Florida. The small woodpecker decomposed, and its bones became scattered in the stream. The currents and eddies of the stream moved one of the upper wing bones, a humerus, until it lodged in the streambed, falling in with bones of other species of the age. A mastodon (*Mammut americanum*), an early horse (*Equus* sp.), a tapir (*Tapirus viroensis*), a camel (*Tanupolama mirifica*), and numerous other bird bones mixed together in the sediments of the streambed. Years passed and the bones filled with minerals from the limestone bedrock.

In 1939 James Gut passed by the now dry and exposed old streambed and noticed some of the fossil bones. The wing bone from this Red-cockaded Woodpecker was eventually sifted from the dry streambed and now resides in the Pierce Brodkorb Collection at the Florida Museum of Natural History in Gainesville. This humerus, the only known fossil bone of the Red-cockaded Woodpecker, was reported in 1959 by Glen Woolfenden, now with the Archbold Biological Station and a professor emeritus at the University of Southern Florida. The presence of the fossil bone in Florida does not completely reveal where Red-cockaded Woodpeckers originated as a species. It does support the view that the species evolved in Florida and spread north and west from Florida after the last glaciation.

Woodpeckers are widely distributed throughout most continents. As a group they demonstrate limited ability to colonize islands, especially oceanic islands. They have, however, managed to spread to many islands in the Malay Archipelago, the West Indies, the Philippines, Japan, Andaman Islands, and a few additional continental islands (Okinawa, Cozumel, Sri Lanka). The limited ability to colonize islands has presumably prevented woodpeckers from reaching New Guinea and Australia.

Sibley and Monroe (Burt Monroe with the University of Louisville) divided the pied woodpeckers, to which the Red-cockaded Woodpecker (currently *Picoides borealis*) belongs, into two genera: *Picoides* (11 species) and *Dendrocopos* (22 species), which collectively have a nearly cosmopolitan or worldwide distribution. *Dendrocopos* occurs exclusively in the Old World, whereas *Picoides* occurs primarily in the New World. In contrast, Short placed all pied woodpeckers within the genus *Picoides* and eliminated the genus *Dendrocopos*. *Picoides* is most diverse in North America, only two species being found in South America. One species ranges from North

America across much of northern Eurasia. The Red-cockaded Woodpecker is sympatric in the southeastern United States with two widespread species of *Picoides*, the Downy (*Picoides pubescens*) and Hairy (*P. villosus*) woodpeckers. Other *Picoides* in North America include Black-backed (*P. arcticus*), Ladder-backed (*P. scalaris*), Nuttall's (*P. nuttallii*), Strickland's (*P. stricklandi*), Arizona (*P. arizonae*), Three-toed (*P. tridactylus*), and White-headed (*P. albolarvatus*) woodpeckers.

Nuttall's and Ladder-backed woodpeckers from the southwestern United States and Mexico are considered by some to be the closest relatives of Red-cockaded Woodpeckers. Short has suggested that Red-cockaded Woodpeckers originated as an eastern isolation of the Ladder-backed Woodpecker line. He suggested that Red-cockaded Woodpeckers became specialized or continued a long line of specialization for foraging and nesting in pure pine forest habitat. Derek Goodwin also supported this view and stated that *borealis* is probably closely allied with *nuttallii* and *scalaris*. K. H. Voous Jr. doubted that *borealis* is a recent (Pleistocene) member of the *scalaris-nuttallii* group, suggesting instead that Red-cockaded Woodpeckers are an old member of the ladder-backed group resulting from an earlier divergence. He further suggested that the southeastern United States was a refuge for the Red-cockaded Woodpecker during the arcto-Tertiary glacial periods.

An alternative view suggests that the Hairy Woodpecker is the closest taxonomic relative of the Red-cockaded Woodpecker. Jerry Jackson presented pro and con arguments for both points of view at the first Red-cockaded Woodpecker symposium, in 1971. He generally favored grouping Red-cockaded Woodpeckers with Hairy Woodpeckers (and their close relative, the Downy Woodpeckers) because of the historical uncommonness of Hairy Woodpeckers in the southeastern United States and the incomplete red nuchal band on the back of the head of male Hairy Woodpeckers, which is perhaps analogous to the separation of the two red cockades on the Red-cockaded Woodpecker's head. Hairy Woodpeckers become more common with increasing frequency of hardwoods in the eastern and northern United States, whereas Red-cockaded Woodpeckers become less common. This suggests that speciation could have occurred as a result of Red-cockaded Woodpecker specialization and adaptation to the open, fire-maintained pine forests and subsequent invasion of such habitats regionwide.

Hairy Woodpeckers occur in pine forests of the western United States and in the Bahamas, according to the well-known ornithologist James Bond (who was also the inspiration for Ian Fleming's 007). Jackson suggested that the rarity of Hairy Woodpeckers in southeastern United States pine forests may result largely from competitive niche segregation with Red-cockaded Woodpeckers. In a personal communication, John Fitzpatrick, then with the Archbold Biological Station in Florida, suggested that in the southeast Hairy Woodpeckers may specialize in patches where severe fires have killed many trees; for example, in sand pine communities and slash pine flatwoods. Habitat preferences of Hairy Woodpeckers are difficult to predict throughout the range of the species. In the southeast in general, however, they appear to prefer recent disturbances, such as storm damage or bark beetle–caused mortality, in forest habitats that contain hardwoods.

Alden Miller, with the Museum of Vertebrate Zoology in Berkeley, suggested that Red-cockaded Woodpeckers may have arisen as the result of hybridization between Ladder-backed and Hairy woodpeckers based on the marked similarity of a known hybrid of *villosus* and *scalaris* to the Red-cockaded Woodpecker. Jackson and Short, however, both examined Miller's hybrid and did not find its similarity to *borealis* as remarkable as Miller did.

Taxonomy

Jackson provided the best summary of information on the taxonomic history of the Red-cockaded Woodpecker. The following is a condensation of Jackson's synthesis. Louis Jean Pierre Vieillot, a French naturalist, is credited with being the first to describe the Red-cockaded Woodpecker as a species (*Picus borealis*) in 1807. Vieillot painted and described the woodpecker as "Le Pic Boreal," the northern woodpecker, quite unaware that the species had primarily a southern distribution in North America (Figure 3.1). Vieillot also erred in his description of the plumage pattern of the woodpecker, stating that it had white outer tail feathers and a band of red on its head, a description better fitting Hairy Woodpeckers than Red-cockaded Woodpeckers.

A description and name far more fitting for the woodpecker came in 1810. Alexander Wilson, considered the father of American ornithology, was unaware of the woodpecker species described by Vieillot in 1807 and described a *Picus querulus* in volume 2 of his *American Ornithology*. The term "querulus" means chirping and was an excellent description of the "chortling" vocalizations given by the small bands of woodpeckers that Wilson observed and obviously took time to watch. Wilson also described the habitat of the species as pine forest and observed the woodpecker in Georgia, North Carolina, and South Carolina.

FIGURE 3.1 An early painting of the Red-cockaded Woodpecker,
Le Pic Boreal, by Louis Jean Pierre Vieillot in 1807.

Jackson delved extensively into the historical origin of Wilson's name
for the Red-cockaded Woodpecker, and an interesting story unfolded. During the American Revolutionary War, the colonial army was poorly
equipped compared to the British military, and officers often used a few
colored feathers stuck in their hats as an insignia of rank in place of the
more expensive brass and gold braids used by the British. Hats were em-
bellished with feathers, ribbons, or leather to signify rank, and such orna-
ments were collectively termed cockades. The ornamentation on fancy
British uniforms was often referred to as "macaroni" by the colonial Ameri-

cans, hence the taunt against the British in the song "Yankee Doodle": "stuck a feather in his hat and called it macaroni." Thus, Wilson named the woodpecker "Red-cockaded" because the red feathers that were present on the heads of the males resembled a cockade, the American insignia of rank in the Colonial Army. Unlike today, the term "cockade" was a familiar word in the early 1800s.

As noted by Jackson in 1971, Carl Illiger, an early American ornithologist, described the Red-cockaded Woodpecker as a new species a third time, being unaware of the descriptions of both Vieillot and Wilson. Illiger called the bird the Red-streaked Headed Woodpecker (*Picus leucotis*), probably basing his description on a specimen collected in Georgia or South Carolina in 1810 by John Abbot. Johann Georg Wagler, another ornithologist of the 1800s, gave the species a fourth name in 1827, when he described the woodpecker as a new species and named it after Vieillot as *Picus vieilloti*.

In 1941 Alexander Wetmore, with the Smithsonian Institution, preferred the generic name *Dryobates*, and indicated that two subspecies, *Dryobates borealis hylonomus* and *D. b. borealis*, existed. The subspecies from southern Florida, *hylonomus*, had shorter wings than the nominate, *borealis*, subspecies. The American Ornithologists' Union (AOU) accepted Wetmore's decision, thus eliminating other specific names, and established two subspecies for the Red-cockaded Woodpecker in 1945. In 1946 W. E. Todd disagreed with this split and noted Robert Ridgway's earlier comments in the literature supporting the same position as Todd's.

Around 1950 *Dendrocopos* became the accepted genus for the Red-cockaded Woodpecker. The 1957 AOU *Check-list of North American Birds* indicated an acceptance of Wetmore's subspecies determination and listed the species as *Dendrocopos borealis borealis* and *D. b. hylonomus*. In 1976 the AOU merged the New World *Dendrocopos* species into the genus *Picoides*, already represented in North America by the three-toed woodpeckers, giving the woodpecker its current scientific nomenclature *Picoides borealis*, and continued to consider the species polymorphic with two subspecies. In 1977 Robert Mengel, with the University of Kansas, and Jackson analyzed wing lengths from a wide variety of northern and southern Red-cockaded Woodpeckers and suggested that wing-length variation is clinal, changing gradually from north to south, indicating a monomorphic species. Similarly, in 1990 Armando Pizzoni-Ardemani, a graduate student working with Walters, found geographic variation in a number of characters to be clinal, birds from cooler northern and drier inland and western locations being larger than more southern and eastern birds. In 1983 the

AOU acknowledged the subspecies concept for bird species in general, but did not include subspecies because of a desire to expedite publication of the 1983 checklist. Whether the Red-cockaded Woodpecker currently is "officially" listed as a monomorphic or polymorphic species is unclear. Regardless, the basis of designating subspecies for Red-cockaded Woodpeckers is questionable.

Morphology

Woodpeckers in general are characterized by their relatively straight and often chisel-tipped bill, which is used for foraging and cavity excavation. The bill is typically shaped like a wedge when viewed from above, an adaptation for excavation. They have a thicker, more bony skull relative to their size than most birds. This is an adaptation to the forces absorbed by the skull as the bird hammers away at woody tissue. Woodpeckers also have an unusually long and hard-tipped tongue, typically barbed with small spines that aid in the extraction of arthropod prey from crevices and chambers within trees (Figure 3.2). The spines, which are absent in young woodpeckers but appear with age, also permit the adherence of sticky saliva to the tip of the woodpecker's tongue, so that it acts much like flypaper in the capture of insects. The sublingual salivary glands of woodpeckers are modified and greatly enlarged to produce the quantities of saliva needed during foraging. F. Lucas, with the U.S. Biological Survey during the early 1900s, reported that the bony hyoid apparatus located in the woodpecker's neck was longer and of greater strength than in most birds to support the additional length and mobility of the woodpecker's tongue. The woodpecker's tongue is so long that portions of the base must loop up around the skull somewhat in order for it to fit inside the head when it is retracted (Figure 3.2). The shaft of the tongue is very flexible and can compress somewhat when the tongue is retracted.

ADULTS

The Red-cockaded Woodpecker is a relatively small, black-and-white bird weighing from 40 to 55 g. David Ligon, then a graduate student at the University of Florida, had observed that five males from Florida averaged 42.4 g and four females averaged 45.3 g. Data reported by Short also suggest sexual dimorphism, with females being larger than males. Additional data, however, indicate that this is not the case, and in fact the converse may be true.

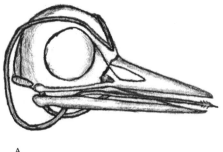

A

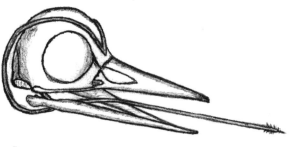

B

FIGURE 3.2 Diagram of the skull of a woodpecker revealing
how the tongue is *(A)* withdrawn and *(B)* extended. Drawing
by R. N. Conner.

Male Red-cockaded Woodpeckers in eastern Texas weigh more, not less,
than females. Average weight of adult males from longleaf pine habitat in
eastern Texas was 47.5 g (sample size, n = 29), whereas adult females averaged 45.3 g (n = 18). In loblolly/shortleaf pine habitat adult males averaged
49.1 g (n = 23), whereas females averaged 46.9 g (n = 16). Pizzoni-Ardemani
also found males to be significantly heavier than females in the North
Carolina Sandhills, with males averaging 49.6 g and females 47.6 (n = 90).

The backs of Red-cockaded Woodpeckers are barred in a horizontal
black-and-white pattern, as are the wings. The woodpecker has a black cap
on its head and a very distinctive large white ear covert or cheek patch
(Plate 8). The adult male, but not female, has several red "cockade" feathers
on each side of the head in the posterior of the black cap near its junction
with the white cheek patch. These feathers are typically covered by black
crown feathers unless the woodpecker becomes excited and exposes them.
The male's cockades make a poor field mark, because they are seldom

visible even when viewed with binoculars. Otherwise, the two sexes are extremely similar in plumage, making sexing difficult in most instances except in the hand. The iris in the eyes of adult Red-cockaded Woodpeckers is reddish. Sides of the breast and body are white or grayish white and streaked with black. The chin, throat, and breast are basically white, and the lower rump and tail coverts are black.

The bills of Red-cockaded Woodpeckers average about 21–23 mm in length, are slightly curved, and have a chisel-like tip. Jackson reported that mean upper bill (culmen) length of 266 males (22.6 mm) was significantly longer than that of 164 females (21.8 mm). Most *Picoides* woodpeckers exhibit a bony swelling of the frontal bones at the base of the upper bill. This region of the skull in woodpeckers is basically an empty space with densely scattered bony struts. Below the base of the upper bill is a similar region of bony struts. Thus a network of tiny bony struts that distributes and cushions the force of a blow when the woodpecker pecks on wood reinforces the area where the upper bill attaches to the skull. The culmen attaches to the skull with a nasofrontal hinge, which is folded somewhat to provide additional cushioning during pecking. Well-developed musculature attached to the woodpecker's bill also absorbs some of the concussion when the woodpecker pecks an object. As the woodpecker hammers away, muscles (most likely the protractor pterygoidei) are stretched to help buffer the force of the blows on the cranium. Feathers cover the nostrils on the bill to help prevent wood chips and dust from entering. The nostrils are linear with slitlike openings, making it doubly difficult for debris to enter with the air as the bird pecks and breathes.

Red-cockaded Woodpecker tail feathers, or rectrices, are about 74–81 mm in length. The tail feathers are acuminate, which means they are pointed and have a very strong central shaft that is stiff all the way to the tip of the feather (Figure 3.3). The stiff tail feathers are pressed against trees and used as a support as the woodpecker forages, excavates a cavity, or climbs on tree boles and branches. They are used very much like a third leg. The pygostyle, or tailbone, of woodpeckers is enlarged to permit the greater muscle attachment necessary to move the tail feathers to support the body. Red-cockaded Woodpeckers have 12 tail feathers; the sixth, or outermost, feather on each side is white and diminutive and often overlooked. The fourth and fifth outer tail feathers are white with black bars on the inner vanes, whereas the inner tail feathers are black. The fourth rectrix is the longest.

Wing length of Red-cockaded Woodpeckers ranges from 108 to 124 mm. Wing and tail length vary geographically, inland and northern birds having

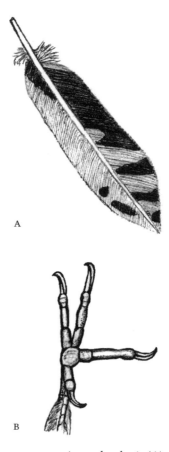

FIGURE 3.3 A woodpecker's *(A)* rectrix (tail feather) and *(B)* zygodactylous, or scansorial, foot. Drawing by R. N. Conner.

longer wings and longer tails. Western birds, although otherwise relatively large like northern and inland ones, have relatively short wings.

The tarsi (unfeathered portion of the legs) of the woodpecker are gray and fairly sturdy for a bird of its size. The musculature of the woodpecker's legs is very strong, enabling the bird to climb around on the boles of trees easily. Natural selection has forged functional modifications in the skeleton and musculature of the legs of woodpeckers. The birds adopt a pecking stance in which the body and head are held as distant as possible from the tree in order to deliver hard blows. Fourteen cervical vertebrae attach the skull to the body. Anatomical changes for the delivery of more forceful blows typically include lengthening of the distal portions (extremities) of the legs and heightening of the spines (vertebral processes) in the cervical vertebrae. However, according to Lowell Spring, with Western Oregon State College, functional modifications for forceful pecking conflict with features that promote smoothness and efficiency of climbing around the bole and branches of trees, constraining modifications of the legs especially.

The feet of woodpeckers have a zygodactyl toe arrangement, meaning that two toes orient toward the front of the bird and two toward the back (Figure 3.3). In actual practice, one of the toes that supposedly orients toward the back is usually placed laterally when a woodpecker grasps a tree, to give the bird support from side to side. This toe orientation has been termed "scansorial" by Walter Bock and Waldron Miller, with the American Museum of Natural History, suggesting that woodpeckers can change the orientation of the toes to match the features of each particular perch. The second and third toes are oriented forward, whereas the fourth toe orients laterally and somewhat backward when needed. The first toe, or hallux, is generally oriented toward the back, and its claw usually does not even hook into the bark when the woodpecker is climbing or perched. The claws on the distal ends of woodpecker toes are deeply curved and quite sharp, permitting easy penetration of bark and wood in order to grasp trees when climbing.

SEXUAL DIMORPHISM

Although very similar in plumage, male and female Red-cockaded Woodpeckers appear to differ morphologically in subtle ways. That males may be slightly larger was discussed above. Although earlier studies produced inconsistent results, Pizzoni-Ardemani's detailed analysis showed that males have longer legs, toes, and bills, and shorter tails, than females. All of these differences may be adaptations related to the well-documented foraging specializations of the two sexes discussed in Chapter 7. Shorter legs and longer tails are advantageous in foraging on trunks, as females do, and larger toes are advantageous in clinging to twigs and limbs, as males do. Longer bills in males may reflect a greater emphasis by males on excavation, as opposed to probing, in acquiring food.

JUVENILE CHARACTERISTICS

Young Red-cockaded Woodpeckers can usually be distinguished from adults by plumage characteristics until September, October, or even November of their first year. The young of both sexes have streaklike white feathers on their foreheads and a grayish tinge to the white cheek patch. Overall, their plumage appears browner than that of adults. Jackson observed that some adult females occasionally retain some white feathers in their forehead. This is evidently rare and perhaps local, as we have not observed it in eastern Texas or North Carolina. Young males have a patch of red feathers on the top of their heads that is about 15 mm in diameter (Plate 9). The red crown patch is molted, as are the white forehead feathers, in late summer to early fall. Juvenile females typically become indistinguishable from adults earlier than juvenile males, because remnants of the red crown patch remain visible longer than remnants of the gray cheek or white forehead. In North Carolina, for example, one can be certain of female age (juvenile or adult) only through August, whereas one can be certain of the age of males captured through September. Young juvenile males lack the red cockades typical of adult males, but as they mature, they can begin to grow their red cockade feathers before they completely lose their red crown patch. Jackson noticed that the tenth primary wing feather of young Red-cockaded Woodpeckers is generally longer, broader, and more rounded on the tip than that of adults. Jackson also observed that the color of the irises in the eyes of young woodpeckers is dark and has a buff-yellow to smoke-gray tinge, rather than the deep chestnut red of adults. Adults generally have a complete, white eye-ring, whereas juveniles often do not.

NESTLINGS

Ligon made detailed observations of the development of nestling Red-cockaded Woodpeckers. Recently hatched nestlings are naked with bright pink skin, and legs and feet that are nearly white. Their eyes and ear openings are closed. At 5 days of age the bill has darkened some, and feather tracts (areas of the skin that produce feathers) are visible on the crown and wings (Figure 3.4). Specifically, femoral, spinal, ventral, and scapular tracts are apparent at 5 days. Five-day-old nestlings can probably hear well, as the ear orifices are open. By 10 days of age, all feather tracts are well developed and the tips of the tail feathers are exposed. Ten-day-old nestlings have opened their eyes and have darkened feet and tarsi (lower legs). By day 18 young woodpeckers are nearly fully feathered and actively peck at intruding biologists. Nestlings are fully feathered and ready to fledge at 24–26 days of age, although some wait a bit longer. Some of the wing and tail feathers may still be partially ensheathed at their bases at this age, even in young that have fledged.

FIGURE 3.4 Red-cockaded Woodpecker nestlings showing early signs of feather tracts. Photo by Sue Daniels.

MOLT

Adult woodpeckers molt once every year, usually after the nesting season and before winter. The central tail feathers, which are the strongest, are usually retained the longest, while other tail feathers are dropped and new ones regrown first. Normally, molt is from the outer feathers inward, enabling the woodpecker to continue to use its tail while climbing and pecking. The primary feathers are molted from the inside out, and the secondaries from the outside in, as in other birds. In North Carolina, molting adults are observed from late June to early October. Red-cockaded Woodpeckers, like other woodpeckers, lack down feathers that serve as insulation in other birds.

Parasites

Conner and coworkers have shown that Red-cockaded Woodpeckers tend to use the newest cavity available in their cavity tree cluster (the breeding male's roost cavity) for their nest cavity (see Chapter 5). A possible selective advantage to this behavior is a reduction in parasite load. A variety of parasites (lice, flies, and mites) are known to occur on woodpeckers in the eastern United States. Lice (*Degeeriella* sp.) have been reported specifically for Red-cockaded Woodpeckers, and mites have been observed while handling woodpeckers in the field by Chuck Hess, working with Fran James on the Apalachicola National Forest, and by Jerry Jackson. However, although parasitic arthropods have the potential to affect avian reproductive success and the fitness of adult breeders, ectoparasites and negative effects from them are rarely observed in Red-cockaded Woodpeckers. Thus, Red-cockaded Woodpecker selection of the newest cavity available for their nest site may be related more to some other benefit than parasite load reduction.

CHAPTER FOUR

Red-cockaded Woodpecker Distribution

Past and Present

Distribution Prior to Recorded History

The distribution of the Red-cockaded Woodpecker prior to the original description by Vieillot in 1807 is virtually unknown. The partial left humerus from the Pleistocene of Orange County, Florida, described in Chapter 3, is the only direct evidence of the distribution of the species prior to 1807. Thus the distribution of Red-cockaded Woodpeckers from their evolution as a distinct species until their description by Vieillot and the subsequent collection of distribution data is speculative. However, once the obligatory relationship between Red-cockaded Woodpeckers and the species of southern pines evolved, the limits of the distribution of the woodpecker were set by the collective distribution of southern pines. The recent changes in pine distribution in the southeastern United States are outlined in Chapter 2. Within this time frame, the range of the Red-cockaded Woodpecker must have expanded and contracted with the available pine habitat. Since the most recent glaciation, the distribution of fire-maintained pine ecosystems of the southeastern United States, which includes the Red-cockaded Woodpecker's ecosystem, likely has expanded from a limited refuge in the extreme Southeast (Florida and Georgia) to the recent historical range. The birds presumably expanded their populations slowly, being a resident species with low rates of new territory establishment (see Chapter 6), but they had many centuries to expand. Certainly their powers of expansion matched those of their equally sedentary hosts, longleaf pines and other pine species. For several thousand years, up through the arrival of Europeans on the continent, suitable pine habitats covered vast areas, and the species was probably abundant regionally.

Through much of this period Red-cockaded Woodpeckers coexisted with Native Americans. The impact of Native Americans on the Red-cockaded Woodpecker and the fire-maintained pine ecosystems the woodpecker required can only be inferred indirectly. Presumably Native Americans had little direct interest in the species, as Red-cockaded Woodpeckers were not of economic value. However, Native Americans developed elaborate agricultural economies throughout the historical range of the woodpecker and in the process impacted the region substantially. Some forest clearing occurred near the larger population centers of the mound-building cultures, but the total at any one point in time would have had only limited impact on the Red-cockaded Woodpecker. Similarly, the growing of corn in close proximity to pine habitat could have provided an additional source of prey for Red-cockaded Woodpeckers. Red-cockaded Woodpeckers are known to extract corn earworms (larvae of the moth *Heliothis zea*) by excavating through the shucks and extracting the larvae. Some groups of woodpeckers have been observed obtaining a major portion of their prey from this source during periods of up to several weeks. It is possible that this unusual foraging behavior has occurred since Native Americans first began to cultivate corn in the region.

It was in the use of fire as an ecological tool, however, that Native Americans had the greatest impact on Red-cockaded Woodpeckers. Fire is one of the defining elements of Red-cockaded Woodpecker habitat, and Native Americans used fire extensively. In the course of burning to improve hunting success and facilitate travel, Native Americans inadvertently practiced sound woodpecker management. The impact of fire due to Native American activities is difficult to separate from that of other sources of ignition. However, as outlined in Chapter 2, the tremendous expansion of fire-maintained pine ecosystems after the last glaciation was coincident with substantial Native American populations in the region. Whether this expansion was due primarily to climatic changes or was significantly influenced by Native American burning practices remains an interesting research topic. We can conclude that Red-cockaded Woodpeckers benefited, perhaps greatly so, from the presence of Native Americans for several thousand years.

Recorded Historical Distribution

Vieillot originally described the distribution of the Red-cockaded Woodpecker as the northern United States in error, which accounts for the inappropriate specific epithet "borealis" (northern). By the 1830s the pioneer American ornithologists Alexander Wilson and John James Audubon had documented the occurrence of the Red-cockaded Woodpecker from New Jersey to Tennessee and south and west to Florida and Texas. Subsequent

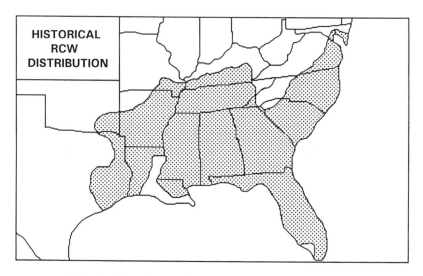

FIGURE 4.1 Historical distribution of the Red-cockaded Woodpecker.

records established the presence of breeding populations in Virginia, Kentucky, Missouri, Arkansas, and eastern Oklahoma. There is even an isolated breeding record from Assateague Island in Maryland. Isolated records from north central Texas, Ohio, and Pennsylvania, as well as Audubon's New Jersey record, are presumably of vagrant or dispersing individuals. Indeed, a vagrant individual appeared in northern Illinois, more than 400 miles from the closest existing population, in August 2000.

The general limits of the distribution of historical breeding populations are shown in Figure 4.1. Within this extensive range, the habitat prior to alteration by Europeans varied; consequently, the population density of Red-cockaded Woodpeckers varied. In prime habitats, which were widespread from the eastern Carolinas south through Florida and west across the Gulf Coastal Plain, Red-cockaded Woodpeckers were apparently common in suitable habitat. Audubon, writing in 1839, stated that Red-cockaded Woodpeckers were "found abundantly from Texas to New Jersey, and as far inland as Tennessee." Obviously Audubon was referring to good habitat, and he further refers to the species as "nowhere more numerous than in the pine barrens of Florida, Georgia, and the Carolinas." In *A History of North American Birds*, published in 1874, the authors (Baird, Brewer, and Ridgway, well-known names in American ornithology) cite Dr. Woodhouse, who found Red-cockaded Woodpeckers common in eastern Texas and the Indian Territory (Oklahoma). The species was apparently

Table 4.1. *Red-cockaded Woodpecker Population Densities*

Population	Birds/km²	Groups/1,000 ha
Sam Houston National Forest (Texas)	6.3	17
Wade Tract (Georgia)	5.7	16
Francis Marion National Forest (South Carolina)	4.3	18
Jones State Forest (Texas)	6.8	20

Sources: Rudolph and Conner, unpublished data (Sam Houston National Forest); Engstrom and Sanders 1997 (Wade Tract); Hooper and Lennartz 1995 (Francis Marion National Forest); Donna Work, personal communication (Jones State Forest).

Note: High densities were determined for limited areas of larger populations in recent decades.

less numerous to the north and west as pine became a less significant part of the forest landscape. Yet even in southern Missouri, near the northern limit of the species range, E. S. Woodruff reported the species as "fairly common" in the early 1900s.

More detailed information on Red-cockaded Woodpecker populations prior to extensive logging is lacking. Some of the early anecdotal accounts are difficult to interpret, because the existence of cooperative breeding was not recognized at the time. It is evident that the species was reasonably common and a characteristic inhabitant of appropriate habitat throughout the southeastern United States. Population densities existing on small tracts at the time of this writing support this interpretation of the early accounts. Table 4.1 presents population densities observed in recent years. Densities of 3.5–5.0 birds per sq. km and 12–13 groups per 1,000 hectares, applied to the area available prior to logging, suggest a substantial population. Current densities occur in second-growth forests, in which loss of old growth and encroaching midstory limit the numbers of potential cavity trees; also, substantial areas of second-growth forests are unsuitable for foraging or nesting due to recent harvesting. The population studied by Todd Engstrom and Felicia Sanders in the vicinity of the Wade Tract in southern Georgia—the only population that has substantial old growth in good condition available to it—has a density of approximately 16 groups per 1,000 hectares.

The Logging Era

Thus, at the time of the arrival of Europeans in North America, the Red-cockaded Woodpecker was a characteristic and common bird of fire-maintained pine ecosystems in the southeastern United States. During the next 500 years, the situation changed drastically. Change was slow at first and limited to the Atlantic seaboard. During the early 1800s Red-cockaded Woodpecker habitat was still intact throughout most of the species' range. The succeeding 150 years saw a progressive alteration of the habitat, interrupted only briefly by the Civil War. The once extensive pine forests of the southeastern United States were logged for timber and converted to agriculture, primarily between 1870 and 1930. The destruction of the forests progressed generally from east to west, culminating in the logging of the pine forests of eastern Texas—the western extreme of the bird's range—by 1930.

Logging and conversion to other types of land use have had a major impact on the landscape of the southeastern United States. Michael Lennartz, with the U.S. Forest Service, Southeastern Experiment Station, and coauthors compiled information on the extent of forested habitats dominated by pines for the second Red-cockaded Woodpecker symposium, in 1983. The data pertained to the situation in the late 1970s and early 1980s. The area they examined was essentially the range of the Red-cockaded Woodpecker, totaling approximately 206 million hectares. Of this total, approximately 25.6 million hectares were in forests dominated by the seven major species of southern pines. Only 0.6 million hectares (2.5% of total pine area) were estimated to be in stands where the age of the dominant trees exceeded 60 years, the average *minimum* age for potential cavity trees (see Chapter 5).

The contrast between the extent of fire-maintained pine ecosystems in the South today and at the time of European contact is striking. The area of longleaf pine at that time has been estimated at 24.3 million hectares, and the area of loblolly pine at 38.9 million hectares. With allowances for the other pine species, the total area of fire-maintained pine ecosystems may have exceeded 100 million hectares, or approximately 50% of the available land area. The majority of the 100 million hectares could have been suitable nesting habitat for Red-cockaded Woodpeckers, compared to the 0.6 million hectares minimally suitable at present. At the density of 16 groups per 1,000 hectares observed in old-growth habitat in southern Georgia (see above), 100 million hectares would theoretically have supported a maximum of 1.6 million groups of woodpeckers. Obviously, catastrophic disturbances, differential abilities of pine species and regions to support woodpeckers, and other factors would reduce this value to an unknown extent.

In retrospect, the logging of the original forests was exceedingly thorough, as witnessed by the extreme rarity of old-growth forest in the southeastern United States at present. The ecological impacts of the disappearance of old-growth forests on a landscape scale were severe. The extent of alterations in avian populations was variable. The loss of forest birds from areas converted to agricultural use was essentially complete. However, in areas where second-growth forest regenerated, forest birds persisted. Due to the lack of detailed observations prior to removal of the original forests, and the rarity of existing old-growth forests, details such as relative abundances are unknown for the avian communities of these forests. Major changes no doubt occurred, but remarkably few avian species became extinct or seriously threatened with extinction due to the cutting of the old-growth forests. The Passenger Pigeon (*Ectopistes migratorius*), Carolina Parakeet (*Conuropsis carolinensis*), and presumably Bachman's Warbler (*Vermivora bachmanii*) are extinct, but factors in addition to forest cutting were responsible. Of these three, only the Passenger Pigeon may have been dependent on vast tracts of old-growth forest. Abundant mast production may have been required for survival of this highly nomadic, massively gregarious species. Even in this case the excessive market hunting, which would have eliminated the species in any event, clouds the role of forest cutting in population declines. Wild Turkeys (*Meleagris gallopavo*) were also once close to extinction, but hunting pressure was the primary cause.

Two additional species suffered devastating impacts due to the logging of the old-growth forests of the southeastern United States, both woodpeckers. The Ivory-billed Woodpecker (*Campephilus principalis*) was apparently dependent on extensive tracts of old-growth forest. It is best known as a species of hardwood bottomland forests, due in part to the classic study of the species by James Tanner, a student at Cornell University working under a National Audubon Society Fellowship along the Tensas River in northeast Louisiana. However, early records document regular use of upland habitats, including pines, indicating that fire-maintained pine communities may have been an important habitat for the Ivory-billed Woodpecker. The species became extinct in the United States with the felling of the last extensive remnants of old-growth forest. The species currently maintains a precarious existence in montane pine habitat in Cuba.

The Red-cockaded Woodpecker also suffered major population declines with logging of the original forests. Dependent on pine habitats, the fate of populations was determined by the fate of the pine forests (Figure 4.2). The dependence of Red-cockaded Woodpeckers on older pines as cavity sites placed unique constraints on the ability of Red-cockaded Woodpeckers to persist following the initial logging of the forests. Total removal of older pines in a forest leads directly to the extirpation of Red-cockaded Wood-

FIGURE 4.2 Cutover pine forest in eastern Texas near the turn of the century, show-ing some of the remnant pines following timber harvest. Courtesy of the East Texas Research Center, Steen Library, Forest History Collections, Thompson Family Lum-ber Enterprises Collection P90T:292, Stephen F. Austin State Univ., Nacogdoches, Texas.

peckers. However, the species is extremely tenacious, and as long as exist-ing cavity trees survive—or potential cavity trees are present—individual groups of woodpeckers can often survive severe habitat loss. In fact, small populations of this species can be remarkably persistent because of the un-usual breeding system, as is discussed in Chapter 6. Although the details were not recorded, Red-cockaded Woodpeckers survived the catastrophic impacts of logging of the original forests in numerous locations.

Several characteristics of the cutting of the forests allowed the survival of Red-cockaded Woodpeckers. Most important was the patchy nature of the early logging activities. Some tracts survived the initial wave of log-ging for various reasons, and the woodpecker populations they contained remained intact. Remnant populations on the quail plantations of south-ern Georgia and the McCurtain County Wilderness Area in Oklahoma still persist in areas that were never extensively logged. Other areas, especially some longleaf pine habitats, were extensively tapped for pine resins (also termed pitch, gum, or oleoresin) to make naval stores such as turpentine and tar, used primarily in the construction and operation of wooden ships (Figure 4.3). While naval stores were produced in many areas, the Carolinas

FIGURE 4.3 Tapping of longleaf pines for the naval stores industry and manufacture of turpentine. Courtesy of U.S. Forest Service Archives.

were particularly prodigious producers, leading to the characterization of North Carolina as the Tarheel State. The value of these stands in the production of naval stores prevented, or at least delayed, the logging of significant areas of Red-cockaded Woodpecker habitat. In most areas, logging was not as thorough as with present practices, and significant numbers of trees were not cut. These were often old trees that had turpentine scars, were infected with red heart fungus, and were of little monetary value—precisely the trees required by Red-cockaded Woodpeckers for cavity excavation. The survival of these crucial trees, combined with minimal foraging opportunities, allowed the survival—at reduced densities—of numerous populations throughout the range of the species.

The conversion of cutover lands to agriculture prevented the survival of many woodpecker populations. However, those lands that remained partially forested and met the minimum requirements of Red-cockaded Woodpeckers, at least for the short term, allowed the survival of the species in numerous areas. Regeneration of forests by natural regeneration and planting, plus the abandonment and reforesting of agricultural lands, resulted in the slow recovery of Red-cockaded Woodpecker habitat through the middle of the century.

The Modern Era

The initial logging of the pine forests of the southeastern United States resulted in a major decrease in the Red-cockaded Woodpecker population. Populations survived in many areas and presumably increased as the habitat slowly recovered. After World War II, however, two factors combined to halt, and then reverse, the recovery of Red-cockaded Woodpeckers. First, logging practices changed from the "cut out and get out" practices of earlier decades to more sustainable approaches. In order to increase regeneration opportunities, the practice of leaving cull pines declined. It became standard practice to employ some type of site preparation following harvesting that removed any remaining trees and prepared the area for planting. Pines infected with red heart and other fungi were also selectively removed in many areas to reduce infection rates. Thus the relict pines, often with red heart infections, that allowed Red-cockaded Woodpeckers to survive the initial wave of logging were increasingly rare in the forest landscape.

Coincident with the intensification of silvicultural practices was the implementation of short rotation ages. The original forests contained variable numbers of old pines, 200–450+ years in age depending on the pine species. In the managed forests, concepts of economic maturity rather than physiological maturity determined the timing of successive harvests, and thus the age structure of the forests. If sawtimber was the end product, rotations of 35–70 years were typical; if pulp for paper and other products was the goal, rotations of 20–40 years were the norm. There was no perceived need for trees older than 70 years, so few were produced. The implementation of these two practices, removal of relict trees and short rotations, had a devastating impact on Red-cockaded Woodpecker populations. As these practices became widespread, Red-cockaded Woodpeckers rapidly disappeared or were reduced to small remnant populations throughout most of their original range.

The second factor that had a major impact on Red-cockaded Woodpecker populations was less obvious but equally devastating. Prior to the arrival of Europeans, Native Americans and natural sources of ignition caused the pine forests of the Southeast to burn frequently. European colonists initially continued these practices with minimal change in the fire regime. During the mid-1900s, changing views on the impact of fire on timber resources and wildlife, increasing development, potential liability considerations, and efficient fire suppression techniques resulted in a major change in the fire regime. Fire frequencies declined, and the seasonal occurrence of fire shifted from a summer peak to a peak during the winter and early spring, when control of prescribed fire was more easily maintained.

The ecological impact of the change in the fire regime was widespread. The reduction in the influence of fire on the pine forests of the region changed the ecological balance almost immediately. The forest structure that prevailed in much of the region with frequent and effective fire, a pine savanna with a diverse herbaceous understory, collapsed. Established pines survived, and a midstory of woody vegetation—previously suppressed by fire—developed rapidly and suppressed the herbaceous understory. In effect, the ecological rules were changed, and vast areas of fire-maintained pine forests began successional changes toward mixed hardwood forests. Many pine forests came to consist of a pine overstory and a mixed hardwood midstory—an unstable forest structure in which dominance of the pines is maintained by periodic harvesting and either intensive site preparation and replanting or the use of herbicides. Even in more xeric sites not suitable for most hardwoods, a dense midstory of scrub oaks developed below the pine canopy. The development of a dense hardwood midstory leads to territory abandonment by the Red-cockaded Woodpecker as discussed in Chapter 9, as well as declines in numerous other species of plants and animals that contribute to the unique biodiversity of the southeastern United States. Combined with the great reduction in area of pine forest, the effect of hardwood encroachment was devastating.

Additional impacts to future Red-cockaded Woodpecker population distribution are on the horizon. The rapidly developing concepts of "new forestry" point to the most severe alteration of natural forest ecosystems yet experienced. New forestry, which is rapidly being implemented on industrial forest lands across the southeastern United States, is economically driven and explicitly treats pine trees as an agricultural crop. Rotations for pulp production, already short, will be reduced even further. In addition, sawtimber rotations will be shortened through a variety of techniques, including pruning of trees to produce "sudden saw logs." The expanded use of genetically improved trees, herbicides to reduce competition from most other vegetation, and fertilization to increase growth and compensate for productivity declines resulting from biomass and nutrient losses in the resulting agricultural system all point to the nearly total loss of fire-maintained southern pine ecosystems on many thousands of hectares where some semblance of natural communities still persists. The important ecological role of fire, the complexity of which is only now beginning to be realized, will also be lost as new forestry expands on the landscape. The role of fire in reducing competing vegetation will increasingly be taken over by the intensive use of herbicides.

The Extent of Population Decline

The decline of Red-cockaded Woodpeckers was recognized early in this century. E. E. Murphey, in a chapter on the species for A. C. Bent's well-known series on the life histories of North American birds, stated that

> from many sections of the South where it was formerly common, the Red-cockaded Woodpecker has disappeared by reason of the ruthless destruction of pine forests by the lumbermen. When the large timber is cut out, the birds leave the locality and apparently do not return.

The continued decline of Red-cockaded Woodpecker populations has been tracked in the literature up to the present. Numerous references to the extirpation or decrease of local populations have been recorded.

It was not until 1971 that the first estimate of rangewide population was published. Jerry Jackson used published information, museum records, and information from numerous individuals familiar with the species in an effort to document the past and present distribution of the species and its current status. Although the data were incomplete, Jackson was able to develop maps showing counties with known records of Red-cockaded Woodpeckers prior to 1960 and during the period 1960–1971. These maps, although tentative, demonstrated that even though the species still existed across most portions of its historical range, it was no longer found in many areas where it once occurred. Jackson also compiled information on numbers of birds throughout its range and conservatively estimated that approximately 3,000 Red-cockaded Woodpecker groups survived. He admitted the limitations of his data, but speculated that probably no more than 10,000 birds actually survived. Based on current group sizes, Jackson's estimate translates into 4,000–4,500 groups, which represents only 0.3% of the 1.6 million groups we speculate may once have existed. Even if the original number was only a tenth of what we estimate, 97% of the original population of this species has been lost (see Figure 1.4).

Jackson's 1971 report on the status of Red-cockaded Woodpeckers documented serious declines and the scattered distribution of the surviving populations, most of which were small. His estimates have proven remarkably accurate, considering the difficulties involved in attempting to precisely estimate numbers of a small bird spread over millions of hectares. Subsequent surveys have raised the estimates somewhat, but the general conclusion remains intact: the Red-cockaded Woodpecker has been extirpated over much of its former range, and surviving populations are mostly small and isolated. A common species once distributed fairly continuously in mostly large populations was now rare and patchily distributed in mostly small populations. Jackson published a revised survey in 1978. He estimated 3,473 groups of woodpeckers throughout the species range. Eighty-four percent were estimated to occur on federal land.

In 1983 Michael Lennartz and several coauthors, including Jackson, published a survey of the status of Red-cockaded Woodpeckers on federal lands (national forests, national wildlife refuges, and military reservations). Their median estimate was 2,677 groups plus approximately 300 groups in areas not specifically included in their survey. This figure was very close to Jackson's 1978 maximum estimate of 2,904 groups on federal lands.

In 1989 Ralph Costa and Ron Escaño, biologists with the U.S. Forest Service, published a compilation of Forest Service records through 1986. Costa and Escaño reported a total of 2,115 groups of Red-cockaded Woodpeckers on national forest lands. This total is very close to the Lennartz estimate of 2,121 groups on national forest lands. Of the 21 populations surveyed, 4 were estimated to be stable, 7 stable or decreasing, and 10 decreasing. Five national forest populations were extirpated between 1970 and 1986, which represents a loss of 19% of the populations that still remained in 1970.

In 1995, Fran James attempted to assess the status of Red-cockaded Woodpeckers across their entire range (Table 4.2). She also estimated that these numbers represented 4,430 breeding groups in the early 1980s, but that the number had declined to 3,420 by 1990, a loss of 23% over about 10 years (Figure 4.4). We attempted to update these population estimates for 1998, which revealed increases in some populations and decreases in others (Table 4.2).

The difficulties of estimating the total population of Red-cockaded Woodpeckers are substantial. A major difficulty that affects all of the above surveys is that significant portions of the data used to compile the estimates are out of date, often by several years. Given the rate of decline observed in many Red-cockaded Woodpecker populations in recent years, this can result in overestimates of total population. The individual population surveys that form the basis of the above studies were often incom-

Table 4.2. *Number of Active Red-cockaded Woodpecker Clusters by Geographic Area*

State	Locality	Early 1980s	1990	1998
		Number of Active Clusters		
Alabama	Talledega National Forest	175	127	126
	Conecuh National Forest	15	13	14
	Bankhead National Forest	8	0	0
	Other	3 + ?	16 + ?	4 + ?
Arkansas	Ouachita National Forest	25	16	14
	Felsenthal National Wildlife Refuge	25	28	15
	Other	239	88 + ?	35 + ?
Florida	Apalachicola National Forest	510	590	630
	Eglin Air Force Base	243	208	280
	Blackwater State Forest	50	29	18
	Osceola National Forest	44	44	54
	Withlacoochee State Forest	21	39	47
	Big Cypress National Preserve	18	36	40
	Ocala National Forest	41	5	13
	Avon Park Bombing Range (DOD)	25	8	21
	Other	184	157	123 + ?
Georgia	Fort Benning (DOD)	111	176	187
	Fort Stewart (DOD)	138	128	189
	Red Hills Hunting Plantations	144	180	177
	Okefenokee National Wildlife Refuge	48	45	26
	Piedmont National Wildlife Refuge	24	23	35
	Oconee National Forest	22	12	18
	Fort Gordon (DOD)	35	3	2
	Other	31	29	8 + ?
Kentucky	Daniel Boone National Forest	8	4	7
Louisiana	Kisatchie National Forest	?	340	314
	Fort Polk/Peason Ridge	?	63	45
	Other	?	61	17 + ?
Mississippi	Bienville National Forest	120	86	106
	Homochitto National Forest	50	27	40
	Noxubee National Wildlife Refuge	23	17	37
	DeSoto National Forest	30	16	16
	Other	7 + ?	8	6 + ?

Table 4.2. *Continued*

State	Locality	Number of Active Clusters		
		Early 1980s	1990	1998
North Carolina	North Carolina Sandhills/ Fort Bragg and vicinity	593	371 + ?	562
	Croatan National Forest	43	47	58
	Camp Lejeune (DOD)	26	34	48
	Sunny Point Military Ocean Terminal	7	5	6
	Other	32	8	?
South Carolina	Francis Marion National Forest	514	315	319
	Carolina Sandhills National Wildlife Refuge	122	125	127
	Fort Jackson (DOD)	18	15	13
	Savannah River Plant (DOE)	17	7	29
	Other	185	156	340 + ?
Oklahoma	McCurtain County Wilderness Area	31	15	11
Tennessee	Other	6	1	0
Texas	Sam Houston National Forest	172	133	168
	Angelina and Sabine National Forests	66	28	49
	Davy Crockett National Forest	46	26	48
	Jones State Forest	?	15	14
	Fairchild State Forest	?	12	4
	Other	16 + ?	51	26 + ?
Virginia	Other	12	5	5

Sources: Data are adapted from Fran James 1995. Recent figures are from U.S. Fish and Wildlife Service, U.S. Forest Service, and the authors' records, unpublished data.

Note: "Other" entries are totals for all areas within a state other than those listed. Incomplete data are indicated by "?".

Abbreviations: DOD, Department of Defense; DOE, Department of Energy.

plete, especially for the larger populations, resulting in underestimates of total population numbers. On some forests in recent years, these sources of error have remained even with the increased attention directed toward the status of Red-cockaded Woodpeckers. Despite the difficulties, these studies have provided biologists with increasingly accurate estimates of total population size. We can reasonably conclude that the remaining population had declined to less than 6,000–7,000 groups by the early 1970s, less than 5,000 by the early 1980s, and less than 4,000 by 1990.

The magnitude of the declines in recent decades, combined with the fragmented nature of the remaining populations and massive habitat degradation, clearly indicate that the Red-cockaded Woodpecker faced extinction within a time frame measured in decades, not centuries. The decision

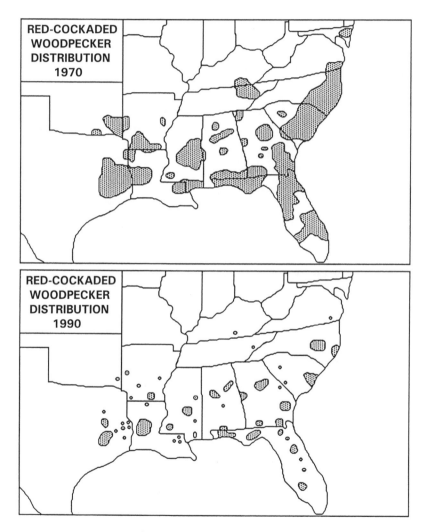

FIGURE 4.4 Approximate distribution of the Red-cockaded Woodpecker in 1970 and 1990, showing a reduction in the range of the woodpecker over the two-decade period.

to list the species was not premature. Timely listing of the species provided an opportunity to recover the species on remaining habitat without resorting to captive breeding and other extremely costly approaches.

The Current Population

The current population of Red-cockaded Woodpeckers is widely scattered over much of the original range. Table 4.2 lists the existing populations by state and provides some supporting information on current population trends. Four populations contain in excess of 300 groups, but 59 of the 73 populations (81%) contain fewer than 75 groups. Populations are generally separated by unsuitable habitat and are isolated from other populations genetically and demographically. Due to a complex of factors detailed in Chapter 9, most populations have decreased, often severely, in recent years. Habitat deterioration and fragmentation and isolation of populations are primarily responsible.

A major dichotomy exists between the extent of Red-cockaded Woodpecker decline on public lands and on private lands. Populations on public lands, the large majority of which are in national forests, have generally fared much better in recent decades. The most important factor leading to this difference was the longer timber rotation age generally practiced on public lands in the recent past, typically 60–80 years on national forests. A large portion of private industrial land in the southeastern United States is managed for pulp production and has very short rotation ages, typically 20–40 years. Consequently, potential cavity trees have been essentially eliminated from most of the private industrial land. On public lands, significant numbers of potential cavity trees still exist, although they are typically of minimum age for cavity trees. Existing cavity trees have also received better protection on public lands, slowing the decline of the woodpecker populations.

Only two substantial populations remain on private land. A significant portion of the North Carolina Sandhills population, which numbers nearly 400 groups, lives on private horse farms, on golf courses, and in residential areas (Figure 4.5). The Red Hills population in southern Georgia, numbering nearly 200 groups, is found on private plantations managed primarily for quail hunting.

This sobering summary of the recent information on Red-cockaded Woodpecker populations clearly supports the decision by the U.S. Fish and Wildlife Service to include the Red-cockaded Woodpecker on the Endangered Species List when it was first initiated in 1973. Although the total

FIGURE 4.5 Red-cockaded Woodpeckers are quite tolerant of human activities, as shown by the close proximity of its cavity trees to buildings in Louisiana. Photo by R. N. Conner.

population—approximately 10,000 individuals—was substantial compared to many other listed species, the reduction from the original population was extreme (we estimate in excess of 99%) and the survival of the species was severely threatened. The distribution of the remaining woodpeckers, which in many small subpopulations were genetically and demographically isolated from each other, was a critical factor. Each subpopulation behaved independently, and as Table 4.2 demonstrates, a large majority of the populations have been declining. Consequently, the risk of extinction was greater than the total population size might indicate.

Recently intensified management efforts have considerably improved the outlook for survival of the Red-cockaded Woodpecker. The primary causes of continued declines are now known. For the short term—the next 25 years, perhaps—intensified management, including hardwood midstory reduction (preferably through aggressive use of growing-season fire), artificial cavities, and translocation of birds, is absolutely essential to stabilize and reverse recent population declines. Such a program has been initiated on national forests in Texas and has reversed severe population declines on all four national forests, and similar successes have occurred in North Carolina (see Chapters 10, 11). Similar results are possible with any remnant population, given a sufficient land base and appropriate management decisions, and efforts to increase woodpecker populations are now under way in many locations. Recent success reintroducing pairs of Red-cockaded Woodpeckers in Texas demonstrates that reestablishment of extirpated populations is also feasible. It thus appears that the population declines that continued through the 1970s, 1980s, and 1990s can be halted or even reversed in the 2000s. This is discussed more thoroughly in Chapters 10 and 11.

Thus the state of Red-cockaded Woodpecker populations at the time of this writing is that they are stable or even increasing where appropriate, intensive management is being practiced, and are declining where it is not. This will no doubt continue to be the case over the short term. For the long term, maintaining viable populations without intensive woodpecker management will require management of the fire-maintained pine ecosystems. The primary requirements are, first, an adequate density of older pines for cavity excavation and foraging, and, next, a sparse hardwood midstory, preferably maintained by growing-season fire. Ideally, this would be achieved in the context of a fire-maintained ecosystem with a diverse and well-developed herbaceous layer. Achieving these requirements will take time, but will greatly reduce the necessity for intensive management of the woodpeckers themselves. Recent changes in management policy in many, but not all, areas are conducive to achieving these requirements. An additional requirement is the establishment of populations of sufficient size to preclude demographic and genetic problems, as detailed in Chapter 6. Our views on future population trends are expressed in Chapter 11.

CHAPTER FIVE

Cavity Trees in Fire-Maintained Southern Pine Ecosystems

Cavity Construction in Live Pines

The frequent fires that swept through the southern pine forests greatly reduced the abundance of hardwoods and dead trees (snags) relative to their abundance in hardwood stands in riparian areas and moist bottomlands. As a result, sites for nesting and roosting in upland areas were scarce for many woodpecker species. Live pines and pine snags became the primary source of potential nest sites for woodpeckers. Because fire also takes a high toll on pine snags, even this resource was likely not abundant for woodpecker use. The tops of some senescent pine trees occasionally die, but these potential cavity sites are much rarer in pines than in hardwoods.

Most species of eastern North American woodpeckers primarily nest in snags and dead portions of live hardwoods, which often have large dead branches and limb stubs that provide an avenue for fungal decay to enter the heartwood of the bole. Only a few species of eastern woodpeckers (mainly Pileated and Hairy woodpeckers) can excavate a cavity into living portions of hardwoods. In such cases, the birds must chisel through 4–8 cm of living, undecayed sapwood (xylem tissue) before reaching decayed heartwood, where the main chamber of the roost or nest cavity is excavated. Fungal decay softens wood tissue and reduces the effort required for cavity excavation. Consequently, in oaks, hickories, and maples of North American forests, the presence of heartwood-decaying fungi (primarily *Spongipellis pachyodon*) is a prerequisite; otherwise the heartwood in the center of hardwoods is too hard for cavity excavation. Many small woodpeckers, such as the Downy Woodpecker, also require dead, decayed sapwood through which to excavate the cavity entrance.

WOOD STRUCTURE AND PINE RESIN

Although wood in southern pines is softer than that of hardwoods, mature southern pines typically have 8–15 cm of undecayed, living sapwood. Pines also produce oleoresins (pine gum) and typically ooze copious amounts of these resins when wounded. Thus the combination of sapwood thickness and pine gum makes cavity excavation in living pines much more difficult than in snags—so difficult, in fact, that the Red-cockaded Woodpecker is the only southern woodpecker that has adapted to the exclusive use of live pines as cavity sites.

The sapwood of a pine is composed of living, undecayed xylem tissue. Excavation into the sapwood immediately produces sticky oleoresins, created in the needles and other differentiated tissue and transported throughout the tree. Wounding caused by pecking, or other injury, stimulates gum flow through resin canals to the injury site. This wounding response has most likely evolved as a defense against bark beetle attack. Pine gum is produced and, if sufficiently abundant, can "pitch" the beetles out, preventing a successful infestation.

Thus when a woodpecker begins to make a cavity (cavity start), resin flowing from the wound typically produces an icicle-like formation extending down the bole of the pine. The creamy white appearance of the "icicle" results when the initially clear resin crystallizes and multiple layers harden on top of older crystallized resin (Figure 5.1).

Copious resin flow at a cavity start can stop the woodpecker from excavating. Resin eventually will stop flowing from the initial wound, however, when the resin saturates the exposed xylem and crystallizes. The hardened resin blocks resin canals at the wound and serves as a scab to prevent disease (fungus) from growing into the exposed xylem tissue. Resinosis, saturation, and crystallization of resin must occur before Red-cockaded Woodpeckers can resume excavation of the cavity entrance tunnel. As the woodpecker excavates crystallized resin and newly exposed sapwood to deepen the entrance tunnel, resin again flows to the wound and impedes progress. These excavate-and-wait cycles continue until the woodpecker reaches the heartwood of the pine. Thus, excavation time is directly affected by the thickness of the sapwood and the ability of the pine to produce resin.

Pine resin poses a clear danger to Red-cockaded Woodpeckers. We have observed several instances where Red-cockaded Woodpeckers became stuck in a pool of resin that formed in a cavity entrance tube. This can happen during excavation or during use of a completed cavity whose entrance tunnel was improperly constructed. The typical cavity entrance tube rises

FIGURE 5.1 Red-cockaded Woodpecker cavity start showing the formation of the resin "icicle" below the excavation
site. Photo by Dan Saenz, U.S. Forest Service.

slightly as it goes through the sapwood toward the nest chamber. In addition, the lower outside lip of the entrance is beveled downward. This construction design, if correctly excavated, prevents resin flow into the cavity
and permits any that oozes from the sapwood during excavation to flow
outward and down the bole of the tree. Droplets of gum that form at the top
of the cavity entrance are regularly removed by the woodpeckers. Roosting
Red-cockaded Woodpeckers frequently will perch in their cavity entrance
tubes prior to emerging in the morning or when they hear a noise that war-
rants visual investigation. If a bowl-like hollow is created in the entrance
tube, resin can accumulate in it. As the birds lie in the entrance, resin can
seep into their feathers. If a bird cannot tear loose, it will struggle until
exhausted and eventually die (Figure 5.2). Birds also have been trapped in
resin within chambers of cavities constructed by humans where the chamber breached the sapwood of the pine. Cavity excavation is thus a dangerous
procedure that requires precise behavioral adaptation.

The heartwood, where the vertical cavity chamber is excavated, is composed of dead xylem tissue and does not actively transport pine resin (Figure 5.3). Consequently, excavation can proceed more quickly. A cavity
chamber must be excavated in heartwood because resin will flow into the
chamber if sapwood is breached. When finished, the cavity chamber is approximately 23 cm from top to bottom, 10–14 cm front to back, and about

FIGURE 5.2 A Red-cockaded Woodpecker that died as a result of getting stuck in the pooled pine resin in its cavity entrance tube. Photo by R. N. Conner.

10 cm from side to side (Figure 5.4). Cavity entrance tunnel length can vary from 3 to 18 cm and is primarily a function of the sapwood thickness. Often the entrance tube extends several centimeters into the heartwood before the cavity chamber is made. Over years of use, the inside dimensions are frequently enlarged. There are reports of auxiliary chambers that extend above the original cavity chamber, which might allow two woodpeckers to roost in such cavities. We have never seen excavated chambers above cavities. Decay occasionally is so extensive above a cavity, however, that a small chamber is produced solely by wood decay.

The length of time required for Red-cockaded Woodpeckers to excavate through the living xylem tissue primarily depends upon the ability of the pine to produce resin and upon sapwood thickness. It also appears to depend on the consistency of the construction effort. Excavation of the entrance tube of a cavity in the sapwood can take several months to a few years, or many years if construction is only intermittent. Some pines can produce copious amounts of resin and "pitch" the woodpecker out in a fashion similar to what they do to bark beetles. Young pines generally produce more resin than older pines. Because a pine's ability to produce resin is less during the colder weather, excavation through sapwood might be accomplished more easily during winter months than during spring, summer, and fall. Most cavity excavation, however, takes place during and after the nesting season, when resin flow is the highest but resin viscosity is the least. This suggests that the woodpeckers prefer to contend with rapidly flowing, runny pine gum rather than thick, rapidly congealing gum.

Heartwood structure can affect cavity excavation. Growth rings produced by the cambial layer of pines during the growing season consist of

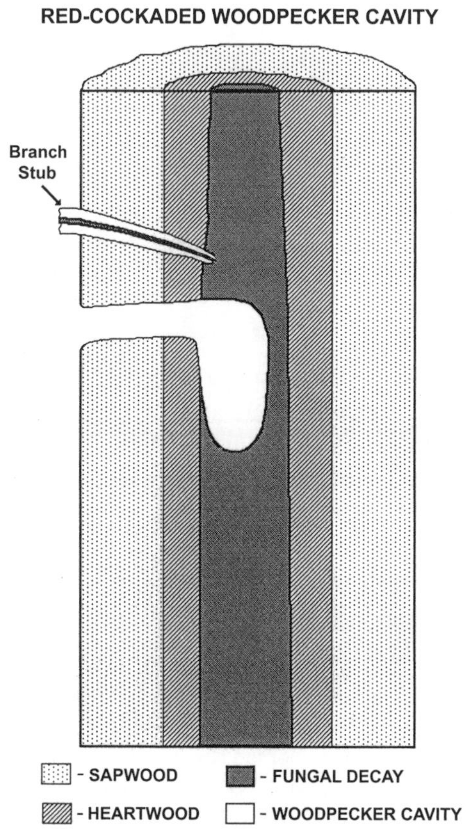

RED-COCKADED WOODPECKER CAVITY

FIGURE 5.3 Diagram of a vertical section of a Red-cockaded Woodpecker cavity showing the heartwood, sapwood, fungal decay column, entrance of fungus through branch stub, and the general shape of the cavity and entrance tube. Computer drawing by R. N. Conner.

FIGURE 5.4 Vertical section of a Red-cockaded Woodpecker cavity in a long-leaf pine tree. Note that there is no fungal decay in this cavity tree, indicating that it may have taken more than 6 years for the woodpeckers to excavate the cavity. Photo by R. N. Conner.

wood of two densities. Xylem tissue produced during the spring (spring-wood), when water is typically more abundant, is less dense and lighter in color than xylem produced during the summer (summerwood). Thus growth each year produces a ring of light-colored, softer springwood and a ring of darker, harder summerwood. If the pine is growing vigorously, both spring and summerwood will be thick. Suppressed growth caused by crowding or prolonged drought produces tightly packed rings, sometimes so thin that magnification is required to count pairs of rings to determine tree age. Excavating horizontally, across the grain of dead xylem tissue, is easier than going with the grain and excavating vertically, downward into the heartwood. As most wood fibers are oriented vertically, vertical excavation in the heartwood tends to spread the wood apart, allowing it to spring back when the bird's bill is withdrawn. Excavating across the grain of the wood horizontally tends to cut or chip the wood.

Conner and coworkers have shown that Red-cockaded Woodpeckers preferentially select pines with thin sapwood and large diameters of heartwood. They measured the diameter of heartwood occupied by cavity chambers by measuring the horizontal distance from 2 cm below the entrance tube to the back of the cavity. Cavity chambers used an average 11.2 cm of the heartwood, and measurements ranged from 9.4 to 17.2 cm. Most cavities used between 10 and 12 cm of the heartwood. The larger cavities had been in use for more than 5 years, and the cavity chambers had been expanded by Red-cockaded Woodpeckers during repeated years of use. The expansion of cavity chambers by Red-cockaded Woodpeckers typically enlarged the back portion of the cavity (horizontally elliptical) and not the lateral portions of the chamber. The chambers had not been enlarged by other species of woodpeckers. Thus, to house the largest cavities, the heartwood must be at least 18–20 cm in diameter.

FUNGAL DECAY OF THE HEARTWOOD

Excavation of the vertically oriented wood fibers within the heartwood can be facilitated by the presence of a heartwood-decaying fungus (heart rot). Presence of heart rots can substantially decrease the time and energy required to excavate a cavity chamber by decreasing wood hardness and causing the wood to fracture when pecked, permitting easy removal of excavated wood. Hence, cavities in pines with thin sapwood, low resin production, and extensive heart rot probably account for cavities being excavated in only a few months. Cavities completed in pines with thick sapwood, copious resin flow, and no heartwood decay may account for the upper limits of construction time.

Although red heart fungus is the most common heart rot found in pines used for cavity trees, other species of wood-decaying fungi have been detected (*Phaeolus schweinitzii, Lenzites saepiaria, Lentinus lepidius,* and *Phlebia radiata*). Red heart fungus enters the heartwood of pines (Figure 5.5) by growing through the heartwood of broken, exposed branch stubs. The branch must be at least 5 cm in diameter to contain sufficient heartwood for successful infection. After growing within the pine's heartwood for a decade or more, the fungus forms a cinnamon-colored, spore-producing conk on the bole of the pine. Red-cockaded Woodpeckers will often excavate a cavity under or near the fungal conk, suggesting that the woodpeckers may visually orient on the conk to find decay. Rudolph and coworkers tested this hypothesis, however, and found no experimental evidence of such behavior. Occasionally the cavity entrance tube is excavated through the dead branch stub that initially allowed fungal access to the heartwood.

Research by Conner and Rudolph, as well as studies by Robert Hooper, with the U.S. Forest Service's Southern Research Station in Charleston, South Carolina, and by Jerry Jackson, all with the collaboration of their colleagues, indicates that Red-cockaded Woodpeckers preferentially select

FIGURE 5.5 Cross section of a loblolly pine showing the decayed heartwood and column of fungal tissue connected to the fungal conk on the exterior of the pine. Photo by R. N. Conner.

pines with heartwood decay for cavity excavation. To a large extent, this may be due to the birds' selecting old pines for cavity excavation (see below), coupled with the high incidence of fungal infection in old trees. But the woodpeckers appear to be able to detect and select pines that are infected with red heart fungus as well. Hooper examined longleaf pine cavity trees on the Francis Marion National Forest, in South Carolina, and the Osceola and Ocala National Forests, in Florida. He compared pines selected by Red-cockaded Woodpeckers with adjacent pines of similar size and age. Pines selected by woodpeckers had a significantly higher red heart fungus infection rate than unused pines. Hooper demonstrated that the woodpeckers can detect a pocket of decay within the bole of the pine and excavate into the small area of decayed heartwood. Previous research by Conner in Virginia had suggested that other species of woodpeckers using hardwoods were able to detect decay by percussing the hardwood's bole or branch and listening for a particular resonance. Hardwoods, however, typically have much thinner sapwood than southern pines, which may make detection of decay easier. Thus it may be more difficult for Red-cockaded Woodpeckers to detect decay within pines. Other means of detection may be necessary.

Fungal decay oxidizes (softens) the springwood component of the heartwood first. Red heart fungus decays the lignin, or dark tissue, producing wood that has a speckled appearance created by off-white cellulose contrasting with the cinnamon-colored fungal tissue. Typically, 10–20 years must pass before summerwood becomes extensively decayed. Eventually the fungus produces a fruiting conk on the bole of the pine to disseminate its spores. Fungal conks can be as small as a quarter or as large as a bread box (20 × 35 cm). Red heart fungus adds a new layer of growth each year on the bottom of the conk (hymenium), which can reach 30 years in age (Plate 10).

Red-cockaded Woodpeckers may excavate in pines lacking decayed heartwood. It has been suggested that the woodpeckers tend to select pines that contain "juvenile" wood at cavity height in such instances. Juvenile wood is produced when young trees grow very rapidly, and it is structurally weaker than wood produced in subsequent years. Only pines on high-quality soils would be able to produce juvenile wood at heights suitable for cavity construction. Hooper has suggested that the Red-cockaded Woodpecker's first preference is for pines with heartwood decay; second, for sound heartwood grown rapidly; and third, for sound heartwood grown at average rates.

TIME REQUIRED FOR CAVITY EXCAVATION

The time used by Red-cockaded Woodpeckers to excavate cavities is significantly longer in longleaf pines than in either loblolly or shortleaf pines. Conner and Rudolph observed that excavation time in Texas averaged 6.3 years in longleaf pines, which was 2–3 times greater than average excavation times required in loblolly and shortleaf pines. Three cavity starts in longleaf pines were active for more than 9 years; these cavities still had not been completed by the end of the ninth year.

Sergio Harding, a graduate student working with Walters, now at Virginia Tech, estimated that in the North Carolina Sandhills excavation required 13 years on average in longleaf (9 years of which is spent excavating the entrance tube) and 4 years in loblolly. Intermittent construction, with intervals as long as 9–12 years between bouts of excavation activity, contributed to the long excavation times in North Carolina. Although some cavities were still unfinished after 16 years, others were completed from start to finish in only 2 years.

As mentioned above, the presence of red heart fungus can significantly reduce the time required for excavation of the heartwood of pines. Fifteen longleaf pines in Texas that had decayed heartwood required an average of 3.7 years to excavate, whereas 10 longleaf pines without decay required an average of 5.0 years.

If cavities are in short supply prior to the nesting season, Red-cockaded Woodpeckers will spend a large portion of the day excavating in order to have a cavity in which their eggs will be laid. In extreme cases, they may excavate the entire cavity chamber in a matter of weeks. At the other extreme, the birds may spend years constructing an entrance tunnel, excavating a little each summer, perhaps skipping some years altogether. The length of time required to construct a cavity is a key factor shaping other aspects of the species' biology (see Chapter 6). The unusually long time required, as well as the variability therein, seems to arise primarily from construction of the entrance tunnel, rather than the cavity chamber. The presence or absence of heartwood decay, however, also has a significant effect on the time required to excavate the cavity chamber, as discussed above.

The control of excavation behavior is one of the remaining mysteries about this species. Certainly, variation in the response of the tree, the internal condition of the sapwood and heartwood, and dangers posed to the birds by the sticky resin are important factors, but there may be social factors as well. Groups that do not produce fledglings appear to devote more time to excavation. Harding found that although several group members may work on the same cavity, there is often a primary excavator. Whether this is the breeding male, breeding female, or a helper is unpredictable. Of course, the birds that begin excavation of a cavity seldom live to see its completion.

Resin Wells

Behavioral adaptations have enabled Red-cockaded Woodpeckers not only to circumvent the difficulties caused by a pine's resin system but to use it to their advantage. As a cavity nears completion, Red-cockaded Woodpeckers begin to make small (2–4 cm diameter) excavations into the cambium around the cavity entrance. As the cavity is used over the years, more of these small excavations will be made, often extending 2–3 m above and below the cavity entrance. These special excavations, termed resin wells (Figure 5.6), are pecked daily by the woodpecker giving them a reddish appearance and causing fresh resin to flow down the bole above, below, and beside the cavity entrance (Figure 5.7). Eventually a large portion of the bole becomes white, as the originally clear pine resin crystallizes.

Red-cockaded Woodpeckers also peck and pry the loose bark off of the

FIGURE 5.6 Close-up of an active resin well excavated by Red-cockaded Woodpeckers. The woodpeckers peck at these wells daily, which results in a continuous flow of clear, fresh pine resin. Photo by R. N. Conner.

boles of their cavity trees, creating a relatively smooth surface. It is likely that natural selection favors pairs that exhibit this behavior, because cavity trees with smooth boles are more difficult for rat snakes to climb than pines with intact bark. Also, resin flowing downward over a smooth bole coats the bark more evenly, increasing the effectiveness of the resin barrier. Resin becomes particularly sticky when direct sunlight warms it. Cavities may be oriented in any direction, but throughout the South there is a significant tendency for cavities to be excavated on the southwestern side of cavity trees. As a result, the entrances receive direct sunlight in the late afternoon, making the gum flow extremely sticky just before evening when the woodpeckers go to roost.

Cavity trees currently used by Red-cockaded Woodpeckers are called active cavity trees and have fresh clear resin flowing from resin wells. Cavity trees that are no longer in use, and therefore lack active resin wells, are termed inactive. In the absence of direct observation of cavity use (dawn or dusk visits to cavity trees), the active/inactive status of a cavity tree is best judged by whether the bark around resin wells has a reddish appear-

FIGURE 5.7 An active Red-cockaded Woodpecker cavity tree. Resin flowing down the
bole of the pine provides a barrier against rat
snakes. Cavity entrances are usually oriented
toward the southwest, where warmth from the
afternoon sun increases the stickiness of the
pine gum. Photo by D. C. Rudolph.

ance, which would indicate recent pecking, and by whether there is fresh
clear resin flowing down the bole of the pine from the resin wells. An overall reddish appearance of the bole resulting from bark scaling is an additional indicator of active cavity trees.

THE EVOLUTION OF RESIN WELLS

How the habit of maintaining resin wells evolved is an interesting question. The woodpeckers probably made cavities and used live pines for nest
and roost trees prior to the evolution of resin well excavation. Once made,
cavities were probably used for many years, as they are today. Woodpeckers using snags for nest sites do not have to contend with attempts by the
tree to grow over and seal wounds, which is what the cavity represents to
the tree. But species nesting in live wood must contend with this response.
Thus, initial pecking at areas around the cavity entrances may have been
required in live pines to prevent the pine from closing the cavity entrance
with annual growth. Pecking primarily during the tree's growing season
in late spring and early summer, which corresponds to the nesting season,
would have produced copious resin flow. Thus a second advantage of this

pecking could have arisen. Through time, woodpeckers that pecked more, and in more areas on the bole of the pine, likely had higher nestling survival rates than other family groups due to the effects of resin on predators. This selective advantage could have been the driving force behind the evolution of the resin well maintenance observed today.

RESIN WELLS AS A BARRIER AGAINST RAT SNAKES

Regardless of how the behavior evolved, its primary advantage today is clear. Although early literature suggests that resin wells and the white appearance of cavity trees serve as territorial markers for the woodpeckers, the majority of data indicate that the combination of resin flow and smooth bark created by Red-cockaded Woodpeckers serves primarily as a barrier against predatory rat snakes, including the various subspecies of *Elaphe obsoleta* and the corn snake (*E. guttata guttata*). Rat snakes are probably the major predator of cavity nesting birds within the range of Red-cockaded Woodpeckers, and numerous reports exist of rat snake predation on avian eggs and young. Resin warmed by the afternoon sun on boles of cavity trees would be particularly effective against rat snakes that hunted at dusk.

Rat snakes are agile tree climbers and can easily climb pines by wedging ventral scales between bark furrows and protrusions while climbing (Plate 11). Rudolph, Conner, and colleagues have shown that small snakes are better climbers than large ones, because they fit into bark furrows more easily and weigh less and thus break off bark protrusions less often. The woodpeckers benefit from their own bark-flaking behavior by creating a smoother tree bole that provides rat snakes smaller furrows and fewer protrusions. The benefits of the sticky resin flow become obvious when snakes attempt to climb cavity trees. As they encounter the fresh resin, they arch their bodies away from the bole of the cavity tree because the resin accumulates between their overlapping ventral scales. Resin rapidly prevents normal scale operation and thus climbing; rat snakes soon fall to the ground after encountering sticky resin, or turn around and climb down. Experiments conducted by Rudolph, Conner, and colleagues demonstrated that snakes contaminated with resin after an initial climbing attempt on an active cavity tree were unable to climb any tree subsequently that day. There is also some indication from observations by Jackson that the phenols in pine resin may have a toxic effect on rat snakes.

The socially dominant, breeding male Red-cockaded Woodpecker typically selects the newest cavity in the cavity tree cluster for his roosting tree. His cavity is typically used as the nest tree during the subsequent breeding season. The pine the breeding male selects as his cavity tree typically produces significantly more resin than active cavity trees used for roosting by other members of the woodpecker group. Thus the cavity selected by the breeding male for his roost and nest site likely has a better resin barrier to protect him and his offspring against rat snakes than cavities used by other group members.

The resin barrier created by Red-cockaded Woodpeckers is extremely effective when working properly and serves well to prevent snake predation. However, if resin wells are not regularly excavated, or if the cavity tree's ability to produce resin dwindles, predation is a distinct possibility. Jackson observed rat snake predation on a nestling Red-cockaded Woodpecker after a large chunk of white crystallized resin had broken off, leaving an area on the bole near the cavity entrance free of fresh resin. We have often seen rat snakes make successful climbs to cavities where there is no longer active pecking at resin wells. Such cases are the exception, however. In many populations, the rate of nest predation is remarkably low for a cavity-nesting bird at this latitude, and we suspect that where the predation rates are higher, the birds are using cavity trees whose resin flow is less than optimal.

The shortleaf pine forest of the Ouachita National Forest in Arkansas is one area that may have unusually high nest predation rates. Joe Neal and others with the University of Arkansas measured how frequently rat snakes attempted to climb cavity trees on this forest. They monitored 13 cavity trees by attaching netting to the trees' bases and intercepted rat snakes climbing 11 (85%) of the trees. Attempts by rat snakes to climb cavity trees coincided with the peak of the Red-cockaded Woodpecker nesting season. Neal also observed that rat snakes were more likely to attempt to climb cavity trees than similarly sized pines that did not contain a cavity.

Red-cockaded Woodpeckers are not the only birds to use pine gum in the vicinity of their nest cavity. Red-breasted Nuthatches (*Sitta canadensis*) are well known for this behavior, which in this species is presumed to be a deterrent to competitors for the nuthatch's cavity. It may instead function to reduce predation. Although the resin barrier caused by Red-cockaded Woodpeckers is effective against rat snakes, data demonstrating that it is a deterrent against other cavity users are absent.

Cavity Competition and Kleptoparasitism

Many species of birds will use Red-cockaded Woodpecker cavities. In fact, secondary and even other primary cavity nesters in fire-maintained pine ecosystems are often highly dependent on cavities constructed by Red-cockaded Woodpeckers, as only they can excavate cavities in live pines. Thus Red-cockaded Woodpeckers are a keystone species within fire-maintained southern pine ecosystems because they are the cavity pathfinder for many other wildlife species. This dependence, however, may have negative effects on Red-cockaded Woodpeckers. Tufted Titmice (*Baeolophus bicolor*), Great Crested Flycatchers, White-breasted Nuthatches (*Sitta carolinensis*), and Eastern Bluebirds are frequently observed using Red-cockaded Woodpecker cavities. These avian species will use cavities that are intact (unenlarged by other woodpeckers) in spite of copious resin flow from resin wells, although typically they use cavities abandoned by the woodpecker. John Dennis reported that an Eastern Bluebird and a Pine Warbler became stuck by resin to the boles of cavity trees in South Carolina, but isolated instances like these probably do not discourage frequent use by such species. Red-cockaded Woodpeckers have some capacity to defend their cavities against species such as these, although they are not highly effective in this, and conflict is presumably reduced by the different and wider range of cavities acceptable to the other species. If these secondary cavity users acquire a Red-cockaded Woodpecker cavity, they do not alter it, and Red-cockaded Woodpeckers may later reclaim it.

INTERACTIONS WITH OTHER WOODPECKERS

Other woodpeckers, like the Red-bellied (*Melanerpes carolinus*) and Red-headed (*M. erythrocephalus*) woodpeckers, Northern Flickers (*Colaptes auratus*), and particularly Pileated Woodpeckers, are a different story. They not only usurp Red-cockaded Woodpecker cavities but typically damage the cavities in the process. These woodpeckers are larger than Red-cockaded Woodpeckers and thus Red-cockaded Woodpeckers cannot defend their cavities from them. In addition, the other species must excavate the entrance tube to a greater diameter before they can gain access and enlarge the cavity chamber. Pileated Woodpeckers are the largest of these species and account for the major portion and extent of the damage (Figure 5.8). Other woodpeckers may damage the cavity irreparably, but often they enlarge the entrance only slightly, so that Red-cockaded Woodpeckers will still use it once the other species abandons it. Such is not the case with Pileated Woodpeckers (Plate 12). They frequently will enlarge only the entrance tunnel, but to such an extent that it renders the cavity unacceptable for Red-cockaded Woodpeckers to use for nesting or roosting. Red-cockaded Woodpeckers, like most primary cavity nesters, prefer the mini-

FIGURE 5.8 Pileated Woodpeckers frequently enlarge Red-cockaded Woodpecker cavity entrances, making them unsuitable for the endangered woodpecker to use. Fully enlarged cavities do serve as nest and roost sites for other large secondary cavity users. Photo by R. N. Conner.

mum entrance diameter that allows them access to the cavity. Cavities usually are abandoned if the enlarged entrance tunnel exceeds 7 cm in diameter. Entrance diameters preferred by Red-cockaded Woodpeckers range from about 4 to 5 cm.

We have observed several instances in old cavity trees where Pileated Woodpeckers have excavated a complete cavity by enlarging a Red-cockaded Woodpecker cavity site and successfully fledged young Pileated Woodpeckers. The heartwood in such trees is extremely large in diameter (>25 cm); otherwise resin would seep into the cavity chamber from the damaged sapwood. Encountering heartwood with an insufficient diameter or an excessive amount of resin flow when enlarging the entrance tube probably deters Pileated Woodpeckers from completely enlarging more Red-cockaded Woodpecker cavities into full cavities. Often Pileated Woodpeckers destroy the cavity entrance by enlarging it and then abandon the cavity. Alternatively, Pileated Woodpeckers may be attracted to Red-cockaded Woodpecker cavities for reasons other than cavity excavation; perhaps they are viewed as potential foraging sites.

Once Red-cockaded Woodpecker cavities are enlarged by Pileated Woodpeckers, many other species of cavity-nesting birds and mammals can use the cavities. Fox and gray squirrels, Eastern Screech-Owls (*Otus asio*), American Kestrels, and Wood Ducks often use these enlarged Red-cockaded Woodpecker cavities. Gray treefrogs (*Hyla versicolor, H. chrysoscelis*), broad-headed skinks (*Eumeces laticeps*), mud dauber wasps (Sphecidae), paper wasps (*Polistes* spp.), and honeybees (*Apis mellifera*) are also often found in both enlarged and unenlarged cavities. Of these species, only paper wasps and honeybees can prevent Red-cockaded Woodpeckers from using the cavities.

INTERACTIONS WITH SOUTHERN FLYING SQUIRRELS

Gum flow from resin wells does not deter cavity use by what is believed by many to be the Red-cockaded Woodpecker's primary rival for cavities, the southern flying squirrel (*Glaucomys volans*) (Plate 13). Similar to Red-cockaded Woodpeckers, flying squirrels prefer unenlarged cavities, which increases the potential for competition. Numerous anecdotal observations indicate that flying squirrels occasionally prey on Red-cockaded Woodpecker eggs and possibly nestlings. However, observations of successful Red-cockaded Woodpecker nests in one cavity while southern flying squirrels were occupying a second cavity in the same tree are also common. It has also been suggested by others that flying squirrel abundance is associated with hardwood midstory and overstory foliage, and that the increased presence of hardwoods in woodpecker cluster areas results in increased woodpecker interaction with flying squirrels. Research by Conner and coworkers indicates that flying squirrels are frequent cavity occupants in loblolly pine forests, but are also frequent occupants in open longleaf pine forests that are nearly devoid of hardwood vegetation at all foliage layers (Figure 5.9). They did not see higher frequencies of Red-cockaded Woodpeckers roosting in the open as a result of an abundance of flying squirrels in loblolly/shortleaf pine habitat. On the contrary, woodpeckers roosting in the open were observed more often in longleaf pine habitat devoid of hardwoods.

Competition among species is very difficult to detect. Many species overlap in their use of food and nest site resources, but overlap by itself does

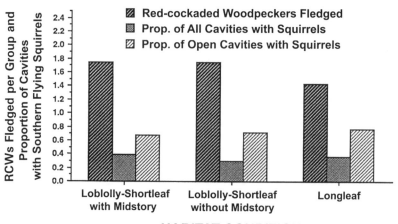

FIGURE 5.9 Red-cockaded Woodpecker fledging success, the proportion of all un-enlarged Red-cockaded Woodpecker cavities occupied by southern flying squirrels (*Glaucomys volans*, G. V.), and available unenlarged cavities (those not used by Red-cockaded Woodpeckers) occupied by southern flying squirrels in loblolly-shortleaf pine habitat with and without hardwood midstory and in longleaf pine habitat.

not necessarily represent actual competition. Competition occurs when species share the same resource and that resource is in limited supply, which results in a negative, population-level impact on one or both of the species involved. Although interactions between Red-cockaded Woodpeckers and other cavity users like flying squirrels traditionally have been called competition, there are two problems with this designation. The first is the lack of evidence of a population-level effect, and the second is the nature of the interactions.

When species compete, the negative impact of one on the other is often first detected as a depression of reproductive success. Thus, if flying squirrels were having a negative effect on Red-cockaded Woodpeckers, we would expect to see decreased fledging success in the woodpecker population as a result of either squirrel predation on eggs and young or inability of the woodpeckers to obtain sufficient cavities for nesting as a result of competition.

Research in Texas by Rick Schaefer, a graduate student at Stephen F. Austin State University working with Conner and Rudolph, indicated that fledging success of Red-cockaded Woodpeckers is highest in loblolly/short-leaf pine habitat, despite the presence of a high midstory, and lowest in longleaf pine habitat, which is relatively devoid of hardwoods. Fledging success was not related to squirrel abundance, regardless of habitat type.

Neither the effect of hardwood density on flying squirrel density nor the effect of squirrel density on nesting success of Red-cockaded Woodpeckers is clear. At most, a few nests may be lost to southern flying squirrels, a few groups may be unable to nest, and a few birds may have to roost in the open, raising their mortality slightly.

Kevin Laves, a graduate at Clemson University, observed some evidence of these effects in the Carolina Sandhills National Wildlife Refuge in South Carolina. Red-cockaded Woodpecker fledging success increased in areas where southern flying squirrels were actively removed, but this increase was not consistently related to nest loss. On the Frances Marion National Forest, Susan Loeb, with the U.S. Forest Service's Southern Research Station, and Hooper observed that Red-cockaded Woodpeckers were more likely to fledge young in clusters provisioned with nest boxes for Eastern Bluebirds and southern flying squirrels. Mitchell and colleagues, with Georgia Southern University, obtained different results on Fort Stewart, Georgia. They observed that neither the removal nor exclusion of flying squirrels affected the probability of fledging young or the number of young fledged.

Flying squirrels and other cavity users may impact nesting success in some populations, but these effects are small, and they clearly are not universal. Given that most populations have surplus nonbreeders (helpers and floaters; see Chapter 6), even where such effects exist, they do not seem to alter populations in terms of number of breeders. Aggressive interactions occur between Red-cockaded Woodpeckers and other species that cause individuals to be adversely affected, but such interactions cause no harm to the overall population. Pileated Woodpeckers are another story, since they destroy cavities, and there is a lot of evidence that permanent cavity loss negatively affects populations.

Therefore, most populations are likely unaffected by use of cavities by species such as Red-bellied Woodpeckers and southern flying squirrels. Routine destruction of southern flying squirrels and rat snakes because of possible adverse effects on Red-cockaded Woodpeckers is a practice that is difficult to justify and should cease in the absence of any evidence of a positive population-level effect.

As for the nature of the interactions, John Kappes, a graduate student at the University of Florida, argues that they differ greatly from the true competition that exists between secondary cavity users, where two or more species vie for a limited supply of nest holes. Instead, several other species attempt to steal cavities excavated by Red-cockaded Woodpeckers, but Kappes suggests the reverse never occurs. Thus these instances are more like cases of nest parasitism, such as practiced by House Sparrows (*Passer domesticus*) and other poor nest-builders (European Starlings, *Sturnus vulgaris*, are famous for this, and are indeed one of the species that usurps Red-cockaded Woodpecker cavities), or even cases in which one species steals food from another, as gulls do from Sanderlings (*Calidris alba*) feeding on the beach. Indeed, Kappes has argued that the term used for the latter interactions is the correct one for interactions between Red-cockaded Woodpeckers and other cavity users. This is cavity kleptoparasitism, not competition, according to Kappes.

Cavity Tree Selection

Red-cockaded Woodpeckers use a variety of tree species for their cavity trees over their range in the southern United States. Longleaf, loblolly, shortleaf, slash, pond, pitch, and Virginia pines are all used, and there are reports of occasional use of baldcypress. Longleaf, loblolly, and shortleaf pines are the species used most often.

OLD PINES ARE REQUIRED FOR CAVITY TREES

Much debate has occurred about the ages of pines needed to provide potential cavity trees for excavation. Potential cavity trees must be old for several important reasons. First, cavities must be excavated in heartwood, and heartwood with a sufficient diameter (14–16 cm) must be present at heights preferred by Red-cockaded Woodpeckers (generally 6–25 m). Heartwood diameter at cavity height is a function of tree age, with older pines having more heartwood at greater heights. If the heartwood is too small in diameter, the sapwood may be breached during excavation. Such a cavity would probably be abandoned before completion, due to pine gum seepage into the cavity chamber. Second, like heartwood diameter, the presence of red heart fungus in a pine is related to the pine's age, as well as to other factors such as soil quality (site index) and soil moisture. Old pines have a higher frequency of heartwood decay, and such decay is more advanced the older the tree becomes. *Phellinus pini* is a very slow-growing fungus. Inoculation studies by Conner indicate that once the fungus infects the heartwood, at least 12 years—possibly as many as 20—may be required before an adequate amount of heartwood has decayed at an inoculation site to house a cavity.

Cavity trees that are sufficiently old (probably 150+ years old) have extensive heartwood development in the trunk well into the pine's crown. As a result, cavities can be excavated higher above the ground than in younger pines. Field observations by Conner and Rudolph from a small (4-hectare) public park in eastern Texas indicate that active longleaf pine cavity trees that have cavities in the upper bole region (n = 11) average 199 years old and range from 146 to 250+ years. Cavities in such trees can be placed 18–27 m above the ground. This height has some potential advantages. Rat snake predation may be reduced and gum flow from resin wells would have less ignition risk during prescribed or natural fires. The oldest longleaf pine cavity trees that are still being used by woodpeckers in Texas (>250 years old) have cavities that are 24–27 m high, all of which are in the crown region of the tree. Thus, in the original old-growth forests, natural and Native American fire probably did not ignite the resin and destroy many cavity trees because cavities would have been well above the fire. In many current forests, however, cavities are placed in second-growth trees or small, stunted old-growth relic pines. In these forests, cavity heights are mostly 5–10 m, so that resin ignition during fires is frequent.

Older pines also have slower resin crystallization rates. Thus, pine gum on older cavity trees stays sticky longer than gum on young cavity trees, providing a better barrier against rat snakes.

A rather shortsighted but popular viewpoint is that the average-aged or even the youngest pines used by the woodpeckers are the maximum-age classes of pines that need to be provided for the birds. Use of young pines (40- to 50-year-old loblolly pines; 40- to 60-year-old longleaf pines) is often cited as evidence in support of this position. This view is often favored by individuals who would like to see economic gain from timber products optimized by growing pines on short rotations, i.e., harvested at young-tree ages.

The ages of pines excavated and used by Red-cockaded Woodpeckers cover a wide range. Generally speaking, the woodpeckers can use cavity trees with an age range of 40–180+ years in loblolly pine, 40–320+ years in shortleaf pine, and 50–450+ years in longleaf pine. Red-cockaded Woodpeckers have excavated cavities in pines as young as 30–40 years old. Reported average ages of cavity trees are 63–196 years for longleaf pine, 70–101 years for loblolly pine, 75–150 years for shortleaf pine, 62–130 years for pond pine, 70–76 years for slash pine, and about 70 years for Virginia pines.

It is difficult to determine the woodpecker's preference for ages of cavity trees, because timber harvesting has severely truncated the available age distribution of pines; most older-age classes have been cut throughout the entire southern United States. Typically, the mean age of cavity trees in a forest is highly dependent on the ages of the oldest pines available, suggesting that age of the trees selected is limited by availability, rather than being a true reflection of inherent preferences. The age range of longleaf pine cavity trees in Texas in 1983 was 49–332 years, with the average about 130 years (Figure 5.10). Sue Zwicker, a graduate student working with Walters, found that in 1993 at Camp Lejeune in North Carolina the age range of longleaf pine cavity trees was 61–221 years and the average was 111 years. In Texas, the age range of loblolly pine cavity trees was 49–156 years, and in North Carolina it was 57–113 years. The age range of Texas shortleaf pine cavity trees was 54–169 years. Research by Rudolph and Conner indicates that Red-cockaded Woodpeckers in Texas preferentially selected the oldest pines available: longleaf pines, 119–162 years old; shortleaf pines, 100–105 years old; and loblolly pines, 87–96 years old.

Studies in some of the oldest longleaf pine habitat in existence, using tree diameters to estimate ages of cavity trees, suggest that Red-cockaded Woodpeckers actually prefer very old cavity trees. On the Wade Tract in southern Georgia, Red-cockaded Woodpeckers selected longleaf pines 180–240 years old for cavity trees, whereas on the Fontainebleau State Park in southern Louisiana longleaf pines 208–374 years old were selected. Preferred loblolly pines on the Fontainebleau had an age range of 156–171 years. Longleaf pine cavity trees on the Oakmulgee Ranger District of the Talladega National Forest averaged 189 years old, with the oldest active cavity tree attaining 368 years of age. The bulk of longleaf pine cavity trees in the Sandhills of North Carolina mostly had an age range of about 150–250+ years, and on the Francis Marion National Forest their age range was about 65–200 years, with an average of about 115 years prior to Hurricane Hugo in September 1989.

A comparison of active and inactive cavity trees in Texas indicates that the age distribution of active cavity trees of each pine species closely matches the age distribution of inactive cavity trees, especially for longleaf and shortleaf pines (Figure 5.11). This suggests that the age of a pine probably does not affect its utility as a cavity tree substantially, although age affects its probability of being selected for excavation.

Over the past 20 years, the woodpeckers in Texas have continued to select the oldest pines available from a pool of trees of increasing age for the cavities they initiate. On the Angelina, Davy Crockett, and Sam Houston National Forests, in general the mean ages of cavity trees were significantly greater (19–68 years) than those of random trees distant from clusters, regardless of tree species or national forest. When pines in the same stand were examined in Texas, cavity trees were still on average significantly older than randomly selected noncavity trees by 3–34 years (see Figure 5.10). Similar results were obtained in North Carolina. However, a comparison by Rudolph and Conner of the ages of 140 recently excavated

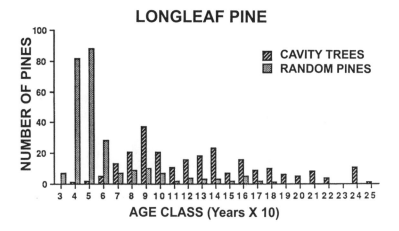

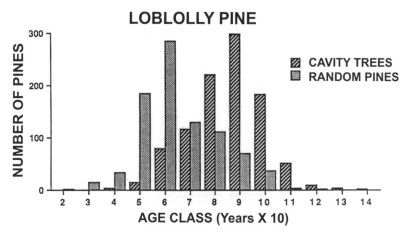

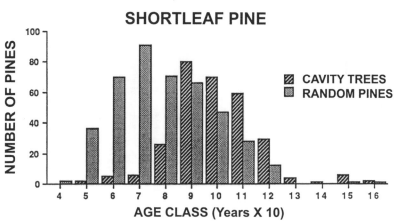

FIGURE 5.10 Ages of Red-cockaded Woodpecker cavity trees versus random pine trees in eastern Texas for longleaf, loblolly, and shortleaf pines. Red-cockaded Woodpeckers typically select the oldest pines available for their cavity trees.

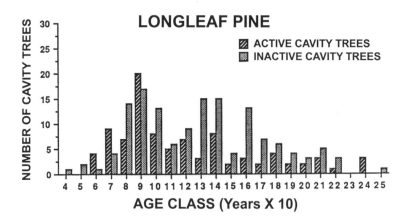

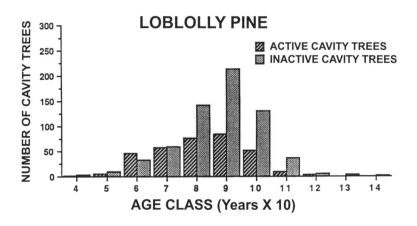

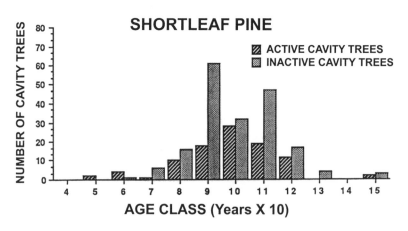

FIGURE 5.11 Ages of active and inactive Red-cockaded Woodpecker cavity trees in longleaf, loblolly, and shortleaf pines in eastern Texas. Active cavity trees tend to be younger than inactive, abandoned trees.

cavity trees in Texas with the ages of trees with older cavities on the Texas national forests revealed that recently initiated cavity trees were typically similar in age to trees with older cavities. That is, the age of pines at the time excavation is initiated continues to rise as the oldest pines available on Texas national forests mature. Thus, the current average ages of pine species in Texas do not provide optimum cavity trees for the endangered woodpecker.

It often is difficult to determine the exact ages of longleaf pines used by Red-cockaded Woodpeckers for cavity trees. If longleaf pines are grown on a plantation after being planted as seedlings, they can grow to breast height (about 1.3 m) in only 5–7 years. Longleaf pines that regenerate naturally, in contrast, may remain in the grass stage for more than 20 years before initiating height growth. Thus, ages of longleaf pines reported in the literature and in this text may actually underestimate longleaf ages by as much as 25+ years. This complicates the determination of timber rotation ages, especially if stands are regenerated through natural reproduction rather than planting containerized seedlings. Containerized seedlings typically remain in the grass stage for only 2–3 years before initiating height growth.

PATTERNS OF CAVITY TREE USE

Red-cockaded Woodpecker cavity trees can be used for extremely long periods of time. Several generations of woodpeckers can use the same cavity for over 30 years if it is excavated into sound longleaf pine heartwood. Although heartwood decay is an asset when a Red-cockaded Woodpecker needs to excavate a complete cavity quickly, presence of the fungus reduces the expected life span of the cavity once it is completed. The softened, decayed wood that initially facilitated the excavation of the cavity chamber eventually decays the heartwood so extensively that the cavity becomes too deep and is abandoned by the woodpecker.

In addition to decay and enlargement by other species (see above), there appears to be another reason for abandonment of a cavity, namely the physiological ability of the cavity tree to synthesize and transport pine resin. The woodpeckers will abandon a tree when pine gum no longer flows from its resin wells. The ability of a pine to produce resin is a function of several factors, some of which may relate to the Red-cockaded Woodpecker's initial reason for selecting a particular tree for cavity excavation. A pine's ability to produce resin is somewhat related to the size of its crown. Oleoresins are produced in pine needles and resin canals and ducts, which compose the tree's resin system throughout the tree. The resin system is separate from the xylem and phloem tissues, which transport water, minerals, and sugars. Ambient air temperature has an obvious effect: pines produce less resin during the colder months. As spring approaches, pines begin their spring growth phase, when most of a tree's energy is put into diameter and height growth. Growth continues as long as water is sufficient and temperatures do not become excessive. During summer, the pine shifts from wood growth to tissue differentiation, and more photosynthates and energy are then directed toward resin production. If drought and heat are too severe, even resin production will acutely diminish. Soil characteristics interact with season (temperature) and affect moisture availability. Thus, on poor soils some trees may be stressed or suppressed earlier, and therefore put less net effort into growth and more energy into resin production during a greater portion of the year. Many of the remaining cavity tree clusters with Red-cockaded Woodpeckers are located on such soils. It is not known whether Red-cockaded Woodpeckers actively select pines for cavity trees on such sites, or use the sites because pines on better soils produced better timber and therefore were cut during the initial harvest.

Recent research in Texas has shown that pines on the edges of forest stands and those in open savanna-like stands produce significantly more resin than pines within interior forests or dense forest stands. Red-cockaded Woodpecker selection of cavity trees in open stands and on the edges of dense stands may be directly related to a potential cavity tree's ability to produce oleoresins.

In summary, the mature pines selected by Red-cockaded Woodpeckers for cavities have some common characteristics (Table 5.1), including old age, sufficient heartwood, presence of red heart fungus, a relatively open bole, and good resin-producing abilities. Studies of growth rings from tree cores help reveal how these characteristics arise. Analyses of tree growth dynamics indicate that most Red-cockaded Woodpecker cavity trees have undergone one or more periods of suppressed growth followed by a release and faster growth. Although suppressed growth and release could be an artifact of the old age of many cavity trees, they may reflect conditions necessary to produce the ideal pine tree for cavity excavation. Suppression causes lower limbs to die and drop off, producing an open bole and a greater accumulation of heartwood. A subsequent release enables the pine to add foliage to its crown. Pines with large crowns are able to produce more resin than pines with small crowns. Since red heart fungus enters the heartwood of the pine's bole through branch stubs, limb pruning by suppression would provide fungal spores greater access to heartwood in the larger limbs and subsequently the pine's bole. Suppression also slows the pine's growth, ex-

Table 5.1. *Characteristics of Red-cockaded Woodpecker Cavity Trees and Habitat Compared to Randomly Selected Mature Pines*

Variable	Cavity Trees (n = 1,828)		Mature Pines[a] (n = 1,682)		t	P
	x̄	(SD)	x̄	(SD)		
Diameter at breast height (cm)	50.2	(8.6)	43.5	(8.8)	22.4	<0.001
Age (yrs)	99.2	(26.3)	72.6	(20.4)	31.1	<0.001
Tree height (m)	27.7	(4.0)	27.2	(3.4)	3.4	<0.001
Bole length to crown (m)	14.9	(3.8)	13.8	(3.8)	8.3	<0.001
Crown volume (m³)	505.8	(409.7)	369.3	(337.2)	10.4	<0.001
Midstory height (m)	4.6	(3.8)	6.7	(3.5)	17.3	<0.001
Basal area pine overstory (m²/ha)	13.7	(4.8)	15.7	(4.3)	12.5	<0.001
Basal area hardwood midstory (m²/ha)	1.9	(2.4)	2.7	(2.4)	9.9	<0.001

Note: A 2-tailed *t*-test was used to compare means.

[a] Mature pines were randomly selected.

posing dead branch stubs to fungal spores for extended periods of time before the pine can overgrow the stub and deny fungal access.

Differences among pine species affect how Red-cockaded Woodpeckers benefit from using them. Longleaf pine is probably the species of choice for a cavity tree. Longleaf pine lives far longer than other species of southern pine (250–450+ years). As a species, they are much more resistant to disease and insect infestation than other southern pines, primarily because of their ability to produce much greater amounts of pine resin. Greater resin yield is of obvious benefit: it improves the effectiveness of the resin barrier against climbing rat snakes. Shortleaf pine's ability to produce resin is less than that of longleaf pine but greater than loblolly pine's. Although loblolly and shortleaf pines do not produce as much resin or live as long as longleaf pine, they generally acquire characteristics required for use as cavity trees at younger ages, because they have greater diameter growth at younger ages and have higher frequency of heartwood decay. Decay can gain access to heartwood more quickly because of the species' lower resistance. Unfortunately, the stress placed on loblolly pine cavity trees by Red-cockaded Woodpeckers during cavity excavation and subsequent maintenance of resin wells may increase mortality from bark beetles. Work by Conner, Rudolph and coworkers indicates that resin production declines more rapidly in loblolly pines, and that cavities in this species are used for shorter periods of time than cavities in longleaf pines (see below).

The Cavity Tree Cluster

The aggregation of cavity trees previously and currently used by a group of woodpeckers in their defended territory is termed a cavity tree cluster (Plate 14). In the past, such groups of cavity trees have been called colonies. The term "colony," however, is inappropriate, because only one breeding pair of Red-cockaded Woodpeckers (a breeding male and breeding female) use each cluster of cavity trees, not several as was once believed. Other individuals in the family group using a cluster of trees typically are non-breeding adult offspring from previous nestings.

In the aboriginal forests of the South, old pines were likely abundant, which presumably permitted relatively close grouping of cavity trees within a cluster. This may have facilitated defense of territory and cluster against other Red-cockaded Woodpecker groups. In many areas today, trees suitable for cavity excavation within a territory are few and widely scattered. Thus trees within a cluster often are not clumped spatially but are linked through their use by a single-family group. Spatial clustering of cavity trees within a cluster appears to be greater where old-growth pines are more widely available, such as on the Wade Tract in Georgia. Although the trees of a cluster are often all within 100 m of each other, active cavity trees used by group members may be as far as 400–800+ m apart if availability of old-growth pines is low. Distant cavity trees are sometimes beyond the defended boundary of the territory.

HABITAT WITHIN CAVITY TREE CLUSTERS

Just as they require certain characteristics of cavity trees, Red-cockaded Woodpeckers prefer particular habitat conditions around their cavity trees. Open, older-growth pine forest habitat, relatively devoid of hardwood midstory foliage with a primarily grass and forb ground cover, is the preferred condition for cluster areas. Ideal open forest conditions are produced by frequent growing-season fires (see Chapter 2), where hardwoods of varying sizes are killed back or eliminated and only a very few overstory hardwoods survive. Pine basal area in clusters of cavity trees can range from less than 2 to 20 m^2/hectare. Preferred pine basal areas probably average from about 5 to 16 m^2/hectare. In most woodpecker populations, hardwood midstory is less prevalent around active cavity trees than inactive cavity trees (Table 5.2). Because loblolly and shortleaf pines grow on more mesic sites, which burn less often than drier longleaf pine sites, there is usually a greater occurrence of hardwood species in loblolly and shortleaf pine clusters, particularly in the midstory and understory.

Table 5.2. *Characteristics of Active and Inactive Red-cockaded Woodpecker Cavity Trees in Eastern Texas*

Variable	Active Cavity Trees (n = 584)		Inactive Cavity Trees (n = 1,148)		t	P
	x̄	(SD)	x̄	(SD)		
Diameter at breast height (cm)	48.5	(7.8)	51.1	(8.1)	6.31	<0.001
Age (yrs)	94.8	(25.2)	101.5	(26.8)	4.75	<0.001
Tree height (m)	27.6	(3.6)	27.7	(4.1)	0.31	0.76
Bole length to crown (m)	15.0	(3.9)	14.8	(3.7)	1.06	0.29
Crown volume (m³)	451.3	(400.6)	538.3	(413.9)	4.10	<0.001
Midstory height (m)	4.1	(3.7)	4.8	(3.9)	3.51	<0.001
Basal area pine overstory (m²/ha)	13.3	(4.9)	13.9	(4.7)	1.94	0.05
Basal area hardwood midstory (m²/ha)	1.5	(2.1)	2.0	(2.5)	4.12	<0.001

Note: A 2-tailed *t*-test was used to compare means.

Throughout the South, many Red-cockaded Woodpecker populations currently are found on soils that are sufficiently poor to stress plant communities. In Texas, clusters in longleaf pine are typically located in deep loamy sands (Tehran and Letney soil types) containing materials of volcanic origin. These soils contain very little organic material and as a result have a very low moisture-holding capacity. Hot summer days produce high soil temperatures that can negatively affect hardwood vegetation. Loblolly and shortleaf pine cavity trees in Texas are typically located on shrink-swell clays of the Woodtel and LaCerda soil types. As the water content of these soils changes, the clay expands and contracts, often stripping root hairs from trees growing on such sites. Basic salts found in these soils further aggravate the poor plant-moisture relationship. With human suppression of natural fire, Red-cockaded Woodpeckers may have persisted longer on these stressful sites than in other areas because less hardwood vegetation was present.

CAVITY TREE MORTALITY WITHIN THE CLUSTER

Under normal conditions a group of Red-cockaded Woodpeckers (formerly termed a clan) excavates a sufficient number of complete cavities to provide all group members with an individual roost cavity. Although one cavity tree can have more than 10 cavities, most cavity trees contain one or two usable cavities. Additional cavities are often excavated in a cavity tree as

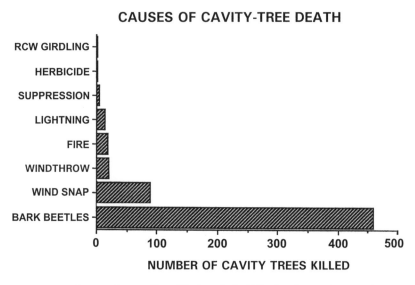

FIGURE 5.12 Causes of mortality of Red-cockaded Woodpecker cavity trees in eastern Texas between the early 1980s and early 1990s. Southern pine beetles were the major cause of cavity tree death in loblolly and shortleaf pines, whereas fire was the major cause of death in longleaf pines.

older cavities become unsuitable. Thus a group of woodpeckers can have from 1 to 13 or more cavity trees that they are currently using, have used in the past, or are currently excavating. If roost cavities are in short supply, woodpeckers occasionally roost in recently dead cavity trees. There are also rare reports of nesting in dead pines. Woodpeckers will also roost in the crotch of forked pines if cavities are limited, pecking resin wells around and below the fork.

The availability of cavities is affected by more than just the frequency of suitably decayed old pines. Excavation of new cavities must at least equal losses of existing cavities to other cavity users, enlargement by Pileated Woodpeckers, and tree mortality. Factors affecting cavity tree mortality are important considerations in the ecology of the woodpecker (Figure 5.12). Studies of cavity tree mortality in Texas by Conner, Rudolph, and colleagues indicated that bark beetles (Scolytidae) were the primary cause of mortality (53%) in loblolly and shortleaf pines (n = 453). Although some cavity tree mortality resulted from infestations of bark engraver beetles (*Ips* spp.) and black turpentine beetles (*Dendroctonus terebrans*), most bark-beetle-induced deaths (95+%) resulted from southern pine beetle infestations. Wind snap, in which pines break at the nest cavity (30%), and fire (7%) also caused considerable mortality. Fire was the primary cause for cavity tree mortality in longleaf pines. Additional causes of cavity tree death are windthrow, in which the tree is uprooted (primarily resulting from damage caused by the root-decaying fungus *Heterobasidium annosum*), and lightning. Each of the last two factors accounted for 4% of the total mortality in these studies.

The dynamics of cavity tree turnover appear to vary from site to site, as well as among pine species. In North Carolina, southern pine beetles do not pose the problem that they do in Texas, nor are they the threat to longleaf pines that they are to loblolly and shortleaf pines. The predominant cause of cavity tree mortality in North Carolina is lightning, whereas in old-growth longleaf pine on the Wade Tract in southern Georgia and on the Frances Marion National Forest in South Carolina, wind damage has been a major factor (see below).

Cavity tree mortality rates are affected by cluster status. In Texas, cavity tree mortality rates in active clusters were about twice those in inactive clusters. During initial studies by Conner and Rudolph, comparisons of active cavity trees with inactive cavity trees in active clusters revealed no difference in mortality rates. This suggested that some factor associated with active clusters might be the cause of their higher mortality rate. Management associated with cluster areas often involves prescribed fires, thinning of overstory pines, and removal of hardwoods. These operations involve the use of heavy equipment within cluster areas, may include plowing of fire lanes, and can result in injury to some trees' trunks and root systems, as well as soil compaction. Thus, increased tree mortality might be associated with disturbance to the cluster site. Disturbance is a well-known cause of increased bark beetle activity (see Chapter 8). Recent research by Conner, Rudolph, and colleagues, however, indicates that active cavity trees within active clusters are infested at higher rates than inactive cavity trees. This suggests that some attribute, perhaps the volatiles from pine resin flowing down the boles, may enhance the attractiveness of active cavity trees to southern pine beetles.

Fire too can stress trees and may contribute to higher mortality in (managed) active clusters. Prescribed growing-season fire is particularly likely to cause stress when first administered after a long period of fire suppression. High fuel loads lead to hot fires and crown scorch, and buildup of duff may lead to root damage. After a few fire cycles, the risk is greatly reduced, but many forests may experience a period of increased mortality of cavity trees and other remnant old growth as fire is reintroduced if managers do not take suitable precautions.

As noted above, wind damage is the second-greatest cause of Red-cockaded Woodpecker cavity tree loss in eastern Texas. The distance between cavity trees and openings in the forest canopy (primarily clear-cuts) appears to affect the rate at which wind damage occurs. In most forests, the distance between cavity trees and openings in the forest canopy is largely affected by management's retention of a 61-m (200-ft) cluster protection buffer of uncut trees around each cluster of cavity trees. However, after buffers are established, woodpeckers often excavate cavity trees closer than 61 m to the edge of the cluster, especially if excessive hardwood midstory foliage is present within the cavity tree cluster area. Studies in Texas and Louisiana by Conner and Rudolph suggest that the 61-m buffer distance is adequate to protect most cavity trees. Wind-damaged cavity trees in Texas averaged 110 m from forest openings, whereas randomly selected undamaged cavity trees averaged 311 m. Most damaged cavity trees were wind-snapped; 35% were within 25 m and 48% were within 50 m of openings in the forest canopy. However, there may be secondary zones of damage between 75 and 200 m of forest edges, perhaps caused by turbulence of wind swirling above trees after passing over the forest edge. Windthrow and wind snap at some rate should be expected to occur to cavity trees as a natural process.

Sustained high winds that occur during hurricanes can have a devastating effect on Red-cockaded Woodpecker cavity trees and the woodpeckers. Hurricane Hugo passed over the Francis Marion National Forest on 21 September 1989, destroying 87% of the woodpecker cavity trees and about 59% of the foraging habitat, and killing 63% of the woodpeckers. Robert Hooper reported that a population of approximately 470 woodpecker groups was reduced to 249 groups (and 83 single woodpecker clusters) by the 1990 breeding season. Although not as devastating as Hugo was in 1989, Hurricanes Kate and Elena and Tropical Storm Juan in 1985 also impacted woodpecker cavity trees on the Wade Tract in southern Georgia, according to Todd Engstrom and Gregory Evans. In 1996, Hurricanes Bertha and Fran combined destroyed 21% of the cavity trees on Camp Lejeune Marine Base in North Carolina (88 trees), as well as a smaller number on the nearby Croatan National Forest. That same year, the population at Eglin Air Force Base in Florida suffered comparable damage from Hurricanes Erin and Opal. Hurricanes represent an important mortality factor for cavity trees in coastal Red-cockaded Woodpecker populations, characterized by regular instances of incidental loss as well as infrequent catastrophic loss.

CAVITY TREE DYNAMICS WITHIN THE CLUSTER

The population dynamics of cavity trees varies among pine species. In Texas, loblolly and shortleaf pine cavity trees die at an annual rate of about

5.3%, whereas longleaf pine cavity tree annual mortality averages signifi-cantly less, about 1%. But Red-cockaded Woodpeckers excavate loblolly and shortleaf pines at only twice the rate they excavate longleaf cavities (1.4 vs. 0.7 newly completed cavity trees per year). In Texas, the average age of loblolly and shortleaf pine cavity trees at death was 87.8 years (n = 29), whereas longleaf pine cavity trees died at an average age of 151 years old (n = 14). Because most of the mortality in the preceding data was caused by prescribed fires and human-related disturbances, the natural age of death for each species is most likely higher. The longer life span, coupled with longleaf pine's greater ability to produce a sustained supply of pine gum, strongly indicates that it is the best pine species for Red-cockaded Wood-pecker cavity trees.

Studies of small, declining populations in Texas by Conner, Rudolph, and coworkers indicate that Red-cockaded Woodpeckers are sometimes unable to excavate adequate numbers of new cavity trees to offset mor-tality. Over a 9-year period on the Angelina National Forest, an average of 260 cavity trees each year in active and inactive clusters were examined. During this time, Red-cockaded Woodpeckers excavated 16 new cavity trees (11 loblolly and shortleaf pine, and 5 longleaf pine). Over the same period, a total of 57 cavity trees died (41 loblolly and shortleaf; 16 long-leaf). This yields a net loss of 41 cavity trees over the 9 years. In addi-tion to this mortality, cavities were destroyed in 55 additional cavity trees by Pileated Woodpecker enlargement of the entrance tubes. The balance between losses and gains was equally negative when only cavity trees in active clusters were examined. This raises some important points about characteristics of pines preferred by Red-cockaded Woodpeckers for cavity trees and what forest management must provide to ensure an adequate supply of potential cavity trees.

Harding's work showed cavity tree dynamics within the Sandhills popu-lation of Red-cockaded Woodpeckers in North Carolina to be similar to those in Texas. From 1980 through 1995, there was a net loss of 91 cavities from an initial population of 506 cavities, which represents a reduction of 18%. Although cavity tree death accounted for many losses (36%), cavity enlargement accounted for more (60%). Initially the rate of loss to cavity enlargement was about twice the rate of loss to tree death, but with im-proved management in the 1990s (see Chapters 10 and 11), the rate of en-largement has been reduced to the point that losses to enlargement and tree death are about equal. As in Texas, turnover of longleaf cavities is slower than turnover of cavities excavated in other pine species, and be-cause the North Carolina population is more dependent on longleaf pine, overall cavity dynamics are more favorable. In this population, most cavities (82%) are in longleaf pine, and only a few are in loblolly (8%) or pond pine (9%). Cavities in longleaf are used on average about 10 years, those in the other species less than 4 years. In nearly all cases the birds cease using a cavity because it is lost.

Harding found that cavity dynamics are even more favorable in the two coastal populations in North Carolina, Camp Lejeune and Croatan National Forest. Between 1986 and 1996, there was a net loss of only 11 cavities at Camp Lejeune from an initial population of 125 cavities. During that period, 189 cavities were lost (75% to cavity enlargement, 25% to tree death), and 178 new cavities were excavated. Thus the loss of 88 cavities to hurricanes between 1996 and 1997 represented a major tree mortality event (see above). In this population, 71% of the cavities are in longleaf pine, 24% in loblolly pine, and 5% in pond pine.

The Croatan population too is heavily dependent on longleaf pine (66%), but use of loblolly pine is also significant (29%). As a result, turnover of cavities is relatively rapid, with loblolly pine accounting for a disproportionate share of losses and gains. Although the initial cavity population in 1988 contained only 22% loblolly pine compared to 74% longleaf pine, loblolly accounted for 34% of the newly excavated cavities (longleaf 61%) and 34% of the cavities lost (longleaf 63%) between 1988 and 1996. Despite this rapid turnover, there was a net gain of 3 cavities during this period. Cavity enlargement accounted for 81% of cavity losses in this population.

The differences between cavity dynamics in Texas and North Carolina are interesting. The processes that result in cavity loss are common to all the populations examined, but the absolute and relative rates at which they operate vary significantly. This may depend on which species of pine is used and the location (i.e., coastal versus inland), but there is also significant variation in the same process operating on the same species in different populations. Cavity enlargement rather than tree death accounts for the majority of cavity losses in North Carolina, but in Texas losses to the two processes are about equal. However, the annual rate of cavity tree mortality is actually similar in the two areas.

The major difference between Texas and North Carolina is in the rate of new cavity excavation, which is much higher in North Carolina. Because of this, the birds in North Carolina have been better able to replace lost cavities than have those in Texas. It is not clear whether this difference is due to relative availability of trees in which to excavate or differences in bird behavior. However, the populations in which cavity dynamics are most favorable, Camp Lejeune and Croatan, both were increasing. Hence the ratio of cavities to birds declined noticeably, even though the absolute number of cavities did not. Furthermore, both suffered extensive losses to hurricanes in the year following the study period. If these studies had lasted one additional year, both populations, like the Texas and North Carolina Sandhills populations, would have exhibited large net losses in number of cavities. It appears that in all populations studied, the birds have difficulty excavating new cavities as rapidly as old ones are lost to cavity tree death and cavity enlargement.

CAVITY TREE DYNAMICS RELATIVE TO FUNGAL DECAY

Cavities can be excavated into pines that have extensive red heart fungus infection and relatively thin sapwood faster than into pines with thick sap-wood and no fungal infection. The ages to which pines are grown in and around Red-cockaded Woodpecker clusters is crucial. They must become old enough for heart rot to develop at a high enough frequency and to a sufficient extent within the heartwood to facilitate cavity excavation. Heart rots must be extensive enough to permit rapid excavation of complete cavities, in order to offset losses when they occur. All of these factors indicate that a large number of old pines need to be available to permit a sustained supply of new cavity trees through time, especially if cavity tree mortality rates and losses of cavities to factors such as enlargement by Pileated Wood-peckers are high, as they appear to be.

There are geographical differences in rates and ages at which red heart forms in longleaf pines. Hooper has proposed that longleaf pines grown only to 100–120 years of age would have a sufficiently high frequency of heartwood decay for Red-cockaded Woodpeckers. On the Francis Marion National Forest, Hooper has shown that 92% of the longleaf pine cavity trees had heart rot infections, and that over 85% of longleaf pines under 80 years old were infected. In Texas, heart rot infection rates were high in loblolly and shortleaf pines averaging 93 years of age, but were much lower in longleaf pines of similar age and older (Figure 5.13). Conner and Locke examined 18 longleaf pine cavity trees (average age = 118 years old) from eastern Texas in detail during the late 1970s by vertically sectioning the trees. Only 39% had heart rot infections. The youngest longleaf pine cavity tree with heartwood decay was over 100 years old. Texas longleaf pine cavity trees with decay averaged 140 years old, whereas ones with-out decay averaged 104 years old, a significant difference. Conner, Rudolph, and colleagues examined 50 longleaf pine cavity trees (average age = 126.4 years old) in the mid-1990s and detected a 46% fungal infection rate. Heart-

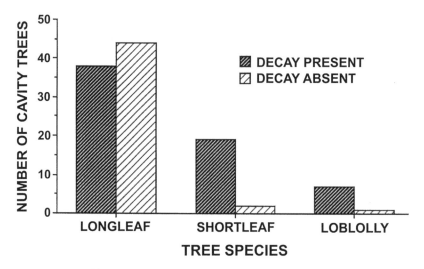

FIGURE 5.13 Red heart fungus infection rates in longleaf, shortleaf, and loblolly pine Red-cockaded Woodpecker cavity trees in eastern Texas. The presence of decay in pines greatly facilitates cavity excavation, reducing the time required to excavate a new cavity.

wood decay in these trees was not regularly detected until they exceeded 128 years old. Soils on the Francis Marion National Forest are much wetter than those of longleaf pine sites in eastern Texas, which may account for some of the difference in fungal infection rates.

If we assume that the woodpeckers are actively selecting pines with heart rot, the data suggest that loblolly forests in eastern Texas that are grown to 100–120 years should provide an adequate number of pines for cavity trees. Such is not the case with longleaf pine. If Red-cockaded Woodpeckers are actively seeking longleaf pines with red heart fungus and can detect the presence of the fungus, longleaf pines with red heart will not be sufficiently abundant to meet their needs in forests grown to less than 120 years in eastern Texas.

The solution to providing sufficient potential cavity trees for Red-cockaded Woodpeckers appears to be directly related to tree age. Ages well beyond the time when decay first appears in pines are needed if mortality is to be balanced with excavation of new cavity trees. Thus whatever timber management system is selected for use in areas managed for Red-cockaded Woodpeckers, the ages of pines produced must be sufficiently high through time to provide adequate numbers of pines with decayed heartwood and suitable sapwood and heartwood characteristics. The spatial availability of these potential cavity trees across the landscape must also be adequate to provide for the creation of new cavity tree clusters and woodpecker groups. Currently the birds in Texas and elsewhere are dependent on a declining pool of remnant old-growth pines. They have presumably been picking the most suitable among these for cavity excavation for some time. As their choices become increasingly limited, it is no wonder that cavity tree dynamics are unfavorable outside of exceptional areas like the Wade Tract. This cannot be expected to change until the second-growth trees germinated around the turn of the century mature sufficiently. It will never change unless old growth is more widely available on future landscapes than it is today.

CHAPTER SIX

Social Behavior and Population Biology

If one had to pick two features of the biology of Red-cockaded Woodpeckers that most distinguish them from all the other bird species in their communities, it would be their habit of excavating cavities in live pine trees, discussed in the preceding chapter, and their social system. The Red-cockaded Woodpecker practices cooperative breeding, a social system in which some mature adults forego reproduction and instead assist in raising the offspring of others. Cooperative breeding is known to occur in over 100 species of mammals, including species as diverse as naked mole-rats (*Heterocephalus glaber*), African hunting dogs (*Lycaon pictus*), and marmosets (family Callitrichidae), and in nearly 300 bird species, ranging from hawks and Ostriches (*Struthio camelus*) to wrens and warblers. This also of course is the social system that characterizes the social insects; the complex societies of bees, ants, and termites with workers and queens are an extreme form of cooperative breeding.

Among birds, cooperative breeding is especially common in certain taxa such as corvids and certain geographical regions, especially Australia. It is quite rare in temperate North America and is uncommon in woodpeckers. No relatives of the Red-cockaded Woodpecker exhibit this social system, indicating that it has evolved in this species independently of its evolution in other species. Among its neighbors, the only other cooperative breeders are the American Crow (*Corvus brachyrhynchos*) and Brown-headed Nuthatch.

Among the cooperatively breeding birds of the world, there is considerable diversity in the mechanics of the social system. In some species the social unit consists of several breeding pairs that nest together. Individuals assist others in raising young, but they also produce their own young. In other species the social unit includes several nonbreeding adults, as well as a number of breeding pairs that nest separately. The nonbreeders, known as helpers, help feed young at one or more nests within the group, and often the breeders feed young at one or more other nests in addition to their own. The most common system, though, is one in which social units consist of a single breeding pair plus one or more helpers. In some of these systems, some helpers attempt to contribute to reproduction, the female helpers by laying eggs in the nest, the males by mating with the breeding female. In other systems, the helpers are strictly nonbreeders and are predominantly previous offspring of the breeding pair. This is the type of social system that characterizes the Red-cockaded Woodpecker.

The Cooperative Breeding System

Compared to other cooperative breeders, Red-cockaded Woodpeckers live in small groups. Although groups as large as seven adults, including five helpers, have been observed, groups of more than three are relatively uncommon, and many groups consist of a breeding pair without helpers. The average group size in all populations studied has been between two and three. In many populations, unpaired males occupy some territories, but even when these groups of one are excluded, average group size is still less than three.

The members of a Red-cockaded Woodpecker group act very much like a unit, traveling through their large territory together, calling back and forth frequently. Group members may even forage on the same tree or branch together, although more typically they spread out among a group of trees. When one bird flies, it gives a contact call. There are occasional displacements from feeding spots, but relations between group members are for the most part tranquil.

GROUP FORMATION AND MALE LIFE HISTORY

Group formation is best understood in terms of the behavior of young birds in the first year after fledging. These birds choose between two possible life pathways that differ greatly in their consequences for the birds' future. Some individuals disperse from their natal territories during their first year in search of a place to settle and breed (Figure 6.1). In most bird species, this is the course that all individuals follow. In cooperative breeders, however, some individuals instead choose not to disperse but to remain in their natal territory as a helper. Among Red-cockaded Woodpeckers, many males choose this option, and this is the source of the vast majority of helpers.

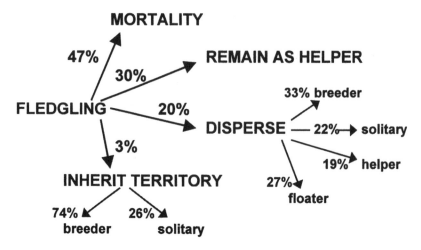

FIGURE 6.1 Life history of male fledglings expressed as annual transition probabilities, measured from fledging to the next breeding season. Based on 2,286 male fledglings from the North Carolina Sandhills.

Helpers assist the breeders in virtually all activities related to territory maintenance and breeding, including defending the territory, constructing cavities, incubating nestlings (helpers have brood patches), feeding nestlings, and feeding fledglings. They do not, however, father any offspring. Copulations between the breeding male and breeding female occur fairly frequently, beginning up to several weeks before egg laying and extending a few days beyond completion of the clutch. These copulations are conspicuous and have been observed many times. Copulations involving helpers have never been observed.

The possibility that copulations involving helpers might occur, but are less conspicuous and thus hard to observe, has been ruled out by studies by Susan Haig (then at Clemson University) and coworkers, in which DNA fingerprints were obtained from all individuals in many groups. In both a small population in South Carolina and a large one in North Carolina, DNA fingerprints indicated that helpers did not father offspring. In fact, in only one of the 101 cases examined was an offspring fathered by anyone other than the breeding male in its group. In this case a helper was present, but he was not the father either. The father evidently was a male from some other group. The three other members of this brood were fathered by the breeding male from the group. No cases in which the eggs were laid by a bird other than the breeding female were discovered in these studies either. Thus the Red-cockaded Woodpecker is among the most strictly monogamous species of any bird studied to date, much more so than most species that live in pairs without helpers.

Although males do not reproduce while serving as helpers, they do not forego reproduction permanently. Helper males attempt to become breeders just as dispersing males do, but they do so by different methods. Essentially, helpers remain in the natal territory and wait for a breeding vacancy to arise in their vicinity. Although by virtue of being sedentary they are unaware of any vacancies distant from them—vacancies that dispersing birds might detect in their wanderings—helpers are able to detect and compete for vacancies that arise nearby, that is, up to two or three territories away from them. Mostly, though, they become breeders in their natal territory or a neighboring territory (Figures 6.2 and 6.3). They obviously are in a good position to detect these vacancies and compete effectively for them, but they may end up waiting years for one to arise. When a breeding male dies, if helpers are present on the territory, the oldest one invariably inherits breeding status. If there are additional helpers, they continue to help the new breeders. Such helpers often end up helping an older brother, an uncle, or some other male relative in such cases.

Although helpers typically become breeders by age 3, 4, or at least 5 years, a few are still helping at age 8 (Figure 6.4). Glenn Woolfenden has suggested that among Florida Scrub-Jays (*Aphelocoma coerulescens*), another

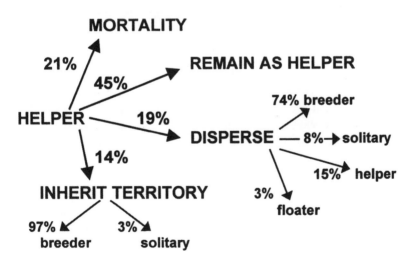

FIGURE 6.2 Life history of helper males expressed as annual transition probabilities, measured from one breeding season until the next. Based on 1,315 male helpers from the North Carolina Sandhills.

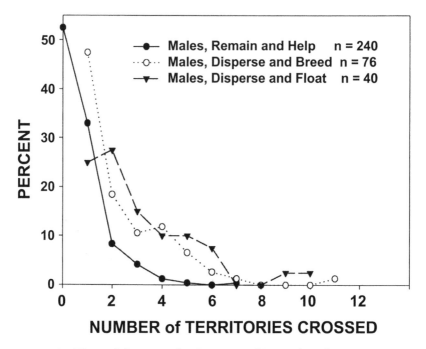

FIGURE 6.3 Dispersal distances of males measured in number of active territories crossed, from the natal territory to the territory on which they acquire breeding status. Individuals that become breeders on the natal territory cross zero territories. The figure contrasts dispersal of males that remain as helpers in their first year (remain and help) with those that disperse after fledging to become either breeders (disperse and breed) or floaters (disperse and float) in their first year. Adapted from S. J. Daniels 1997.

cooperative breeder, certain males are less disposed to attempt to become breeders, and it is these males that remain helpers in old age. Whether there are less-competitive males that choose to remain helpers among Red-cockaded Woodpeckers is not known, but there does seem to be a tendency for helpers to be less likely to become breeders as they increase in age.

Once a male becomes a breeder, almost invariably he remains a breeder until his death. Sometimes an unpaired male will move to an adjacent unoccupied territory, but breeding males do not switch groups as breeding females do (see below). Rarely, a breeding male may be displaced by a new male. In these instances the former breeder typically remains in the vicinity as a floater until it dies, rather than moving elsewhere to breed on another territory. These replaced males almost never seem to survive beyond the first year they are replaced.

Thus the typical life history of a male who chooses the helper path is to remain in its natal territory for some period as a helper, then become a breeder in the natal territory or an adjacent one, remaining in that role until its death.

What about the males that choose to disperse in their first year instead of help? If they are lucky, they acquire a territory, obtain a mate, and begin nesting at age 1. Most are not so fortunate. Many acquire a territory but

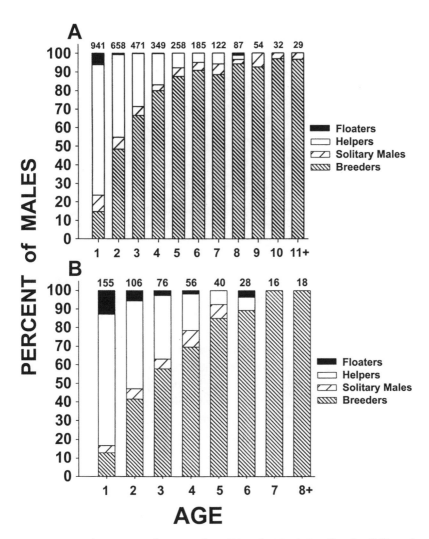

FIGURE 6.4 Male status as a function of age (A) in the North Carolina Sandhills and (B) at Camp Lejeune. Sample sizes are indicated above each bar.

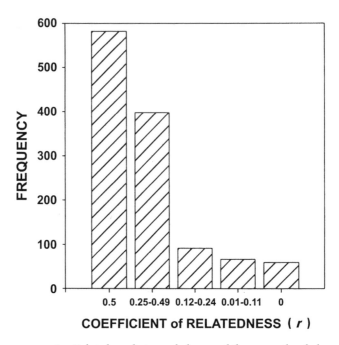

FIGURE 6.5 Relatedness between helpers and the young they help raise in the North Carolina Sandhills. Full siblings have a coefficient of relatedness of 0.5, half-siblings 0.25. A coefficient of 0 indicates helpers were unrelated to the young they helped raise.

do not obtain a mate, and thus are solitary males at age 1. Others do not even acquire a territory and are floaters at age 1. One-year-olds account for a large proportion of solitary males and floater males. These males, if they survive, usually become breeders by age 2 or at least age 3. Some of these departing males join other groups as helpers. Thus, although most helper males are related to the offspring they help raise, a small number are unrelated to the young (Figure 6.5). These unrelated helpers become breeders in the same ways that related helpers do.

Hence both males that depart in their first year and males that remain as helpers become breeders but with some delay, the average delay being longer in helpers than departers. The shorter delay for departers is balanced by a higher mortality rate, as the survival of helpers, living with their family in a familiar territory, is much higher than that of males wandering alone through new habitat in search of a place to live.

Why do some males choose the path of helping and others the path of departing? They do not seem predestined toward a particular path: analysis of heritability indicates no genetic basis for this choice. Individuals apparently somehow choose which life path to follow, based on unknown criteria. One might suspect that high-ranking individuals would be more likely to stay as helpers because they would be higher in the queue waiting to inherit the territory than their lower-ranking brood mates. Of course, even the highest-ranking member of the brood would rank below all the older helpers on the territory. Since it is rare for more than one male from the same brood to stay, high rank may be of little advantage when it comes to inheriting the territory. There are some indications that males on especially good territories may be more likely to stay as helpers, and that males may be more likely to stay when competition for territorial vacancies is more intense, but these are untested ideas. What determines which path a male chooses is yet to be determined.

FEMALE LIFE HISTORY

Although in Red-cockaded Woodpeckers most helpers are male, a few are female. Most females adopt the departing strategy, but some adopt the helping strategy. The departer and helper strategies are executed similarly by the two sexes, with one major exception. Female helpers rarely inherit breeding status in their natal territory and therefore rely much more heavily on moving to nearby territories to become breeders than do male helpers. This difference is related to incest avoidance.

A breeding system in which the male inherited his territory from his father would frequently result in incestuous mating if no mechanism to avoid incest existed. A behavioral mechanism for avoiding incest occurs among Red-cockaded Woodpeckers. If a female's mate dies and her son inherits the territory, she invariably leaves. Presumably the mother rather than the son leaves because males are dominant to females. However, the limited observations available suggest that sons do not employ frequent, intense aggression to force their mothers out. Sons may put subtle pressure on their mothers, or mothers may leave of their own volition. Some females have been observed initially greatly expanding their daily ranges without actually leaving their original territory, presumably in an attempt to contact more neighboring groups to which they might move. Other females depart and search for a new breeding position without preliminary forays, much like birds dispersing in their first year do.

If a female dies or moves and is replaced by a new female before the original breeding male dies, the new female will be unrelated to the helpers present on the territory. In these cases, when the breeding male dies the female remains and mates with the helper who inherits the breeding position. This indicates some ability to distinguish relatives from nonrelatives. The best hypothesis to explain this is that the birds treat as relatives those individuals that were members of their group when they were nestlings or fledglings. One observation that supports this hypothesis is that when relatives that independently disperse from the same natal territory end up together on some other territory, invariably these are individuals that did not know each other previously. That is, one had already left the original territory before the second was hatched.

Male dominance, coupled with incest avoidance, makes it unlikely for a female helper to inherit breeding status. For the highest-ranking male helper to inherit breeding status, all that is required is for the breeding male to perish. For the top female helper to inherit breeding status, not only must the female breeder perish but also the male breeder and all the male helpers if they are related to the female, as they typically are.

Although inheritance of breeding status by female helpers is much less frequent than by male helpers, movement of breeders between groups is much more common among females. About one-fourth of such movements can be related to incest avoidance, as just discussed. Many of the remainder can be related to mate choice, in that the movement follows the death of the female's mate on territories with no helpers. About half the time, the female remains on the territory and pairs with a new male, but the other half of the time she moves. Susan Daniels, a graduate student working with Walters at Virginia Tech, has shown that in deciding whether to remain or leave, females discriminate against young replacement males. Thus many of these movements may represent attempts to avoid pairing with young males, who are rather inept at reproduction (see below).

The remaining female movements, which account for about one-third of the total, involve the dissolution of a pair-bond. It is not unusual among birds for such "divorces" to occur. In other species divorce typically follows unsuccessful reproduction or results in a move to a higher-quality territory. Daniels has shown that young Red-cockaded Woodpecker females (age 1 or 2) are more likely to divorce their mates if reproduction is unsuccessful, but older females are not. Older females sometimes move to another group even though their mates and perhaps helper sons remain, regardless of their prior reproductive success. Daniels has shown that these movements are not related to territory quality either. Patricia Gowaty (then at Clemson University) and Michael Lennartz suggested that there is reproductive competition between mothers and their helper sons, and that this affects the sex ratio of offspring produced (see below). Roy DeLotelle, a biologist working in Florida, has observed that females are more likely to disappear if they have a helper son than if they do not, suggesting that subtle competition between mother and son could be a factor in female movements. The possibility that a female might lose her breeding position if her mate dies might make females with helper sons more likely to take advantage of a chance to switch groups than females with no helpers or unrelated helpers. However, when Daniels tested this possibility, she found that movement was unrelated to the presence of helper sons.

Another possibility is that new females force the old ones out. There are many female floaters, and they often stay in the same territory for long periods, interacting frequently with the resident birds. Some females manage to work their way into the group and may even help feed the young. There is a continuous gradation between females that rarely fight with other group members and that help tend the young (and thus can be appropriately called helpers) and females that skulk within the territory and are attacked by group members whenever discovered (true floaters). It is difficult to know what to call those whose behavior is intermediate, that is, those who are with the group often, fight with others infrequently, and offer no help.

Whether they are called unrelated helpers, floaters, or something in between (we call them helper-floaters), the presence of such females is a source of instability. The helper-floater and the breeding female may coexist through one breeding season but not two. By the next year, either the helper-floater is gone and the breeder remains or the helper-floater has become the breeder and the former breeder has moved to another group or died. How the winner is determined in such cases is unknown. DeLotelle has linked conflicts between breeding females and helper-floaters with movement of breeding females. However, Daniels showed that replacement females tend to be younger than the departing females: if females are driven out, they are driven out by one- and two-year-olds. Nevertheless, aggressive displacement remains one of the few possible explanations of the occasional divorces involving older females. Sophisticated mate choice may be another.

One exception to the usually competitive relationship between established breeding females and immigrant helper-floaters was observed in the North Carolina Sandhills. In this case the new female and the old female coexisted for several years, and eventually both laid eggs in the same nest. The two females raised their big brood together, along with the breeding male and a helper who was the son of the breeding male and one of the females. Several more instances of such co-nesting by two females have been observed recently at Camp Lejeune in coastal North Carolina.

In a complex social system such as that of the Red-cockaded Woodpecker, one may witness everything at least once if one observes long enough. There are isolated cases of males dispersing long distances like fledglings after spending a year or more as helpers, a case of two nests within the same group (reported by Reed Rossell and Jackie Britcher, biologists at Fort Bragg in North Carolina), and even one of a breeding male attempting to copulate with his helper male (after a couple of years without a breeding female in the group). But these rare aberrations have an insignificant impact on the dynamics of the system, which operates according to the rules we have described.

Evolution of the Social System

To this point we have described the unusual social system of Red-cockaded Woodpeckers without addressing why they behave in this unusual manner. What are the advantages of staying as a helper that have led to the evolution of the helping strategy? Why do these advantages, whatever they may be, exist only among Red-cockaded Woodpeckers, American Crows, and Brown-headed Nuthatches? Why do they not exist among their close relatives or the other species in their community? The evolution of cooperative breeding is an issue that has attracted much attention from biologists intrigued by the apparent altruism of helpers, who for a time seem to sacrifice their own reproduction in order to help others. This sort of behavior runs counter to the usual self-interest that is the conceptual core of Darwinian evolution.

After more than two decades of intensive research, it is now clear that helpers are not sacrificing themselves at all. They are staying at home, not in order to help but because this is an effective way for them to become a breeder. The question of altruism exists, but it must be rephrased. Instead of asking why they stay to help, we ask why, given that it is to their advantage to stay, do they bother to help raise offspring? It is obviously to their advantage to help defend the territory on which they too depend, but when it comes to raising offspring, why not let the parents do all the work? Readers whose children have entered adulthood but are still living at home may find it easy to believe that young Red-cockaded Woodpeckers are not staying at home in order to help the parents. They are staying for other reasons, as do some young human adults, but the woodpeckers consistently perform their share of the work, unlike their human counterparts! Thus there are two questions to address: Why stay home? and Why help?

WHY STAY HOME?

Let us begin with the question of why it is advantageous for young male woodpeckers to remain at home rather than depart to search for a place to breed. Our answer is that it has to do with the value of cavities. The great expenditure of time required for cavity construction and the longevity of cavities once constructed were detailed in the previous chapter. Because of these features of cavities, a young bird may be better off competing for territories that already have cavities than moving to a vacant area of otherwise suitable habitat and constructing its own new cavities. This creates a social environment in which competition for breeding vacancies, since it is limited almost entirely to existing territories with existing cavities, is intense. Staying at home as a helper enables an individual to compete effectively for a limited number of territories, namely the natal territory and neighboring ones, a strategy that works well only under conditions of intense competition.

Little is known about the dynamics of competition over breeding vacancies. In a study of another woodpecker that is a cooperative breeder, the Acorn Woodpecker (*Melanerpes formicivorus*), Susan Hannon, Ronald Mumme, Walter Koenig, and Frank Pitelka, at the University of California at Berkeley, have described intense, complex power struggles over vacancies. In this species, related helpers from the same territory may fight as a team to acquire a vacancy, and the largest teams generally win. Proximity to the territory with the vacancy is also a factor. Proximity is likely to be important for Red-cockaded Woodpeckers as well, but fighting as a team is surely less important for them than for Acorn Woodpeckers, although a pair of helpers will sometimes acquire a vacancy together. There are hints of complicated struggles but few good observations of the behavior involved.

The hypothesis that the high value of cavities leads to cooperative breeding relates well to general theory. Natal helpers occur in species in which competition for territories is intense. A leading theory is that these species are characterized by having unusually great variation in their territory quality, and that individuals are better off competing for high-quality territories even if reproduction is thereby delayed, rather than accepting whatever opening they can find. This is because the reproductive success of an

FIGURE 6.6 Theoretical relationship between territory quality and fitness of individuals inhabiting territories. The left half of the line represents territories without cavities, the right half territories with cavities. The slope indicates that factors other than presence of cavities affect territory quality, but the worst territories with cavities are substantially better than the best territories without cavities, by the fitness increment corresponding to the connecting line.

individual with a good territory is much greater than that of an individual with a poor territory. Thus, over its lifetime the individual with the good territory will produce more young, even if it begins reproduction several years later than the one with the poor territory. To fit this idea to Red-cockaded Woodpeckers, simply think of territories with existing cavities as the high-quality ones, and potential territories without cavities as the low-quality ones (Figure 6.6).

This hypothesis explains the low frequency with which new territories are created by Red-cockaded Woodpeckers. Generally the same cavity tree clusters are used year after year. In the absence of artificial cavity construction (see below), new territories are created by two natural processes, which have been well described by Robert Hooper. The first process, territorial budding, does not require the excavation of new cavities. Budding involves the splitting of one territory into two, with a corresponding division of cavity trees. Budding may be carried out by a helper, but it is sometimes achieved by an immigrant male who wrests part of the territory and a cavity or two away from the previous owners.

The second process, called pioneering, occurs when an individual moves into totally vacant habitat and constructs a new set of cavities. Budding is much more widely reported than pioneering, which seems to be exceedingly rare in nearly all populations. In the North Carolina Sandhills (200 groups) the rate of population growth (measured in number of groups) through budding over 16 years was 0.6% per year, whereas the rate of growth through pioneering was only 0.06% per year. On Croatan National Forest (50 groups) in coastal North Carolina, the rate of population growth by budding was 2.1% per year over 7 years, whereas the growth rate by pioneering was 0.3% per year. On nearby Camp Lejeune (35 groups) the rate of population growth through budding over 10 years was 0.6% per year, and the growth rate by pioneering was 1.5% per year. Higher rates of new group formation by pioneering have been reported from an unmarked population on the Francis Marion National Forest by Hooper and colleagues. They estimated a population growth rate by budding of 0.4% per year and a growth rate by pioneering of 2.1% per year.

Although all these rates are low, there is substantial variation among populations in the overall rate of new group formation, and the tendency for one mechanism or the other to predominate. The causes of this variation, as well as the factors that trigger these events, are unknown. This is an important gap in our knowledge of the species. Possibly the presence of large numbers of helpers and floaters stimulates new group formation. Also, the frequency of budding and pioneering might be affected by the distribution of cavities and potential cavity trees. It might be easier for new groups to form by budding when cavity trees are scattered than when they are tightly clustered.

The best evidence supporting the hypothesis about the evolution of cooperative breeding in Red-cockaded Woodpeckers that we have presented is not the lack of pioneering but the positive response to artificial cavity construction. If the difference between good and bad territories is the presence of cavities, then one ought to be able to turn bad territories into good ones simply by adding cavities. Walters and coworkers conducted this experiment in the Sandhills of North Carolina in 1989. Using a technique developed by a graduate student member of the research team, Carole Copeyon, two cavities each were drilled in trees in 10 vacant areas, and also in 10 abandoned territories that contained old but possibly unsuitable cavities (see Figure 10.1). The first year 18 of the 20 sites were occupied, and eventually 19 were. None of the 20 control areas, which were like the experimental sites except that they lacked drilled cavities, had yet been occupied in 1994, 5 years later. Although some of the drilled sites were occupied by previously existing groups, a minimum of 13 new groups was added to the population in response to cavity drilling. This represents a minimum population growth rate of 6.0% per year, nearly an order of magnitude higher than the growth rate observed in the same population (0.7% per year, combining pioneering and budding) in the absence of stimulation by artificial cavity construction. This dramatically illustrates the key role that cavities play in the population dynamics of this species and its use of habitat.

The individuals that occupied the drilled sites were young dispersing birds and helpers from nearby groups. The territories instantly became suitable when cavities were added, and they have functioned as good territories since. In some, the birds have even begun to excavate new cavities in other trees, indicating that potential cavity trees that would have allowed pioneering in these areas existed. The birds could have occupied them, but they chose not to. We think it is because of the influence of cavities on territory quality, which we think has driven the evolution of cooperative breeding in this species. Thus the two key characteristics of the species, use of live pines for cavities and cooperative breeding, are intimately linked in a way that drives the population dynamics of the species.

Cavity construction has been widely and successfully employed since 1989, the most spectacular example being its use in recovering the population on the Francis Marion National Forest from the ravages of Hurricane Hugo. Thus research on cooperative breeding has produced a new management technique that has been highly successful. More important, it has led to a new understanding of population dynamics that radically improves our ability to manage and recover the species by rationally employing a variety of management techniques. These issues are explored in Chapters 9, 10, and 11.

This is a classic example of the value of basic research. It seemed foolish to many only a few years ago to construct cavities for a bird that was perfectly capable of making its own. Approaching the problem purely from a management perspective, it would make sense to build homes for Eastern Bluebirds to increase their population, since they cannot make their own, but not for woodpeckers. The rationale for even attempting the experiment came from a basic evolutionary question: why Red-cockaded Woodpeckers are cooperative breeders.

WHY HELP?

Let us now return to the question of why young woodpeckers help. There are several hypotheses about this (Table 6.1). Perhaps the leading hypothesis is that helping behavior is favored by kin selection. Since the offspring they help raise usually are relatives—generally full siblings or half-siblings or nieces and nephews (see Figure 6.5)—by assisting them, helpers are promoting their own genes, expressed not in their own offspring but in other relatives. After all, full siblings share as many genes as parents and offspring. If the degree of relatedness between offspring and helper is close enough, and the survival of the offspring is raised sufficiently to offset the time and energy invested in helping, helping behavior can evolve through kin selection.

Observations of contributions of helpers of varying degrees of relatedness to the offspring they help raise, however, suggest that although most helpers may derive fitness benefits from helping through kin selection, such benefits are not required for helping behavior to occur. Memuna Khan, a graduate student working with Walters at Virginia Tech, observed in the North Carolina Sandhills that helpers unrelated to the offspring they assisted made variable but substantial contributions to incubating eggs and feeding young, just as related helpers did (Figure 6.7).

The behavior of only a small sample of helpers, in Khan and Walters' study and in others, has been observed. It is clear that helper contributions are substantial (Figure 6.8), exceeding that of either breeder for both incubation and feeding young in some groups. But helper contributions, like the contributions of male breeders and female breeders, vary from group to group. The sample currently is too small to reveal what factors determine whether a particular helper, or breeder, makes a relatively large or small contribution to incubation or feeding young. Roy DeLotelle and Robert Epting (then with the University of Florida) have proposed that helper age may be one such factor; specifically they suggested that yearling helpers feed recently fledged young less than older helpers. Kinship may be a factor as well, but if it is, it has subtle effects on the relative size of the contribution rather than determining whether or not help occurs.

Helpers may also receive indirect fitness benefits if their activities increase the survival of breeders related to them, or if the young they help raise become future helpers of breeders related to the original helper (Table 6.1). The second benefit certainly occurs in Red-cockaded Woodpeckers, but its magnitude is relatively small compared to that of other cooperative breeders. Khan and Walters found in North Carolina that if effects of breeder age and territory quality (see below) were controlled, breeders did indeed enjoy higher survival in the presence of helpers.

A final possible indirect benefit is that helpers, through their assistance, may reduce the effort that related breeders must expend on parental care. Khan and Walters found that indeed breeders spend less time incubating and feed nestlings less often when assisted by helpers (Figure 6.8). This may account in part for the positive effects of helpers on the survival of breed-

Table 6.1. Benefits of Helping Behavior in Red-cockaded Woodpeckers

Type of Benefit	Hypothesis	Results for Red-cockaded Woodpecker
Current Direct Benefits		
Enhanced survivorship of helper	1. Benefits of group living	Helper survival decreases at larger group sizes (Khan and Walters, unpublished data).
Pay-to-stay	2. Breeder demands help in return for tolerating helper.	No information available.
Future Direct Benefits		
Increased future reproductive success of helper	3. Helping experience increases reproductive success when helper becomes a breeder.	Helping experience does not significantly enhance age-specific reproductive success (Khan and Walters 1997).
	4. Original recipients of help may assist helper once the helper attains breeding status.	Reciprocity does not occur more often than expected at random (Khan and Walters 2000).
Increased future probability of breeding	5. Original recipients of help may assist helper in acquiring a breeding position through territorial budding.	Unlikely mechanism, because territorial budding occurs rarely.

6. Original recipients of help may disperse with helper and assist in competition for a territory.	Unlikely, because joint dispersal occurs rarely.
7. Helping may increase a helper's access to mates, through mating with the breeder of the opposite sex.	Unreasonable, because this species is monogamous (Haig et al. 1994).
8. Helping increases access to mates because helper pairs with opposite-sex breeder when the same-sex breeder dies.	Unlikely, because resident breeders are often parents of helper, and individuals who acquire territories through dispersal usually obtain mates also.
Current Indirect Benefits Increased inclusive fitness	
9. Helper increases survival probability of the recipient nestlings.	Helpers increase the number of fledglings by 0.39 (Heppell et al. 1994).
Future Indirect Benefits Increased inclusive fitness	
10. Helper increases the annual survival probability of breeder by (1) group size effects and/or (2) reducing breeder workload, thereby increasing the probability of the breeder producing young in subsequent years.	(1) Helpers significantly increase the annual survival probability of breeders (Khan and Walters, in press). (2) Helpers significantly reduce breeder workload during incubation and nestling feeding stages of the breeding season (Khan and Walters, in press).

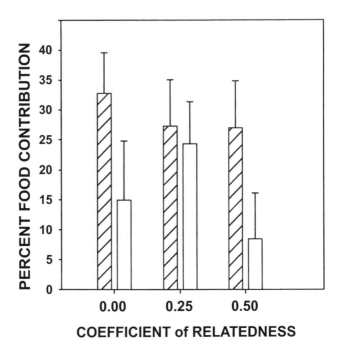

FIGURE 6.7 Percentage of feeding of nestlings (hatched bars) and fledglings (clear bars) by the helper in groups of three as a function of relatedness between the helper and the young. Based on two groups observed in each category in the nestling stage, and three groups in each category in the fledgling stage. Data from the North Carolina Sandhills, courtesy of Memuna Khan.

ers, and also breeders may use their extra time to somehow enhance their future reproduction. For example, they may renest more readily because they have more time for foraging, or they may be better able to complete new cavities that raise productivity because they have more time for cavity excavation.

It is also possible that helpers receive direct benefits from their helping behavior (Table 6.1). The offspring they help raise may stay on the territory and someday become the helpers of the helper, when the latter inherits breeding status. Khan and Walters have shown that this is only a trivial benefit of helping in Red-cockaded Woodpeckers. Another popular notion is that by helping, young birds acquire skills that benefit them when they become breeders; that is, helping is good practice of parenting skills. This hypothesis has been ruled out for Red-cockaded Woodpeckers by Khan and Walters, who found that in the North Carolina Sandhills former helpers do no better when they first attempt reproduction than individuals of the same age that were floaters or solitary males previously. A final possible direct benefit is that by increasing reproduction, helpers raise group size, which increases their own survival. Larger groups may be more effective at spotting predators or finding food than smaller ones, and thus are better ones in which to live. Again, however, data from the Sandhills do not support this. Khan and Walters found that helper survival is in fact lower, not higher, in larger groups. This result suggests that there may be a cost to living in groups, perhaps due to competition over critical resources such as food or cavities. It is interesting that helpers suffer this cost but breeders do not.

The benefits to helpers resulting from their actions thus are becoming increasingly clear. Indirect benefits through kin selection occur, and reduced breeder workload may provide additional indirect benefits. Reciprocity through future assistance from the young a helper helps raise is likely inconsequential, and helpers do not improve their own survival or parenting skills by helping. Another hypothesis proposed by Stephen Em-

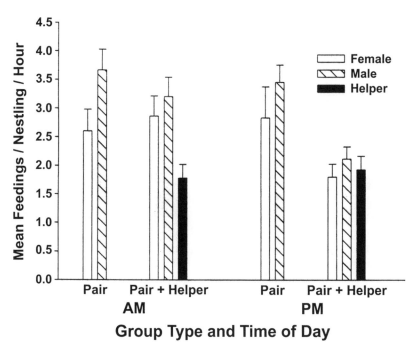

FIGURE 6.8 Individual contributions to nestling feeding for pairs and pairs with one male helper as a function of time of day. Data from the North Carolina Sandhills, courtesy of Khan and Walters, in press.

len of Cornell University, one that appeals to human intuition, is that the breeders will not allow the helpers to stay and enjoy the benefits of their protection and their territory unless the helpers help. Finally, researchers in New Zealand, Ian Jamieson and John Craig, have proposed that there are no advantages to helping behavior at all, that such behavior is an inevitable result of having adult individuals around nestlings and fledglings. According to this hypothesis, helping behavior is merely an accidental byproduct of instincts related to parental behavior. It is highly likely, actually, that the initial inclination of the first helpers was to help. However, one would expect that selection would have eliminated such behavior by now if it had no advantages.

Although it is not entirely clear why helpers help, it is quite clear that such behavior is useful. Most of the hypotheses just discussed depend on the assumption that reproduction is improved through the activities of helpers, and the evidence indicates that it is. In all but one of the populations studied, the number of fledglings produced by groups with helpers was higher than the number produced by pairs without such assistance (by 0.6–0.7 fledglings per group, Figure 6.9). The exception was the small population in central Florida studied by DeLotelle and Epting, in which reproduction in groups with helpers was actually worse than that of groups without helpers.

Since helpers arise as natal offspring among cooperative breeders, one generally finds more helpers on high-quality territories, on which conditions for reproduction are better, than on low-quality territories. Young may be more likely to stay rather than depart on high-quality territories as well. Therefore the correlation between reproductive success and helpers often is due in part to territory quality, with high reproductive success causing the presence of helpers rather than the other way around. Lennartz and coworkers showed this to be the case for the Red-cockaded Woodpeckers on the Francis Marion National Forest. Selina Heppell, a graduate student working with Walters and Larry Crowder at North Carolina State University, subsequently found the same result in the North Carolina Sandhills. By comparing the same pair of breeders on the same territory for years with helpers and years without helpers, they were able to show that part of the correlation between reproductive success and presence of helpers was indeed due to the positive effects of helpers on reproduction. The same pair on the same territory produced on average 0.3–0.4 more fledglings in years when they had help than in years when they did not.

How helpers increase reproductive success in Red-cockaded Woodpeck-

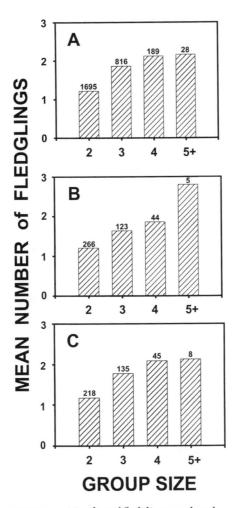

FIGURE 6.9 Number of fledglings produced as a function of group size in populations on *(A)* the North Carolina Sandhills, *(B)* Camp Lejeune, and *(C)* the Croatan National Forest. Sample sizes are indicated above each bar.

ers is not yet clear. One might think the answer is obvious: Since helpers assist in feeding, more food is brought to the nest, and hence more young can be raised. Yet it is often the case among cooperative breeders that assistance from helpers simply enables the breeders to do less provisioning of nestlings rather than resulting in any difference in the total food provided at the nest. Similarly, in species such as Florida Scrub-Jays, helping has less effect on loss of parts of broods to starvation than on loss of whole broods to predation. It seems, surprisingly, that helpers often improve nest defense rather than enhancing feeding of young.

The situation appears to be similar for Red-cockaded Woodpeckers. Khan and Walters found that breeders assisted by helpers reduced their feeding efforts (Figure 6.8) and that their young therefore were fed no more often than young in groups lacking helpers. Both Melinda LaBranche (a graduate student working with Walters at North Carolina State) in North Carolina and Lennartz and coworkers in South Carolina have shown that whole brood loss is less in groups with helpers than in groups without helpers (16% versus 27% in North Carolina). An effect of helpers on partial brood loss has yet to be demonstrated. LaBranche also showed that clutch size is larger in groups with helpers, and therefore that clutch size differences also contribute to higher reproductive success in groups with helpers.

It is not yet clear whether differences in whole brood loss or clutch size are part of the helper portion or the territory quality portion of the correlation between group size and reproductive success. That is, it may be either the high quality of territories on which helpers reside or actions of the helpers themselves (or both) that cause clutch size to be higher and whole brood loss to be lower in larger groups.

Population Dynamics

One of the ironies of the biology of Red-cockaded Woodpeckers is that a species with such unusually great potential for population stability should become endangered. The unusual population dynamics of the species follow from its breeding system, and they can be one of the species' greatest assets or one of its greatest problems, depending on the circumstances. Stability arises from the presence of a large nonbreeding class, the helpers. These provide a pool from which to draw replacement breeders. Thus, unusually low breeder survival and unusually low reproductive output will reduce the breeding population little, if at all, in the following year. The size of the helper class will fluctuate, with variation in mortality and reproduction, but the number of breeders will not (Figure 6.10). Thus, the number of breeding groups in a population is extremely stable. This makes isolated populations, especially small ones, less vulnerable to the adverse effects of random fluctuations in demography, which threaten persistence over long periods of time in more typical species.

There may be some effect of fluctuations in demography on subsequent

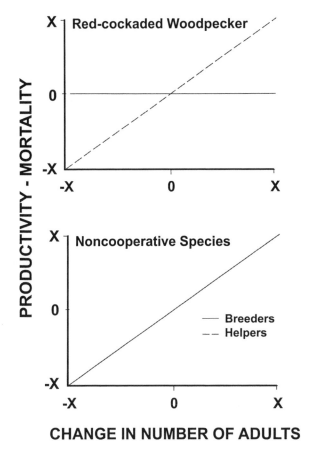

FIGURE 6.10 Effects of production and mortality on breeder and helper numbers. In Red-cockaded Woodpeckers and other cooperative breeders, excess production results in an increase in the helper class, and excess mortality a decrease, whereas breeder number stays constant. In noncooperative species, differences between production and mortality translate directly into changes in breeder numbers. (The relationship may not be direct if breeding space is somewhat limited, so that some birds become floaters instead of breeders.)

reproduction, because less of a buffer is provided by helpers and floaters for females than for males. Therefore poor reproduction or high mortality may be followed by an increase in the number of groups without females, including solitary males and males with helpers the next year. But the number of occupied territories is unchanged, and the population can quickly recover from such events.

The population dynamics of Red-cockaded Woodpeckers also are characterized by high survival rates and low reproductive rates relative to other birds of similar size living in the temperate zone. The demography of Red-cockaded Woodpeckers is like that of tropical species of similar size or temperate species of larger size, and like that of other temperate cooperative breeders like Florida Scrub-Jays and Acorn Woodpeckers. Annual survival rates of adults vary from 75% to 85% for males and 70% to 80% for females, depending on the population, and survival in the first year varies from 45% to 60% for males and 35% to 45% for females. Annual productivity ranges from one to two fledglings per year, depending on the population and the year (see below). Individuals have been known to live 17 years or more, although signs of senescence (i.e., greatly reduced annual survival and productivity) become evident at about age ten. The primary predators of Red-cockaded Woodpeckers are Cooper's (*Accipiter cooperii*) and Sharp-shinned (*A. striatus*) hawks, and rat snakes. High survival and low productivity also contribute to population stability.

The unusual biology of the Red-cockaded Woodpecker reduces the sensitivity of populations to fluctuations in demography, but it heightens sensitivity to habitat quality. If a territory remains a good one, one would expect it to remain occupied continuously over long periods of time. However, a territory can quickly turn from good to poor if cavities are lost (through tree death, hurricanes, cavity enlargement, or hardwood encroachment; see Chapter 5), and the birds will then abandon it. Therefore populations can decline in terms of numbers of breeding groups without noticeable changes in survival or reproductive success. Populations are resistant to threats that impact individual birds, but they are vulnerable to threats that impact habitat.

For small populations, stability is both a blessing and a curse. The population is unusually resistant to further decline, but for the same reasons it has unusually low potential for growth. The number of groups may increase through budding and pioneering, but only very slowly. Otherwise, population growth depends on managers converting poor territories to good ones, primarily through cavity construction and midstory control.

One factor that can reduce population stability is spatial isolation of groups from one another. Again, this is a consequence of the dynamics of the breeding system. One can categorize the dispersal of Red-cockaded Woodpeckers into two types: the short-distance dispersal practiced by helpers and to a lesser extent by breeding females, and the long-distance dispersal practiced by first-year birds. The former type is limited to neighboring territories. If cavity tree clusters are too far from one another, this type of dispersal is precluded. Where territories are grouped together, when a

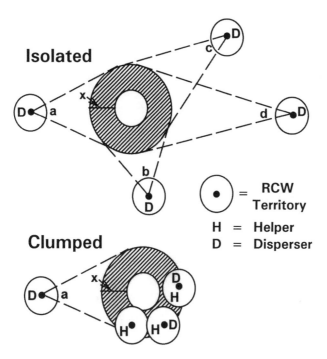

FIGURE 6.11 Differences in dispersal dynamics between isolated and clumped territories. The hatched area indicates the zone (of radius x) in which passing birds are close enough to the central territory to detect a breeding vacancy within it. For helpers, this zone must overlap their territory. None of the other territories are close enough in the isolated case for helpers from them to fill breeding vacancies, but 3 of 4 territories are in the clumped case. Dispersing fledglings must move in a direction that brings them within the zone. The range of directions over which this will occur is indicated by the angles a, b, c, and d, and is a function of distance from the central territory. The proportion of fledglings that will move in the proper direction will be, for example, $a/360$. However, if the natal territory is within the zone, any dispersing fledgling will detect the vacancy if it exists at the time of dispersal. Thus in the clumped case, all helpers and all dispersing fledglings from 3 territories will detect the vacancy, and $(a/360)$ of the dispersing fledglings from the fourth territory will detect the vacancy. In the isolated case, only $(a + b + c + d)/(4 \times 360)$ of the dispersing fledglings from the 4 territories, and no helpers, will detect the vacancy.

breeding vacancy arises there will be a number of helpers in the vicinity to compete for it, along with whatever floaters and dispersing young birds are in the area, and the vacancy will be filled quickly (Figure 6.11). Groups that are clumped spatially enjoy the buffering effect of helpers.

In contrast, if a vacancy arises in an isolated cluster, none of the helpers in the population will be aware of it. The vacancy will be filled only if a long-distance disperser happens to pass through the area (Figure 6.11). Wandering females are no less common among Red-cockaded Woodpeckers than among other species, since so few females are helpers. But wandering males are relatively rare, because only a small fraction of first-year male Red-cockaded Woodpeckers are long-distance dispersers, whereas all first-year males in birds that are not cooperative breeders are dispersers. The odds that a dispersing young male will head from another cluster in the right direction to locate the vacancy may be very slim if the cluster is sufficiently isolated. Thus, the buffering effects of helpers do not extend as much to isolated groups, and they suffer from the small number of males dispersing long distances. For females, the problem of isolation is a simple matter of the reduced efficiency of long-distance dispersal.

GEOGRAPHIC VARIATION

Populations are not uniform in their dynamics across the range of the species. Some differences between populations are related to the condition of the habitat in which the populations reside. In particular, a relatively high proportion of solitary males is an indicator of a population in poor condition. Solitary males result when territories are isolated from one another so that dispersing females have difficulty locating males who have lost their mates, as well as when territory quality deteriorates so that females no longer accept them. Average group size, which reflects the prevalence of both solitary males and helpers, can change greatly within a population as habitat deteriorates or improves. It can also differ greatly in adjacent areas if those areas differ in habitat quality, as Fran James has shown on the two districts of the Apalachicola National Forest in Florida.

It appears that territory occupancy, both in terms of the extent to which males are able to attract mates and the extent to which territories are completely abandoned, is the primary aspect of population dynamics that differs between populations in high- and low-quality habitat. Variation in habitat quality is not, on the other hand, consistently associated with variation in productivity of breeders or mortality. Significant differences between populations in fecundity and mortality exist but appear to be evolved features associated with geographic location rather than responses to habitat conditions.

Red-cockaded Woodpeckers appear to exhibit a common geographic pattern in demography. More northern and more inland populations tend to have higher productivity and lower survival rates than more southern and coastal populations. For example, Mike Lennartz and David Heckel of Clemson University reported a breeding male retention rate (roughly equivalent to annual survival) of 78% for a coastal South Carolina population and 72% for an inland population in northern Georgia. A similar difference between the inland Sandhills population and two coastal populations has been observed in North Carolina. Near the southern end of the range in Florida, DeLotelle and Epting reported a breeding male retention rate of 86% for the small population they studied. Productivity rates at these sites, measured as fledglings produced per successful nest, followed the same pattern, being lowest in Florida and lower in coastal than inland regions in the Carolinas. Differences in productivity are largely a product of differences in partial brood loss (see below).

These variations in demography are not large, but they appear to be consistent. Why such differences have evolved among Red-cockaded Woodpecker populations has not been investigated. Conditions may be better for reproduction farther north and inland, due to higher seasonal peaks of food abundance, but worse for survival because winters are more harsh. Alternatively, these differences may represent variations in life history strategies. In animals, the association of higher survival rates with lower reproductive rates is generally thought to represent increased investment in future reproduction at the expense of current reproduction. That is, less reproductive effort results in lower productivity but higher survival to breed in subsequent years. Higher survival and lower reproduction often are associated with more competitive conditions. Where conditions are more competitive among Red-cockaded Woodpeckers, one might also expect to see more emphasis on the helping, rather than dispersing, strategy. It is not clear whether this is true for male birds. The percentage of males of age 1 that remained in their natal territory, rather than dispersing, is about 70% in all three North Carolina populations, despite the survival differences between them. This figure represents surviving birds, however, and thus is not necessarily equivalent to the proportion that followed the helping strategy during their first year.

Among females, it is clear that there is a greater tendency to adopt the helping strategy in populations where mortality is lower. The proportion of females that remained as helpers is 4% and 7% in the two coastal North Carolina populations, compared to only 1% in the inland Sandhills population. DeLotelle and Epting reported even higher frequencies of female helpers in Florida. Presumably, higher survival results in fewer breeding vacancies and thus more intense competition for those vacancies, increasing the viability of the helping strategy for females.

Population Viability

A critical concern about the population dynamics of endangered species is the long-term persistence of populations. Persistence typically is assessed through the techniques of population viability analysis. The persistence of populations is threatened by several factors, and the effects of all of them must be evaluated. Catastrophes are one form of threat, and the devastation wrought by Hurricane Hugo in 1989 on the Red-cockaded Woodpecker population on the Francis Marion National Forest illustrates that the threats posed by such infrequent events are real (Figure 6.12). One can do little about predicting the occurrence of catastrophes other than determining their likely frequency over large areas for long periods of time. Hooper, working with Colin McAdie from the National Hurricane Center in Florida, conducted such an analysis with respect to hurricanes within the range of Red-cockaded Woodpeckers, and showed that over the time interval of hundreds of years appropriate for viability analysis, such events are almost certain to occur (Figure 6.13). The devastation caused by Hugo represents an extreme, however. The impact of major hurricanes striking the Red-cockaded Woodpecker populations on Eglin Air Force Base in Florida and Camp Lejeune and the Croatan National Forest in North Carolina in 1996 was much less (see Chapter 5). In fact, the impact of these storms at the population level was negligible.

We cannot predict which populations will be affected and when, but we can be sure that what happened on the Francis Marion will happen again somewhere. We can also expect that events such as those at Eglin, Lejeune, and Croatan will be regular. The appropriate response to demonstrated threats from catastrophe is to maintain multiple populations, so that the species as a whole is less vulnerable. This is part of the management strategy for Red-cockaded Woodpeckers. Furthermore, we have the ability to help a population recover quickly from a hurricane (see Chapter 10). This was a factor in minimizing the decline of the Francis Marion population after Hugo and in its subsequent increase, and in preventing the storms in Florida and North Carolina from reducing the number of groups in those populations.

It also is possible to reduce the vulnerability of populations at risk to wind damage. Hooper and McAdie have offered a number of suggestions, such as reducing the access of wind to stands and creating conditions for tree growth that favor the development of greater wind resistance. The presence of large openings near Red-cockaded Woodpecker clusters makes them especially vulnerable to wind damage, as discussed in Chapter 5. The

FIGURE 6.12 Some of the devastation caused by Hurricane Hugo in 1989 at the Santee Experimental Forest. Similar damage occurred on the Francis Marion National Forest. Photo by R. N. Conner.

devastation wrought by Hugo may have been exacerbated by the fact that sections of the forest resembled checkerboards, consisting of openings created by clear-cutting timber that alternated with stands occupied by woodpeckers. This style of timber management can be avoided.

With proper management before and after storms, the impact of hurricanes on Red-cockaded Woodpecker populations can be reduced but not eliminated. By maintaining a sufficient number of populations spread over its entire range, risk to the species as a whole can be eliminated. As long as we can maintain populations against threats other than catastrophes, we can feel fairly confident that the persistence of the species is not seriously threatened by the kinds of catastrophic events it is likely to encounter over the next few hundred years.

Other threats to viability are related to the size of the population. These are divided into three types: (1) demographic fluctuations, (2) environmental fluctuations, and (3) loss of genetic variability. Demographic fluctuations are chance events that affect a particular individual. *Accipiter* hawks may catch woodpeckers every winter, but an element of luck is involved in determining whether a particular attempt is successful. Survival will be higher in a year when luck is more often on the side of the wood-

Hurricane Landfall (1889 – 1989)
In Relation to 1985 Recovery Populations of Red-cockaded Woodpeckers

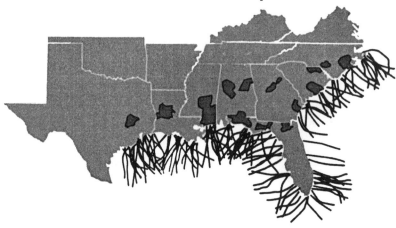

FIGURE 6.13 Hurricane landfall relative to recovery populations of Red-cockaded Woodpeckers between 1889 and 1989 in the southeastern United States. Lines indicate hurricane paths during the 100-year period. Adapted from map provided by R. G. Hooper, U.S. Forest Service.

peckers than in a year when it is more often on the side of the hawks. Environmental fluctuations refer to events that affect all individuals in the population. There are good years and bad years—for example, survival will be higher in a winter in which hawk density is low than in a winter in which it is high. Viability analysis involves determining the extent to which these fluctuations will cause a population to fluctuate in size, and evaluating the probability that over an extended period of time populations of various initial sizes will fluctuate to such low levels that they will go extinct.

The final threat, loss of genetic variability, although just as real, is much harder to assess than demographic or environmental fluctuations. In each generation, each gene present in a population is expressed in a limited number of individuals. Some of these individuals may be very successful and pass the gene on to many members of the next generation. Others may die young and pass the gene on to no one. There is some possibility that all the individuals carrying a particular gene may suffer the latter fate, so that the gene is lost to the population. The smaller the population, the fewer the individuals carrying a particular gene, and the greater the chance that a gene will be lost through this random process. Mutation may regenerate some of the lost genetic variation, but below some population size the rate of loss exceeds the rate of mutation, and a net loss occurs.

GENETIC CONCERNS

Clearly, if all genetic variation is lost, there is no potential for evolution in response to environmental change. The relationship between amount of variation and potential for evolution is unknown, however, as is the relationship between potential for evolution and probability of extinction (Figure 6.14). That is, the relationship between loss of genetic variability and population viability is unknown. We know that this relationship will differ among species, but we do not have the capability of determining the rate of loss that is acceptable for any particular species. We recognize that loss of variation is bad, but we do not know precisely how bad it is. Because

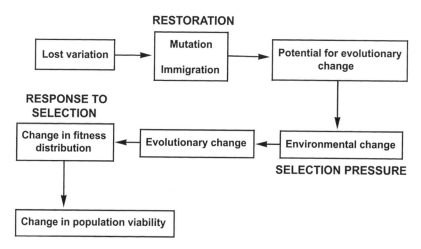

FIGURE 6.14 The complex relationship between loss of genetic variation and loss of population viability. The rate at which variation is lost due to genetic drift is a function of population size, and is readily determined. However, it is difficult to determine how much of the lost variation is restored through mutation and immigration, and thus how much evolutionary potential remains. How much lost potential matters depends on how much the environment changes. Resulting evolution depends on both the capacity for change inherent in the remaining genetic variation and the magnitude and direction of new selection pressures. Population viability depends on how well or poorly the organism's adaptation to its new environment in response to selection equips it to survive and reproduce in that environment. The pathway from potential for change to altered population viability is unpredictable and dependent on the nature of the future environment.

we do not know how much loss is acceptable, the strategy conservationists have adopted is to not allow any.

Determining how large a population must be to avoid any loss of genetic variation is also plagued by uncertainties (Figure 6.14). Population genetics theory provides sophisticated means to calculate the rate of loss for a particular population, but we can only guess what the rate of loss would be that would keep the population in equilibrium with the mutation rate. It is loss above the mutation rate that must be avoided. Conservation biologists, in the absence of precise information, have resorted to general standards. For vertebrates the standard is an effective size of 500 breeding adults, or 250 pairs.

An effective size is the number of breeders that, when all contribute equally to reproduction, corresponds to a particular rate of loss of genetic variability. Thus 500 is the effective size that corresponds to a rate of loss equal to the mutation rate in the standard vertebrate. Of course, not all individuals in a population contribute equally to reproduction. Therefore the effective size of a population is never exactly equal to the actual number of adults in the population. The actual population size is almost always higher than the effective size, often much higher, because nonbreeders (such as helpers) contribute nothing to reproduction, and variation in reproductive success among breeders makes effective size less than actual size.

It is possible to calculate the effective size of a population of any actual size if the demography of the species is known. The demography of Red-cockaded Woodpeckers is well known, enabling, presumably, relatively accurate calculations of effective size. This problem has been thoroughly investigated, especially by Michael Reed, who did so as a graduate student working with Walters at North Carolina State. The population size required to achieve an effective size of 250 pairs was established as roughly 310–390 actual pairs, depending on the details of the demography of particular populations.

Currently, population objectives for the conservation of Red-cockaded Woodpeckers are based solely on these values. Populations are considered viable if they meet the standards for an effective size of 500. Population size objectives thus rest on very shaky ground. We can have some confidence that a population of 310–390 pairs has an effective size in the neighborhood of 500, but we have no idea whether such a population will actually retain all its genetic variability, or if it does not, whether whatever loss occurs will threaten its viability (Figure 6.14). Recent work by Russell Lande at the University of Oregon indicates that mutations that restore lost variability occur at lower rates than those used to determine population size standards. The required effective population size to avoid loss of genetic variability is actually on the order of 5,000 rather than 500. This translates into 3,100–3,900 pairs of woodpeckers.

Lack of precision in genetic standards has led most conservation biologists to advocate avoiding their use in setting population goals and substituting instead goals based on analysis of effects of demographic and environmental fluctuations. For Red-cockaded Woodpeckers, goals based on genetic standards are unachievable anyway. The only realistic way to counter loss of genetic variability from a population is through gene flow into it from other populations, whether by natural immigration or artificial translocation. How much gene flow is required is a complex problem that has yet to be solved. The best strategy currently is to promote gene flow by preserving as many populations within a region as possible.

A more immediate threat posed by loss of genetic variability is inbreeding depression. It is well known that mating of individuals related by common ancestry can result in reduced viability of offspring. Inbreeding depression results from segregation of partially recessive, deleterious alleles. Daniels and Walters found that in Red-cockaded Woodpeckers inbreeding depression occurs and is manifested by reduced hatching rates of eggs and reduced survival of fledglings to age 1 year. As a result, inbred pairs produce 44% fewer offspring surviving to age 1 year than do unrelated pairs.

The probability that inbreeding will occur even if individuals mate at random is greatly increased in a small population. The relevant population sizes are two orders of magnitude smaller than for loss of genetic variability, that is, 50 instead of 5,000. Thus Red-cockaded Woodpecker populations should contain at least 31–39 pairs in order to avoid adverse effects of inbreeding depression. Daniels and coworkers used the complex population model developed at North Carolina State University (see below) to show that in populations as large as 49 groups, close inbreeding will reach levels high enough to threaten population viability. Immigration can ameliorate this effect, and populations of 100 groups are not significantly affected even in the absence of immigration.

Daniels' work suggests that many populations of Red-cockaded Woodpeckers have become vulnerable to the effect of inbreeding depression. Peter Stangel, then with the Savannah River Ecology Laboratory, and coworkers examined genetic variability in several populations of Red-cockaded Woodpeckers and found that levels of variability were still high, although some small populations appeared to have experienced reductions. Populations differ more than is typical of birds, as is expected from the species' unusual social structure, but populations are not genetically distinct. Susan Haig and colleagues confirmed this finding using a second technique. These results are consistent with the idea that Red-cockaded Woodpeckers were formerly distributed fairly continuously and have only recently, in evolutionary terms, experienced fragmentation and isolation of populations. One might therefore expect Red-cockaded Woodpeckers to be unusually susceptible to problems related to loss of genetic variability and inbreeding depression, since they have no history of the conditions that produce these losses. Daniels and Walters' work supports this contention with respect to inbreeding depression. Stangel and coworkers found that adverse effects of inbreeding are not yet evident in small populations, presumably because they have only recently (in evolutionary terms) been reduced to small size. Currently, populations probably are becoming inbred, losing variability, and diverging genetically.

DEMOGRAPHIC CONCERNS

Unfortunately, demographic analyses, although the cornerstone of viability analysis, are inordinately difficult to apply to Red-cockaded Woodpeckers. Use of standard methods requires simplifying assumptions to the point that the complexities of the species' social system are eliminated, making results suspect. Several researchers have modeled population dynamics using simple life tables for females, an approach that fails to capture the buffering effect helpers have on population fluctuations. Because territory occupancy is so critical to population dynamics in this species, and because territory occupancy reflects fluctuations in the male rather than the female population, population dynamics are best modeled including males.

Heppell, Walters, and Crowder used a more sophisticated life history matrix that includes the helper stage to model population dynamics of males. They thus modeled life history not just in terms of births and deaths, as the simpler models do, but also in terms of transitions between the fledgling, helper, and breeder stages. Still, this approach fails to capture the spatial structure in population dynamics. For example, in the model helpers fill any breeding vacancy that occurs in the population, not just those on the natal and neighboring territories. Accordingly, Heppell and coworkers used their matrix to demonstrate the degree to which population numbers would be affected by alterations in different components of the life history, and not to model population behavior over time. They found that population size was especially sensitive to the fecundity of older breeders and the probability of transition from helper to breeder.

Lynn Maguire and her students at Duke University developed a model of interacting groups that includes births, deaths, transitions in status, and even dispersal. From this they derived a stage-based matrix model similar to that employed by Heppell and coworkers. This model also lacks spatial structure and thus suffered from the same shortcomings when used in viability analysis.

The simulation technique known as spatially explicit, individual-based modeling is another alternative to the analytical matrix models traditionally used in viability analysis. This technique allows spatial structure, as well as complexities of the social system, to be incorporated into a population dynamics model. Ben Letcher, Jeffrey Priddy, Walters, and Crowder developed such a model at North Carolina State University (and later Duke University). Their analyses indicate that spatial structure of populations has a large impact on their viability.

Typically, demographic fluctuations are a threat only to populations of about 50 or fewer individuals. Environmental fluctuations are a threat at much larger (hundreds or thousands) and much less predictable population sizes. The magnitude of these threats depends greatly on annual variation in birth and death rates. Mortality rates appear to vary relatively little from year to year in this species (Figure 6.15), which, coupled with the buffering effect of helpers, is likely to result in reduced impacts of demographic and environmental fluctuations compared to those of most species. Therefore, populations potentially are viable at relatively small sizes. On the other hand, annual variation in reproduction is considerable (see below) and hence is likely to be a primary determinant of population viability.

Letcher and colleagues reported that in the face of demographic fluctuations alone, populations of even 15 groups were highly persistent if territories were clumped spatially, but populations as large as 169 groups declined if territories were scattered. Walters, Crowder and Priddy found that when both environmental and demographic fluctuations were included, populations of 100 or fewer groups were vulnerable to extinction, even when territories were maximally clumped. Populations of 250 or more groups were not vulnerable, even when territories were scattered. Viability of populations between 100 and 250 groups depended on spatial configuration as well as population size.

Thus, how large Red-cockaded Woodpecker populations must be to persist over long periods of time is only beginning to become clear. The current standard of 310–390 pairs is based solely on avoiding loss of genetic variability, and is no better than a poor guess. We can expect this standard to be replaced in the near future by a more precise one based on demographic viability analysis that uses individually based, spatially-explicit simulation

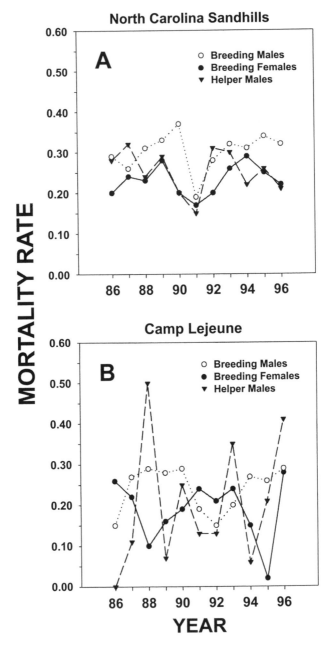

FIGURE 6.15 Variation in annual mortality rates for breeding males, breeding females, and helper males on *(A)* the North Carolina Sandhills and *(B)* Camp Lejeune. Based on monitoring of roughly 200 groups in the Sandhills and 30–50 groups at Camp Lejeune.

modeling. Spatial distribution is as critical as population size. A standard of 250 groups may be reasonable for long-term viability in the face of environmental fluctuations, but populations as small as 100 groups may be viable if properly configured spatially. Even smaller populations may be unusually persistent demographically, but likely will suffer eventually from the effects of inbreeding depression. Over extremely long time frames, populations that meet the viability standard will suffer from loss of genetic variability. A solution to this problem through gene flow is required to ensure viability.

Reproduction

Like nearly all resident species in the temperate zone of North America, Red-cockaded Woodpeckers initiate breeding activities in early spring. In March or perhaps late February, the breeders begin to exhibit courtship behavior, including copulations. Soon the group begins to work intensively on the resin wells around the cavity that is to be used for nesting, which typically is the roost cavity of the breeding male. Egg-laying may begin from mid-April to early May; it starts in late April in most locations. This is somewhat later than for many resident species—for example, chickadees and titmice—but earlier than for migratory species. Perhaps significantly, in many areas Red-cockaded Woodpeckers nest earlier than the other woodpecker species that sometimes usurp their cavities—Red-bellied and Red-headed woodpeckers. Occasionally Red-cockaded Woodpeckers fledge young from a cavity that is later used for nesting by one of these other species.

The basic parameters of reproduction were revealed by the studies of David Ligon in the late 1960s and early 1970s, although the extensive research conducted over the next 20 years has added to knowledge of variation in these parameters. The incubation period for Red-cockaded Woodpeckers is only 10–11 days, measured from the laying of the last egg to the hatching of the first. This is one of the shortest incubation periods of any bird species. Nestlings fledge 24–29 days after hatching (Plate 9). When the young first fledge, they remain stationary on a trunk or horizontal limb for long periods, as the adults fly back and forth bringing them food from wherever they are foraging. Gradually the fledglings begin to move with the group, and adult feeding visits cover shorter and shorter distances.

Eventually the fledglings begin to follow individual adults as they forage, begging for food. During this time, the fledglings may fight frequently over access to adults. One may fly in and displace another to whom an adult was

bringing food, so that the adult feeds the newcomer instead. Or the dominant fledgling may displace another fledgling who is positioned, begging,
near an adult. The fledglings establish dominance hierarchies among themselves. Males dominate females more often than not, but not invariably.
Hatching order may affect dominance status, and thus hierarchies may be
established at the nestling stage, but this has not yet been demonstrated.
The squabbling and begging of fledglings make groups even noisier than
usual during the period 2–6 weeks after fledging.

The rate at which fledglings are fed drops off over time, more dramatically in some years and some groups than in others, and the fledglings increasingly forage for themselves. Still, occasional feedings are observed for
an unusually long period in this species, up to 5 months after fledging. The
fledglings may remain with their natal group for an unusually long time.
Those that become helpers, of course, remain into the next breeding season. But even among those that disperse in their first year, many do not
do so until the next breeding season is about to begin (March or April).
Another peak in dispersal occurs a few weeks after fledging (July or August), but a few fledglings disperse each month between these two peaks. A
late summer peak in natal dispersal is not unusual among birds, whereas
an early spring one is highly atypical.

VARIATION IN REPRODUCTION

Reproduction varies both across populations and across years within populations. This variation may be partitioned into four components: (1) nesting
effort, (2) clutch size, (3) partial brood loss, and (4) whole brood loss.

Nesting effort has two components, initiation of first nests and renesting after nest loss. Nesting effort varies considerably across years but little
across populations. The two components of nesting effort are highly correlated; that is, years in which many groups do not nest are usually years in
which few groups renest. Annual values for proportion of groups not nesting as low as 0% and as high as 25% have been reported, values of 5–15%
being typical (Table 6.2). Groups whose nests fail early and quickly, before
observers can detect the nest, may be recorded as not nesting. Adjusting for
this produces typical values of 5–10% for the proportion of groups not nesting. Groups in which the breeding male is age 1 year are especially likely
to fail to nest, and social conflict within groups is often associated with
failure to nest.

Frequency of renesting averages about 30% but is highly variable across
years. In some years in some populations no renesting occurs, whereas as

many as 75% of nest failures are followed by renesting in others. Typical values are 10–40% for most populations (Table 6.2), although unpublished data provided by Mike Lennartz and Ted Stevens indicate an average value of under 5% was observed on the Francis Marion prior to Hurricane Hugo. Renesting is most likely when nests are lost early in the breeding season, and is much more likely for nests lost at the egg stage than for those lost at the nestling stage. Productivity of renests is much lower than that of first nests, due to greater rates of both whole and partial brood loss. This is consistent with the general trend for nesting success to be highest early in the breeding season, a common trend among birds.

Years in which nesting effort is high are characterized by unusually long nesting seasons, with young being fledged as late as August in some cases. In years when nesting effort is low, nearly all young may be fledged by the end of June (Figure 6.16). Overall, the nesting season is relatively short compared to that of other resident birds.

In a year of high nesting effort, a few groups may attempt a second brood following a successful first nesting, although this is very rare. This has been reported for two populations in North Carolina, one in South Carolina, and three in Florida. Many of these cases occurred in the same year, 1991. In most years no such events occur, and in years when they do the incidence is quite small (<5% of nesting groups). Equally rare are instances in which a group attempts a third nest after two nest failures.

The second component of variance in reproduction is clutch size. Females typically lay 3 or 4 eggs, although complete clutches of 2, 5, and even 1 egg sometimes occur. There are consistent differences across populations in average clutch size, which ranges from 2.9 to 3.5 eggs. Clutch size is one of the components of reproduction that varies among years, but is not one of those involved in geographic variation in productivity. This is somewhat surprising, since in many birds clutch size increases from south to north.

Red-cockaded Woodpeckers typically suffer partial brood loss early in the nestling phase. Broods hatch asynchronously, so that at any point in time the last young to hatch are smaller than those that hatched first, at least early in the nestling phase. Often one or two young die during the first few days after hatching, and some eggs do not hatch at all. The extent of partial brood loss is the major determinant of the geographic variation in productivity described earlier, partial brood loss being greater in populations with low adult mortality. Partial brood loss is reflected in the proportion of eggs that become fledglings in successful nests, that is, nests that produce one or more fledglings. Among inland and northern populations, the average proportion of eggs that become fledglings in successful nests

Table 6.2. Reproductive Parameters for Six Red-cockaded Woodpecker Populations

	Sandhills	Lejeune	Croatan	Marion	Piedmont	Eglin
Years observed	18	12	9	7	10	6
	(1980–1997)	(1986–1997)	(1989–1997)	(1976–1992)	(1983–1992)	(1993–1998)
Clutch size[a]	3.25	3.45	3.25	3.08	3.48	3.18
	(3.08–3.47)	(3.12–3.76)	(3.22–3.63)	(2.69–3.26)	(3.17–3.77)	(3.06–3.32)
Groups that nest	86%	88%	89%	94%	95%	91%
	(75–96)	(79–98)	(84–93)	(88–100)	(89–100)	(84–96)
Nests that fail[a]	22%	19%	22%	32%	22%	24%
	(13–27)	(8–32)	(12–32)	(14–50)	(7–43)	(14–30)
Groups that renest	33%	26%	38%	4%	42%	41%
	(7–66)	(0–64)	(0–67)	(0–10)	(0–75)	(18–67)
Eggs that fledge/successful nest[b]	63%	57%	59%	67%	64%	54%
	(57–68)	(46–67)	(51–69)	(61–82)	(56–71)	(51–60)

Fledglings/successful nest[b]	2.08 (1.88–2.27)	1.99 (1.58–2.20)	1.96 (1.63–2.29)	2.09 (2.00–2.23)	2.31 (2.12–2.65)	1.78 (1.68–1.95)
Fledglings/group[c]	1.48 (1.20–1.66)	1.48 (1.03–1.84)	1.41 (1.10–1.66)	1.50 (1.15–1.89)	1.67 (1.11–2.12)	1.34 (1.09–1.58)

Sources: Unpublished data for Marion and Piedmont were provided by M. R. Lennartz and E. E. Stevens, U.S. Forest Service. Other data are from the authors' unpublished records.

Note: Mean and range among years are provided in parentheses.
    The populations are the North Carolina Sandhills (Sandhills), Camp Lejeune (Lejeune), and Croatan National Forest (Croatan) in the North Carolina Coastal Plain; Francis Marion National Forest (Marion) in the South Carolina Coastal Plain (pre-Hugo); the Piedmont National Wildlife Refuge and Hitchiti Experimental Forest (Piedmont) in the Georgia Piedmont; and Eglin Air Force Base (Eglin) in the Gulf Coastal Plain in the western panhandle of Florida.

[a] Data on clutch sizes and nest failure rates are from first nests only.
[b] The percentage of eggs that hatched and produced fledged young in successful nests and the number of fledglings per successful nest provide measures of partial nest loss and brood size at fledgling, respectively.
[c] "Fledglings/group" refers to all potential nesting groups, and thus provides an overall measure of productivity: fledglings produced per potential nesting group.

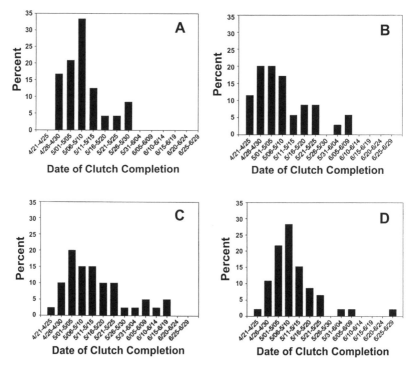

FIGURE 6.16 Dates of clutch completion in the *(A)* 1990, *(B)* 1995, *(C)* 1996, and *(D)* 1997 breeding seasons at Camp Lejeune, indicating the range of variation in breeding synchrony and breeding season length.

ranges as high as 70%, whereas values for Florida populations are lower (Table 6.2), ranging as low as only 40–50% in central Florida. In the North Carolina Sandhills, nests commonly fledge three young, four are not unusual, and five have occurred, whereas in Florida it is rare for more than two young to be fledged.

A large number of hypotheses have been proposed to explain the reproductive complex that includes early onset of incubation, asynchronous hatching, size hierarchies among nestlings, and partial brood loss early in the nestling phase. This complex is seen in many birds, for example, many seabirds and raptors. According to some hypotheses, there is an advantage to having size hierarchies, and early onset of incubation and asynchronous hatching are adaptations to create these hierarchies. According to others, early incubation is advantageous and size hierarchies are an incidental byproduct—perhaps even a disadvantageous one—of early incubation. Some hypotheses involve nestling death by starvation; others, death due to attacks from nest mates. Which hypothesis applies to Red-cockaded Wood-peckers, and even the mechanism producing the mortality in this species, remains to be determined.

One particularly prominent hypothesis is brood reduction. Brood reduction is a hypothesis developed specifically to explain systematic loss of the smallest, last-hatching young early in the nestling period. Brood reduction is thought to be an adaptation whereby breeders match the number of young to the amount of food available after hatching. The last-hatching young are lost early, so that little effort is invested in them when food is insufficient (Figure 6.17).

Melinda LaBranche investigated nestling mortality in Red-cockaded Woodpeckers in the North Carolina Sandhills. Although she found that the extent of partial brood loss varied greatly among years and that losses occurred early in the nestling period—patterns consistent with brood reduction—her evidence did not indicate that the last young to hatch were nec-

FIGURE 6.17 An adult Red-cockaded Woodpecker brings food to its young. Photo by R. N. Conner.

essarily the ones to die. Therefore it is not clear that the brood reduction hypothesis can explain partial brood loss in Red-cockaded Woodpeckers. LaBranche favored another hypothesis, that asynchronous hatching may reduce peak feeding demands on adults, because the peak requirements of nestlings occur on different days.

Although partial brood loss is unusually great in Red-cockaded Woodpeckers, whole brood loss is unusually low, as is typical of cavity nesting species. Fledging success of most bird species is quite low. Species such as Northern Cardinals (*Cardinalis cardinalis*) that build cup nests within understory and midstory foliage typically fledge young from less than 30% of their nests. Cavity nesters in general enjoy a much higher nesting success rate and regularly fledge young from more than 60% of their nests. The proportion of Red-cockaded Woodpecker nests that are successful averages about 80% in most populations (Table 6.2), a value that is not unusual for a primary cavity nester. Most nest losses occur early, a pattern that implicates nest desertion as a factor rather than nest predation. Both Lennartz and coworkers in South Carolina and DeLotelle and coworkers in Florida report that social conflict among group members or with intruders is often associated with nest desertion.

Thus the unusually complex social system may lead to elevated levels of nest desertion compared to that of other cavity nesters. On the other hand, the protection provided by nesting in live pines and producing a resin barrier around the cavity results in unusually low levels of nest predation, even for a cavity nester. Red-cockaded Woodpecker cavities are not constructed in the soft, decayed snags that many cavity nesters use. Raccoons (*Procyon lotor*) and American black bears (*Ursus americanus*) are unable to chew through the wood to prey on Red-cockaded Woodpecker nestlings as they can do with some other woodpecker species and secondary cavity users. Some nests that are lost are destroyed by other cavity users, notably flying squirrels and other woodpecker species, who at least occasionally act as opportunistic nest predators. Loss to major nest predators such as rat snakes is a rather infrequent event, certainly much less frequent than it is in most other species of birds. The resin wells the birds maintain around their cavities are highly effective in thwarting predators. Indeed, the two species of rat snakes appear to be the only major nest predators capable of taking eggs and nestlings of Red-cockaded Woodpeckers, and even they have a very difficult time reaching nest cavities (see Chapter 5).

The general patterns characterizing reproduction do not appear to apply to populations in Arkansas and perhaps Oklahoma at the northwestern edge of the species' range. These populations are less productive than

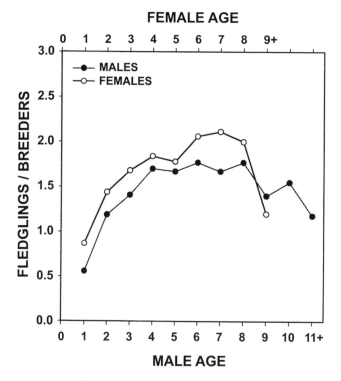

FIGURE 6.18 Reproductive success as a function of age and sex. Only individuals that had acquired breeding status are included. Data from the North Carolina Sandhills.

others, and work by Joseph Neal, Douglas James, and coworkers from the University of Arkansas indicates this to be due to relatively high levels of whole brood loss. Starvation of chicks during periods of poor weather, as well as nest predation, is associated with whole brood loss in this region.

The causes of the considerable variation in reproduction are poorly understood. Helpers account for some variation within years within populations, and life history evolution may account for some differences between populations (see above). Age of breeders also accounts for some of the variation within populations, as reproductive success increases considerably with age through age 4 in both sexes (Figure 6.18). Some of this increase can be attributed to breeding experience, as individuals that have bred before do better than naive breeders (Figure 6.19) and fewer birds are inexperienced at older ages (Figure 6.4). Since helping experience does not provide similar benefits (see above), the important skills gained through breeding experience presumably are related to interaction between mates or to ter-

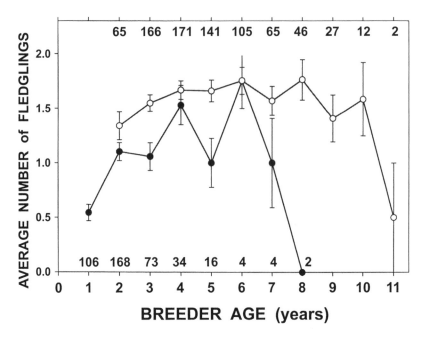

FIGURE 6.19 Fledgling production as a function of age for males breeding for the first time (filled circles) and males that have bred during at least 1 previous year (open circles). Sample sizes for first-time breeders are given at the bottom of the graph, those for experienced breeders at the top.

ritory defense rather than to caring for young. Some of the improvement with age, however, is independent of experience (Figure 6.19), suggesting that birds improve some general skill with age that is relevant to breeding. Foraging ability is an obvious candidate.

Much of the variation in reproductive success, particularly that across years within populations, remains unexplained. One presumes that variation in weather and food supply is somehow involved, but the specific relationships are not obvious.

SEX RATIO

The sex ratio among adults is not even, due to the fact that males adopt the helping strategy more often than females and the survival of helpers is much better than that of departers. There are many more male helpers than there are female helpers and floaters, so that there are about 1.6 or 1.7 adult males per adult female in most populations. Such a sex ratio bias is commonplace among cooperative breeders.

In contrast, sex ratios that differ from 1:1 at hatching are unusual. The Red-cockaded Woodpecker is one of the few birds for which such a bias has been reported. Patricia Gowaty and Michael Lennartz first reported a biased sex ratio among nestlings in the Francis Marion population. They found that among females with helpers, those that had not previously bred in a particular group produced more sons than expected, whereas females that had bred previously in the group produced equal numbers of sons and daughters. Females without helpers produced more sons than daughters as well. Overall, 59% of the young produced were male. Gowaty and Lennartz proposed that females that had bred previously in a group overproduced daughters to reduce the potential for reproductive competition between them and their sons.

Stephen Emlen and his colleagues, stimulated by this report of a biased sex ratio in a cooperative breeder, developed a general model of sex ratio evolution for cooperative breeders. They pointed out that because helpers frequently assist their parents in raising subsequent broods, and because helpers are predominantly male in species such as Red-cockaded Woodpeckers, males in a sense "repay" the cost of their production, whereas females do not. In evolutionary models, the cost of reproduction is measured in terms of lost future reproduction due to energy expended and risks taken during current reproduction. If parents invest equally in sons and daughters, but sons pay back some of the future reproduction sacrificed by increasing their parents' future reproductive success through helping, then males are cheaper to produce than females. Sex ratio theory, developed originally by R. A. Fisher, the noted biologist and statistician, predicts that in this scenario more sons than daughters should be produced. This is a form of the more general mechanism for producing uneven sex ratios known as local resource enhancement.

The observations of Gowaty and Lennartz are based on a small sample, and sex ratio biases of the sort they describe have not been found subsequently. Thousands of fledglings have been sexed in the North Carolina Sandhills population, and hundreds in the North Carolina Coastal Plain populations, and in both cases the sex ratio is even. Walters found that at Camp Lejeune, 49% of 518 fledglings were male, and at the Croatan National Forest 50% of 525 were. LaBranche found that the sex ratio in the Sandhills population was not significantly different from even, but was significantly different from the ratio predicted by Emlen's model, which predicted 54% of young would be male (49% were). The relationship between female tenure and presence of helpers observed by Gowaty and Lennartz did not occur in the much larger Sandhills sample. These results suggest that there is nothing unusual about sex ratios of the young in Red-cockaded Woodpeckers, and that Emlen's model does not apply to this species. Walter Koenig and Walters made some corrections to Emlen's model, but even with these corrections it predicted a male-biased sex ratio for the Sandhills population rather than the even sex ratio that exists.

Questions about sex ratio in Red-cockaded Woodpeckers nevertheless persist. Epting and DeLotelle reported that sex of offspring in central Florida was related to hatch order, with first-hatched eggs more likely to produce females and last-hatched to produce males. Again, the evidence is based on a small sample. If hatch order is a determinant of sex, differences between populations in sex ratio of fledglings could exist because of differences in extent of partial brood loss. The Sandhills population, which exhibits no bias, experiences much less partial brood loss than the populations in which sex ratio biases have been reported. Perhaps whatever mechanisms produce sex-ratio bias operate in some populations but not others. This aspect of the biology of the species will no doubt draw further attention from evolutionary biologists.

General Behavior

The day of a Red-cockaded Woodpecker typically begins around sunrise, or slightly later on overcast days, when it leaves its roost cavity and gathers with other group members in the cavity tree cluster. The birds are highly vocal as they greet one another each day, and they often cling close together on the trunk of a large pine or snag. If the cavity trees are widely scattered, there may be much flying about and calling until the birds are all assembled. Birds often work on their resin wells a little at this time. Then the group is off to spend most of the day (60–90%, depending on season) foraging within their territory (Chapter 7). The birds make short flights from tree to tree, exhibiting the undulating flight characteristic of woodpeckers and producing a whirring of the wings as they take off. Within the trees they move by hitching up the trunks and hopping along the limbs, and the males frequently cling upside down to twigs and limbs. Most flights are short and within or below the canopy, except at the end of the day when the birds may fly all the way back to the cavity tree cluster well above the canopy (see Chapter 7). The birds may not return to the cluster until it is nearly time to roost, or they may return somewhat earlier and work on their resin wells prior to roosting. Birds usually roost about sunset, although roost times are somewhat variable and are earlier on overcast days or in inclement weather.

Although foraging consumes the bulk of the day, along with breeding behavior during the nesting season, the birds devote some time to various other activities as they travel through their territory. Jerry Jackson has been able to provide detailed descriptions of some of these behaviors, which he observed in an injured, captive bird that he maintained for years. Red-cockaded Woodpeckers drink regularly, typically from pools of rainwater located in trees. Birds have been observed drinking from water held in tree crotches, knotholes, and cavities, including flooded Red-cockaded Woodpecker cavities. Lacking these sources, they will also come to the ground to drink from puddles. However, Red-cockaded Woodpeckers obtain much of their water from their food, most of which has a high water content. Bathing is rarely observed, the drinking pools in the trees being unsuitable for this purpose. Evidently the birds' reluctance to come to the ground extends to bathing as well as drinking, although Jackson's captive bird bathed regularly. The birds do sun themselves regularly, and they stretch frequently.

The birds often cease foraging for a period in order to rest and preen. Resting bouts are longer and more frequent in the afternoon than in the morning. Typically, activity of group members is highly synchronized, so that when one bird begins to preen the rest of the group is soon preening as well. The birds usually cling to a trunk when resting and preening, and these two activities are tightly linked temporally and presumably motivationally. The birds are quieter when resting, but once the rest period is over they begin to noisily move and forage again.

When not feeding young, the group typically forages intensely for several hours after dawn. There is then a tendency to have a midday lull, during which the birds are relatively quiet and change trees only occasionally. This period is often spent in the cavity tree cluster, and individuals may spend considerable time excavating cavities or working resin wells, as well as resting and preening. Active foraging then resumes in the afternoon and continues until the birds return to the cluster before going to roost. There is, however, much variation in the daily routine.

Territories and Aggression

Red-cockaded Woodpeckers reside in large, permanent territories. These are "all-purpose" territories; that is, all activities take place within their boundaries. Territorial boundaries are defended by all group members through visual displays and physical aggression. Defense behaviors have been described by David Ligon, by Jerry Jackson, and by Lester Short. Typical defense behaviors include wing-flipping displays, spiraling chases around the trunks of trees, and long looping flights that Jackson has termed "flutter aerial display." Wing-flipping involves a slow extension of the wings over the back and retraction. Ligon called wing-flipping the "open wing display" and believed it to function as a means of intrapair recognition and pair-bond reinforcement. Aggression occasionally escalates to grappling fights. The birds are quite vocal during territorial disputes, employing especially what Jackson labels the "she-u" call (see below).

Similar to Downy and Hairy woodpeckers, Red-cockaded Woodpeckers can make a relatively loud ruffling sound (wing-fluttering described by Ligon) with their wings as they fly (Figure 6.20). This behavior is probably a sign of agitation in all species that do it and may serve as a threat to humans or other intruders at cavity tree clusters. Other signs of agitation are wing-flipping and raised crown feathers.

Territorial defense may take the form of boundary disputes with neighboring groups, or disputes with individual intruders deep within the territory. Boundary disputes between neighboring groups can appear chaotic, with several birds from one group chasing another, or multiple pairs of birds involved in individual battles, or the members of each group clustered together, flipping their wings and calling excitedly. Wing-flipping and spiraling chases are common during boundary disputes, but actual physical fighting between adults is rare. Fledglings are often victimized in one-sided physical fights with adults from other groups during boundary disputes. The intensity of boundary disputes varies with the season. During the breeding season they can be quite intense, but during winter two groups that meet along their territorial boundary, after an initial bout of wing-flipping and excited calling, may settle down and forage peacefully side by side for a prolonged period. Neither group ventures far into the other's area during such interactions, however.

Disputes with intruders are often quite intense. Flutter aerial display and other faster and closer aerial chasing are frequent, as is wing-flipping. Although all group members may participate in an attack on an intruder, the breeding pair plays the dominant role. An individual breeder is especially aggressive against same-sex intruders, but males may be quite aggressive against female intruders in some instances. Outside of the breeding season, intruders are sometimes tolerated after an initial display of aggression. The young of the year frequently are tolerated outside of the breeding season, and adult females may also be, depending on the reaction of the breeding female, but adult males rarely are. This sex difference is consistent with the fact that breeding females, but not breeding males, sometimes switch groups.

An auditory signal conspicuously absent from the defense repertoire is drumming. Drumming substitutes for song as a long-distance territorial defense advertisement in other woodpeckers. For Red-cockaded Woodpeckers, however, it is a softer signal employed over short distances whose function seems to have shifted away from territorial defense to courtship (see below).

Many investigators have examined home ranges of Red-cockaded Woodpecker groups. As one would expect, the core of the home range is the territory, but one of the unusual attributes of this species is that extraterritorial forays are fairly common. Such forays may take the birds into the territories of neighboring birds or into unoccupied space that they do not defend. Use of a neighboring territory, of course, depends on a group being undetected by the territory owners. That such forays occur is likely a consequence of the large size of territories (see Chapter 7). Territory owners presumably cannot defend their boundary with the efficiency of species that have much smaller territories, which provides neighbors with frequent opportunities for intrusion. Indeed, groups may travel completely through a neighboring territory without being detected.

The significance of extraterritorial range is unclear. Groups range outside their territory most often in late summer, when fledglings are present, and in midwinter, when foraging conditions presumably are poorest. This suggests that extraterritorial range may be important in meeting foraging needs (see Chapter 7). On the other hand, much extraterritorial range consists of portions of territories of other groups or areas devoid of foraging habitat, suggesting that extraterritorial forays have a social function. Alternatively, they may have no important function at all—they may simply be a manifestation of low densities and imperfect territorial defense. Territory and home range sizes, and their relationship to critical resources such as cavity trees and foraging habitat, are explored in detail in Chapter 7.

Vocalizations

Vocalizations have been described by Hans Winkler, with the Austrian Academy of Sciences, and Lester Short, as well as by Ligon and by Jackson. These authors have broken the vocal repertoire of the species into a number of discrete categories, but the repertoire could also be described in terms of graded series along several dimensions. The vocal repertoire consists of an unmusical assortment of chirps and squawks. Red-cockaded Woodpeckers are highly vocal compared to their close relatives, a by-product no doubt of their more social nature. Indeed, the most commonly heard calls are the

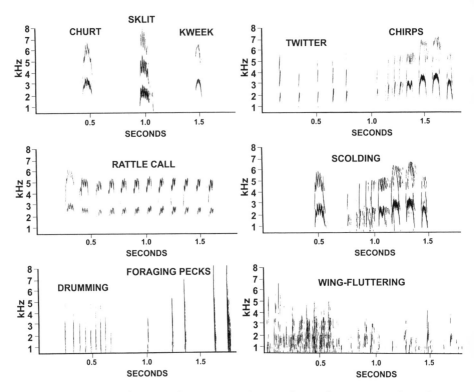

FIGURE 6.20 Vocalizations, drumming, pecking, and wing-fluttering sounds made by Red-cockaded Woodpeckers.

contact calls exchanged by group members. As the group travels, birds call frequently, and it seems they always call when they move from tree to tree. Group members frequently come very near one another, sometimes within centimeters, and often they chortle and chirp when they do.

Red-cockaded Woodpeckers possess a syrinx (vocal apparatus) and one pair of tracheobronchial muscles. The various calls differ little in frequency (Figures 6.20 and 6.21). Gradations instead are in the duration and spacing of notes and perhaps in frequency modulation. Generally, contact notes are relatively long and are spaced far apart, those at the end of the continuum being given singly. These grade into calls used in intense social situations such as territorial disputes, in which the notes become shorter (i.e., sharper) and closer together. In some calls the notes are given rapidly to produce a rattle.

Variation in the physical characteristics of calls, together with their corresponding variation in function, is a subject worthy of further investigation. The functions of the calls of Red-cockaded Woodpeckers generally are poorly documented. Evidence of function consists entirely of contextual information, rather than measures of effects of calls on recipients. Without recordings it is difficult to determine the degree of similarity of calls reported by different investigators, or even of calls heard in similar contexts by the same investigator. The extent to which calls are graded rather than discrete is not yet well documented either. Calls have been classified according to where they lie along a continuum, but whether most sounds produced fall on a restricted few points along the continuum or whether sounds characteristic of all points are produced is not clear, let alone how meaning changes along the continuum. Generally, graded signaling is poorly understood in birds, and Red-cockaded Woodpeckers are not an exception in this respect. Variations on a particular theme may be meaningless, or each slight change could alter the meaning of the call.

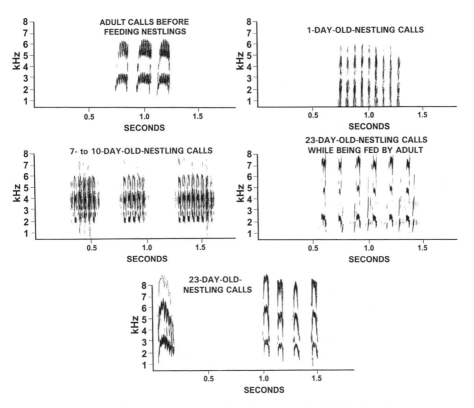

FIGURE 6.21 Vocalizations of adult and nestling Red-cockaded Woodpeckers associated with adult woodpecker visits to the nest cavity to feed nestlings.

CALLS

Ligon described a total of 13 vocalizations given by Red-cockaded Wood-peckers, whereas Jackson described approximately 20, but many vocalizations in both cases were heard in only a few instances. The most commonly heard calls of adult Red-cockaded Woodpeckers are the "churt" and "sklit" (possibly "szrek" or "shrit," as described by Ligon) (Figure 6.20). Jackson suggests that the "sklit" (when birds are excited) is the most commonly heard call and is used when humans are detected within the vicinity of the cluster area as the woodpecker approaches, whereas the "churt" (when birds are not excited) is given when humans are not detected. Several days prior to fledging, nestlings often give both "sklits" and "churts" as they peer out of their nest cavities. Nestlings are extremely vocal while begging for food inside nest cavities at a variety of ages (Figure 6.21).

Scolding calls are given when the woodpeckers are disturbed and are usually heard during human visits to clusters during the nesting season (Figure 6.20). The "kweek" is a shrill call similar to and sometimes transitioning into the "churt" and is often given as adults leave a nest after feeding the young (Figure 6.20). Both Ligon and Jackson also described a "she-u" or "che-u" ("tsi-voo," as described by Winkler and Short) given in rapid succession by adults when another Red-cockaded Woodpecker enters their territory; this call is often accompanied by an open wing display and erratic flight. A bird gives the "shurz-u" call, first described by Ligon, when hawks approach. Usually the woodpecker hitches to the opposite side of the tree as it gives the call.

Undisturbed adults in close proximity to other adults give a somewhat melodious chortle or "whu-whu" call during or after feeding young. "Twitter" calls occasionally are associated with "kweeks" and are given frequently as adults approach their young (Figures 6.20, 6.21). "Wick-a wick-a" is another call indicating excitement, which Ligon heard once when a male called to his mate while defending their nest cavity from a Red-bellied Woodpecker. Red-cockaded Woodpeckers also give a rattle-like call (Figure 6.20) reminiscent of rattle calls given by Hairy Woodpeckers and the whinny given by Downy Woodpeckers. Rattles are often vocalized as adult woodpeckers approach the cluster area or nest tree. Jackson, as well as Winkler and Short, described various chirps, squeaks, contentment peeping, and distress cries, including "wad," "zrips," "deedle-deedle," and "chit." The complete vocal behavior of this social woodpecker remains largely unstudied.

DRUMMING

Similar to other woodpeckers, Red-cockaded Woodpeckers seek out locations on trees to drum (Figure 6.20). Drumming is a rapid pecking that produces a loud sound in most woodpecker species and substitutes for song as a long-distance territorial defense advertisement. Red-cockaded Woodpeckers do not drum as often as most other woodpecker species and do not employ drumming in territorial defense (see above). The majority of the drumming is done on the trunk or limb of live pine trees, which lack the resonance found in the dead, hardened snags (or rain gutters) that many other woodpecker species use as drum sites. This species' evolution in fire-maintained pine ecosystems, which are relatively devoid of snags, may have led to lessening the importance of drumming in long-distance territorial advertisement. The function of drumming seems instead to have shifted to short-distance communication related to courtship. Both sexes drum, but the rate of drumming is by far the greatest for solitary males.

CHAPTER SEVEN

Foraging Ecology of Red-cockaded Woodpeckers

Woodpeckers as a group are supremely adapted for foraging on the boles and limbs of trees. Their relatively large, muscular legs and feet, as well as the zygodactylous arrangement of the toes (see Figure 3.3), allow efficient movement on vertical bark surfaces. This is a very specialized ability in the avian world, and allows access to an abundant foraging substrate generally unavailable to most birds. Of the nearly 10,000 species of birds currently recognized, only the woodcreepers (Dendrocolaptidae—49 species of the Neotropics), nuthatches and creepers (*Sitta* and *Certhia*—30 species primarily of Holarctic distribution), and a small number of additional species scattered among several other groups forage with any efficiency on vertical bark surfaces and occur within the nearly worldwide range of woodpeckers. In addition, two small families (Australian treecreepers [Climacteridae] and sittellas [Neosittidae]) with a total of nine species occur outside the range of woodpeckers in Australia and New Guinea. All of these non-woodpecker species basically glean arthropods from bark surfaces or probe beneath loose bark to obtain prey. They lack the ability, well developed in woodpeckers, to excavate bark and wood, although nuthatches have limited excavating capabilities.

Nearly all species of woodpeckers excavate wood, both to obtain prey and to construct nest and roost cavities. The ability to excavate provides foraging opportunities unavailable to other birds, even those few that can forage and glean on vertical bark surfaces. An entire community of arthropods exists beneath bark surfaces and within the wood of many trees, especially dead and diseased trees. Woodpeckers possess many structural modifications that allow efficient excavation and provide access to this food source. These modifications involve essentially all parts of a woodpecker's anatomy and are discussed in Chapter 3.

With access to a large prey base unavailable to most avian species, woodpeckers have diversified and spread throughout most habitats supporting woody plants. Several species (e.g., flickers [*Colaptes*]) have even become primarily or entirely terrestrial, foraging on and, in some cases, even nesting in the ground. Other species, such as Lewis's Woodpeckers (*Melanerpes lewis*) and Red-headed Woodpeckers, use aerial hawking techniques extensively to obtain flying prey. Most, if not all, species feed at least occasionally on fruits and seeds. However, most species, including the Red-cockaded Woodpecker, forage primarily on arthropods obtained by gleaning and excavating on woody plants.

Foraging Substrates

Studies of the foraging behavior of Red-cockaded Woodpeckers have consistently found that they forage primarily on pines. The percent of foraging on species other than pines has ranged from 0% to 22% in published studies, typical values being 5–10%. Although foraging may occur on a variety of nonpine substrates (even corn plants) depending on location, most occurs on hardwood trees. The extent of foraging on hardwoods varies with location, abundance of hardwoods, season, and sex of the woodpecker. More extensive use of hardwoods tends to occur in mixed pine-and-hardwood habitats where the pine species are typically loblolly or shortleaf. Use of hardwoods is generally less in longleaf habitats where hardwoods are smaller or less abundant due to the greater influence of fire. The highest overall use of hardwoods by both sexes reported to date is for a population studied by Don Wood in the McCurtain County Wilderness Area in the Ouachita Mountains of Oklahoma. This old-growth habitat contains a variety of hardwood species with a variable amount of shortleaf pine primarily on the ridges and south-facing slopes. Due to fire suppression in recent decades, succession to hardwoods is advanced and pine reproduction rare. In this habitat, 15% of total Red-cockaded Woodpecker foraging was on hardwoods.

Seasonal differences in the use of hardwoods are frequently reported. James Skorupa and Robert McFarlane (then with the Savannah River Ecology Laboratory) found no evidence of hardwood use in summer in a population in South Carolina. In winter, however, use of hardwoods increased to 10% of foraging time. In a number of other studies in which seasonal differences have been examined, hardwood use increased in winter.

Several studies have reported males foraging on hardwoods more frequently than females. For example, Patricia Ramey (then a graduate student working with Jackson at Mississippi State University) found males foraging

22% of the time on hardwoods, compared to 6% for females in Mississippi. The difference between sexes was generally less in other studies, but more often than not males foraged more frequently in hardwoods.

Among the species of pines found in the range of the Red-cockaded Woodpecker, most species are acceptable as foraging sites. Only sand pine seems to be avoided, presumably because of its small stature and propensity to grow in dense stands. Sue Zwicker, a graduate student working with Walters at North Carolina State University, found that in coastal North Carolina the birds did not exhibit a strong preference among longleaf, loblolly, and pond pine in habitats in which these species were fairly common. Zwicker did, however, find that males, in contrast to females, tended to favor longleaf over other pine species. She suggested that this may be related to the larger diameter of terminal twigs in longleaf pine, as well as the propensity of males to spend more time than females foraging on twigs and small limbs (see below).

Dying pines are an important foraging substrate of Red-cockaded Woodpeckers in some areas. Pines die from many causes, such as drought, lightning, and disease, but throughout much of the range of Red-cockaded Woodpeckers, southern pine beetle infestation causes or accompanies most tree mortality. Woodpeckers eat many species of beetles and other arthropods, and infested pines can provide superabundant prey. Most, if not all, species of woodpeckers of the southeastern United States, including Red-cockaded Woodpeckers, take advantage of this abundance. When foraging on infested trees, Red-cockaded Woodpeckers generally concentrate on pupae and recently emerged adults that are located closer to the surface than larvae. They feed not only on the southern pine beetles, but also on other species characteristically associated with pine beetle infestations. The availability and importance of foraging on trees infested with southern pine beetles varies geographically, temporally, and with pine species. Infestations and the foraging of Red-cockaded Woodpeckers thereon are discussed in detail in Chapter 8.

Foraging Behavior

When foraging, Red-cockaded Woodpeckers generally fly to a point on a tree, forage upward, and then fly to a different tree to repeat the process. Occasionally they drop to a lower point on the same tree and again forage upward. This last behavior is most often observed on particularly rich foraging trees. Males may make many movements within a tree when foraging on twigs and small limbs (see below). How the birds decide when to leave a tree and what tree to visit next has not been studied. In other species, this often is a function of how successful an individual is in a particular spot, with longer stays and shorter moves indicating greater success. How birds decide where to forage within their territory is discussed later in this chapter.

Red-cockaded Woodpecker foraging behaviors are typical of those of most species of woodpeckers, especially other *Picoides*. They fairly rapidly hitch up the bole of a tree or along limbs, often spiraling as they progress. They intently survey the bark surface, often peering and probing into crevices and under loose flakes of bark. Arthropods discovered are rapidly captured in the bill. When foraging on pines, Red-cockaded Woodpeckers make extensive use of a scaling technique in which loose bark is pried off with the bill or feet. Arthropods exposed on the surface are then captured. An observer following a foraging group experiences a steady, light rain of scaled bark resulting from this activity. The birds also excavate to expose arthropods under bark, within deadwood, or in cones. This foraging behavior is most commonly used on recently dead trees with dense beetle populations, or on small dead limbs containing ant colonies and other prey. Excavating on cones is concentrated in seasons when fresh, green cones are present. Red-cockaded Woodpeckers, unlike many of their relatives, rarely excavate deeply into large-diameter stems.

David Ligon in Florida and naturalist Ted Beckett in South Carolina observed Red-cockaded Woodpeckers fly-catching, which is rarely reported for species of *Picoides*. Foraging for plant material occurs regularly, but provides a minimal portion of the food intake. Seeds and fruits of a wide variety of species are eaten occasionally (see below).

Diet

The composition of the diet of Red-cockaded Woodpeckers is not well known. The first extensive quantitative data come from a study published by F. E. L. Beal and coworkers in 1916. Beal was a biologist with the United States Biological Survey and was involved in compiling information on the food of birds in relation to agriculture. Woodpeckers were of considerable interest to the Department of Agriculture because of crop depredation and the potential for control of insect populations by woodpeckers and other birds in an era preceding the widespread use of pesticides. Based on the analysis of the contents of 99 Red-cockaded Woodpeckers' stomachs, Beal determined that 86% of the food was arthropods and 14% was plant material, mostly seeds of conifers. Ants comprised nearly 52% of the diet and beetles over 15%. The remaining 19% of the diet was composed of a wide variety of arthropods, including cockroaches, crickets, grasshoppers, termites, caterpillars, moths, hemipterans, spiders, centipedes, and millipedes. Unfortunately, Beal did not state if his percentages referred to number of items, weight, or volume of food, preventing accurate assessment of his results.

The endangered status of the species has inhibited in recent times further studies of diet using analysis of stomach contents. Stomach contents can be difficult to interpret in any case, since some items are digested faster than others, so that what is found in the stomach does not necessarily reflect precisely what has been eaten. Much subsequent information on diet comes from incidental observations of foraging. Studies by Dave Ligon, Wilson Baker (with Tall Timbers Research Station), Robert Hooper and Michael Lennartz, James Hanula and Kay Franzreb, and our own observations have basically confirmed Beal's general conclusions. Red-cockaded Woodpeckers prey on a wide variety of arthropods, and probably any taxon of suitable size is taken. Ants, beetles and beetle larvae, cockroaches, centipedes, and spiders probably comprise the major portion of their arthropod prey. Recently, Chuck Hess and Fran James obtained additional quantitative data by flushing the stomachs of adult birds. They, like Beal, found the dominant prey to be ants (58% of biomass), especially the arboreal ant *Crematogaster ashmeadi*. This ant species is very abundant and widespread throughout the range of the Red-cockaded Woodpecker. The stomachs contained a wide variety of other arthropods, beetles and their larvae being most prominent (7%), as well as fruits and seeds (16%) and wood (9%).

The diet of nestlings is easier to determine by direct observation than that of adults, because the adults carry food in their beaks for a while, instead of consuming it quickly, and bring it to a predictable location, the nest cavity. Nest watches indicate the diet of nestlings contains a greater proportion of large prey items than that of adults. Richard Harlow (U.S. Forest Service, Southeastern Forest Experiment Station) and Mike Lennartz observed that insect larvae, centipedes, millipedes, cockroaches, and spiders constituted 95% of the identified prey brought to nestlings in South Carolina. However, they were only able to identify the prey item for 36% of the deliveries, because in many instances the items were too small to be identified. Rick Schaefer in Texas and Louisiana, subject to the same limitations, found that the same types of prey were delivered to nestlings, and that moths and caterpillars were also common prey. Progressively larger items were brought to nestlings as they grew older. Hanula and Franzreb (also with the Forest Service's Southeastern Forest Experiment Station) used cameras mounted at the nest cavity to monitor the prey delivered to nestlings in South Carolina. Wood roaches (*Paracoblatta* sp.) composed 69% of the identifiable prey delivered to the nestlings. The remainder of the prey was composed primarily of the taxa noted above. Hess and James flushed stomachs of nestlings and found that they consumed a more varied diet than the adults in their study population. Ants accounted for only 15% of the biomass consumed by nestlings; beetles and beetle larvae (14%), spiders (15%), and centipedes (12%) accounted for comparable amounts, as did wood (18%!).

Based on these studies, it appears that Red-cockaded Woodpeckers are quite catholic in their choice of arthropod prey. Most taxa occurring on boles and limbs of pines are eaten. Major differences in composition of prey at various times and places are presumably due to differences in availability. High percentages of beetle larvae in the diet are associated with a great abundance of bark beetles, especially during epidemics. High percentages of ants and roaches may also be associated with large populations of these taxa. The exploitation of corn worms, which the woodpeckers obtained by excavating through the husks of corn ears, demonstrates considerable ability to take advantage of novel prey sources.

Seeds and fruits of a wide variety of plants have been recorded in the diet. Fruits of poison ivy (*Rhus toxicodendron*), wax myrtles (*Myrica cerifera* and *M. inodora*), blueberries (*Vaccinium* spp.), magnolias (*Magnolia* spp.), black cherry (*Prunus serotina*), grapes (*Vitis* spp.), pokeberry (*Phytolacca americana*), and black gum (*Nyssa sylvatica*) have all been reported. Fruits have occasionally been reported as food of nestlings. Conifer seeds have been reported in most studies of diet, and in all cases where a specific pine species was identified the species was longleaf pine. In Texas, longleaf pine seeds are eaten extensively in the fall, during the period when seeds are dispersing from the cones. The birds do not forage on loblolly or shortleaf pine seeds, although they have ample opportunity to do so. Presumably the birds eat longleaf seeds rather than those of other pine species because longleaf seeds are much larger. A bird will land directly on a cone or the adjacent branch, pluck a seed from between the cone scales, and fly to the trunk of the tree. On the vertical trunk, the bird will place the seed and attached wing in the angle between the trunk and its breast. It will then peck and manipulate the seed until the wing is detached and then proceed to eat the seed.

Red-cockaded Woodpeckers also forage extensively on trees infested with southern pine beetles when these are available. They do not begin to concentrate their foraging activities on infested trees until the pupae and emerging adults are present. Generally, this is about the stage of the infestation when the pine needles begin to turn red (see Chapter 8). For a period of several weeks such trees are extremely attractive to Red-cockaded Woodpeckers. Two foraging methods are most commonly used on beetle-infested trees. When actually foraging on southern pine beetles, Red-cockaded Woodpeckers and other woodpecker species flake off the outer layers of bark, exposing the pupae and recently emerged adults (tenerals). In the process, the surface layers of bark may be removed from the entire bole, which gives the tree a bright reddish cast. Red-cockaded Woodpeckers also excavate through the bark to expose other arthropods that are abundant at this stage. The large larvae of cerambycid and buprestid beetles are commonly captured. When the needles begin dropping and the bark begins to loosen and slough from the beetle-killed tree, Red-cockaded Woodpeckers cease their foraging activities. The intensive use of a beetle-infested tree, important though it is, persists for only a few weeks.

Red-cockaded Woodpeckers obtain most of their water (see Chapter 6) and minerals from their food. As with most birds, required amounts of most minerals are apparently obtained from the normal diet. A major exception is calcium, which is required in large amounts by females for the formation of eggshells. Various species of birds obtain needed calcium from the ingestion of mollusk shell fragments, calcareous grit, or bones. Red-cockaded Woodpeckers have been observed by Dick Repasky and co-workers at North Carolina State University ingesting bones found in raptor pellets. These pellets are the undigested prey remains regurgitated by hawks and owls, and they typically contain bone fragments. In the observed cases, female Red-cockaded Woodpeckers ingested bone fragments contained in the pellets, and even cached bone fragments in nearby trees for later use.

Sexual Dimorphism in Foraging Behavior

The foraging behavior of woodpeckers, especially in relation to prey size and choice of foraging substrate, is related to size. In general, female woodpeckers are smaller than males. This sets the stage for sexual differences in foraging behavior, which itself may influence the evolution of size differences. Sexual differences in foraging behavior have been observed in a number of woodpecker genera, but most actual data pertain to members of the genus *Picoides*.

Among various *Picoides*, sexual differences account for some variations in foraging behavior. In a study by Olav Hogstad on Three-toed Woodpeckers in Norway, females chose smaller trees than males. They also foraged at lower heights and on smaller stems than males, partly as a consequence of foraging on smaller trees. A similar study of Nuttall's Woodpeckers in California by Mark Jenkins, of the Hastings Natural History Reservation, found no difference in size of foraging trees but demonstrated that females foraged on smaller-diameter stems and at greater heights in the trees. Females were also more likely to glean arthropods from surfaces than were males. A number of studies have documented sexual differences in foraging by Downy Woodpeckers. In this species, females typically forage at greater heights but on larger-diameter stems than males.

Several hypotheses have been suggested to account for the sexual differences in foraging behavior among woodpeckers and other taxa. Size differences between females and males may influence foraging behavior directly, for example, smaller females foraging on smaller stems (but see Downy Woodpecker above). Alternatively, foraging differences could reduce competition for prey. Partial separation of foraging niches could allow specialization by each sex in mated pairs, resulting in more efficient use of the combined foraging habitat. Finally, social dominance of one sex over the other might lead to the dominant sex foraging in the more productive sites and forcing the subordinate sex to forage in less productive sites. Different factors could be involved in different species, or more than one factor could be involved in each case.

Thomas Grubb, of Ohio State University, attempted to directly test the hypothesis that dominant male Downy Woodpeckers force females to forage in less productive microhabitats. Grubb used the developing technique of ptilochronology to evaluate the nutritional status of individual woodpeckers. Ptilochronology involves the examination of daily growth bars on feathers, somewhat analogous to the examination of annual rings of trees (Figure 7.1). Wider growth bars indicate faster growth of the feather, reflecting in part better nutritional status. In the Downy Woodpecker population examined, males exhibited wider growth bars than females, suggesting that they enjoyed a better nutritional status. Grubb then provided supplemental food to the population and monitored the regrowth of previously plucked feathers. He found that with supplemental food provided, there was no statistical difference between feather growth rates of males and females. The growth rate of male feathers had not increased compared to the normal, unsupplemented rate, but that of the females had.

Red-cockaded Woodpeckers also exhibit differences in foraging behavior in relation to sex, but these are unlike those of other *Picoides* species. In general, females tend to forage at lower heights, often low on the bole of the

FIGURE 7.1 Rectrix (tail feather) of a Red-cockaded Woodpecker under low angle incident light, showing the daily growth bars used in ptilochronology studies. Photo by D. C. Rudolph.

tree, a region avoided by males. Females forage mostly on the bole, rarely on small branches and twigs, a favorite foraging area of males (Figure 7.2). Both sexes forage commonly on larger limbs. Females tend to use more superficial foraging techniques (gleaning, probing, and scaling of bark) than males, which excavate more frequently. Jerry Jackson suggested to the authors that because of these differences females, but not males, might experience nutritional stress if dense midstory vegetation develops due to lack of fire and prevents easy access to the lower boles.

Sexual dimorphism in foraging has been reported in a variety of studies from various parts of the species' range. Our own unpublished observations are consistent with these studies. Rudolph and coworkers in Texas, as well as Walters and Armando Pizzoni-Ardemani (a graduate student at North Carolina State University) in North Carolina, examined foraging behavior of Red-cockaded Woodpeckers using color-banded birds, which allowed us to identify known individuals. The differences in foraging position and behavior between sexes were obvious (Figure 7.2); often an observer needed to follow a group only briefly to pick out the breeding female, rather than needing sophisticated statistical analyses to show a sexual difference. Breeding males showed a strong tendency to forage at greater heights than breeding females, and they also spent more time foraging on branches than females, which concentrated more heavily on the boles. A consequence of this pattern was that males generally foraged on smaller-diameter stems than females. Differences in foraging techniques were also evident. Males excavated for prey significantly more often than females, much of the excavation occurring on dead lateral branches that females visited less frequently than males. Females, in contrast, used more superficial techniques such as gleaning and scaling of bark to a greater degree than males.

In Texas, differences between breeding males and helper males were also observed. Helper males foraged at greater heights than females, like the breeding males. However, helper males foraged to a greater extent

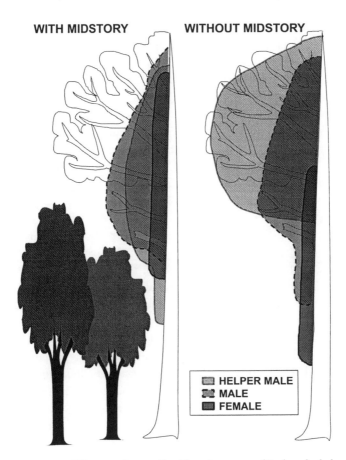

WITH MIDSTORY   **WITHOUT MIDSTORY**

HELPER MALE
MALE
FEMALE

FIGURE 7.2 Diagram of generalized foraging zones of Red-cockaded Woodpeckers of different social status. "With midstory" (left) represents the situation in habitats with abundant hardwood midstory, typical of many loblolly and shortleaf pine stands. "Without midstory" (right) represents the situation in habitats without significant hardwood midstory, characteristic of frequently burned longleaf pine stands.

on lateral branches than the breeding males. Consequently, helper males foraged on smaller-diameter stems than breeding males. Observations of helper females were too few to allow detailed conclusions. It was obvious, however, that their behavior was basically similar to that of the breeding females.

Juvenile Red-cockaded Woodpeckers foraged much as did the adults in both Texas and North Carolina. During the period from fledging until at least October, they followed the adults closely while foraging, often visiting specific foraging sites as the adults vacated them. Presumably juveniles learn valuable foraging skills from such behavior. Adults, including helpers, frequently catch prey and feed it to the juveniles during this period, as described in Chapter 6. The foraging behavior of juveniles is thus influenced by the sex of parent that they follow during foraging. We had too few observations to determine if juveniles tended to follow adults of the same sex.

It is clear from these observations that there is a remarkable degree of separation in foraging position among group members based on sex and social status. At the present time, the selective forces leading to sexual dimorphism in foraging behavior in Red-cockaded Woodpeckers are unknown. Unlike other *Picoides* species, the differences apparently cannot be explained as a consequence of differences in body size, since the larger sex forages on smaller-diameter stems. As detailed in Chapter 3, the sexes differ morphologically in ways that reflect adaptation to different foraging niches. Morphological differences may have evolved in conjunction with foraging differences to enable more efficient utilization of the available habitat and to reduce competition within groups. Alternatively, foraging differences due to dominance behavior may have resulted in selection pressures that led to morphological differences. Dominance of males over females may be a factor: dominance could account for the observed separation of breeding and helper males as well. The separation could also be genetically determined, especially in the case of sex differences.

The role of dominance relationships in producing foraging differences was investigated in Texas. The foraging heights of breeding males and females were compared with simultaneous estimates of their horizontal separation. It was hypothesized that if dominance was a factor, the disparity in foraging heights between males and females should decrease when the two were furthest apart in the horizontal plane. No significant relationship was detected, and very few agonistic interactions between foraging birds were observed. In North Carolina Pizzoni-Ardemani and Walters observed no differences in foraging position between females foraging with their group and females foraging alone. These observations suggest that dominance is not responsible for sexual differences in foraging behavior in Red-cockaded Woodpeckers, at least in the proximate sense, although additional studies obviously need to be performed. It may still be that interference from males in the past led to the evolution of foraging specializations in females that are now inflexible.

Foraging Requirements

TERRITORY SIZE

Each group forages throughout an extensive area which, as discussed in Chapter 6, consists largely of a defended territory but may also include extraterritorial range. The size of the area used for foraging varies greatly depending on season, habitat, and the presence of other groups. Typically the foraging area expands slightly in winter—or at least additional areas are used—unless adjacent groups prevent expansion. This is presumably a response to decreased arthropod availability. Habitat quality is also a factor. The largest foraging areas are generally where foraging is in some way limited. This can result from forest fragmentation, excessive midstory, or sparse trees in savanna habitats. Social interactions also appear to influence the area used. In dense populations, territorial boundaries are generally set by interactions with adjacent groups, and the foraging area is equivalent to the territory. In sparse populations, adjacent groups often do not exist, and the groups engage in extraterritorial forays into the vacant habitat, greatly expanding their foraging area.

The occurrence of extraterritorial forays hinders our ability to determine territory size. Most investigators report home range sizes and do not distinguish between territorial (i.e., defended) and nonterritorial range. Two things, nevertheless, are clear: territories are enormous, considering the size of the bird, and territory sizes are extremely variable. Robert McFarlane has examined the territory sizes of woodpeckers generally and has shown that territory size increases predictably with body size with few exceptions. The exceptions are the Red-cockaded Woodpecker and the two species of three-toed woodpeckers, which have territories an order of magnitude larger than other species of similar size (Figure 7.3).

Home ranges as small as 34 hectares and as large as 400+ hectares have been reported, with population averages varying from about 70 to about 150 hectares (Table 7.1). Within a population, the largest home ranges typically are about twice the size of the smallest ones. The measured home ranges often include extraterritorial range that is part of another group's

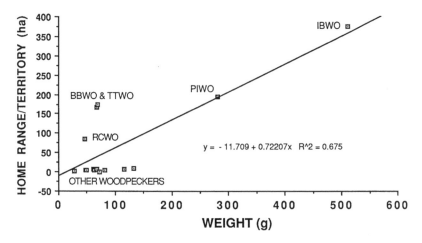

FIGURE 7.3 The relationship of woodpecker body weight to home range/territory size. The graph illustrates the larger home ranges of Red-cockaded Woodpeckers and Black-backed and Three-toed woodpeckers in relation to body weight, compared to other North American woodpeckers. Abbreviations: RCWO, Red-cockaded; TTWO, Three-toed; BBWO, Black-backed; PIWO, Pileated; and IBWO, Ivory-billed woodpeckers. Other woodpeckers are Hairy, Downy, White-headed, Red-bellied, Red-headed, Lewis's, and Acorn woodpeckers; Northern Flicker; and Yellow-bellied and Williamson's sapsuckers. Adapted from McFarlane 1995.

defended area. In the only study in which these extraterritorial areas were distinguished from the rest of the home range, Robert Hooper and colleagues determined that they constituted about 10% of the measured home ranges. Home ranges also include infrequently visited areas between the core ranges of the groups, and usually it is not clear whether or not these are defended. The amount of infrequently visited area is no doubt underestimated in most studies, because sampling is insufficient to detect all such areas. In the study by Hooper's group, such areas constituted about 12% of the home range, with the remaining 78% being defended territory. The latter is probably a higher percentage than in most populations, where open spaces between groups are more numerous. Thus the core areas, defended areas on which groups depend for the bulk of their resources, are considerably smaller than reported home ranges but likely are still in the range of 40–100 hectares. In the North Carolina Sandhills, 30 home ranges averaged 126 hectares, but the core areas, defined as the area in which a group spent 95% of its time, averaged only 81 hectares (64% of the total home range).

Home range size is related to habitat quality, as measured by the density and age of pines. When one compares populations, it is clear that where pines are few and/or young, home range size is larger (Table 7.1). For example, home ranges are especially large at the edge of the species' range in central and southern Florida, where habitat is especially poor.

Among the largest home ranges documented are those in the Curtis H. Stanton Energy Center in Florida, studied by Roy DeLotelle and coworkers. The habitat is a slash pine savanna with sparse tree densities, and home range size averages approximately 150 hectares. In contrast, on the intensively managed Francis Marion National Forest prior to Hurricane Hugo, in the densest areas average territory size was approximately 87 hectares. The habitat there was dense longleaf and loblolly pine with minimal midstory encroachment, and most groups had adjacent groups to restrict movements. Todd Engstrom and Felicia Sanders (a graduate student working with Engstrom) reported an even smaller average home range size (63 hectares) on the old-growth Wade Tract in southern Georgia. They suggested that the small size of territories at the Wade Tract was due to the availability of very old, high-quality trees for foraging.

On the other hand, within populations home range sizes may vary considerably within fairly uniform habitat, indicating that factors other than habitat quality are important. One obvious pattern is that groups with no close neighbors on one or more sides have larger home ranges. Why home ranges would be larger in the absence of neighbors is obvious, but why the neighbors are absent is not. The latter question holds the key to understanding the foraging ecology and current spatial distribution of Red-cockaded Woodpeckers. In some cases it may be that the adjacent habitat is too poor to support neighboring groups, and thus the home range size is a direct reflection of foraging habitat quality. This is clearly not the case in some instances, however. Where apparently suitable foraging habitat is

Table 7.1. *Home Range Size Estimates of Several Red-cockaded Woodpecker Populations*

Location	Mean Size (ha)	Range (ha)
South Florida	144	78–213
Central Florida	116	93–155
	148	116–199
North Florida	129	85–157
	70	58–92
Coastal South Carolina	87	34–225
Coastal North Carolina	153	63–199

not occupied, it may be that cavity trees are lacking (see Chapter 6). Given the demonstrated importance of cavity trees to habitat quality and habitat occupancy, it would not be surprising to find that the distribution of groups is to a considerable extent a function of the distribution of cavity trees, a scarce resource with a sparse, nonuniform distribution. Where this is true, an element of variability unrelated to foraging needs is introduced into the territory size and home range size distribution. Territory size in these instances is only a poor reflection of foraging needs, and limited-use areas may represent available foraging habitat that remains little exploited because of a lack of cavity trees. If this is true, it should be possible to compress some of the larger territories in order to add new groups to the population, without adverse effect on already existing groups. It is clear that Red-cockaded Woodpeckers require large territories, and even larger ones in poorer habitat, but they may not require as much area as some groups currently occupy.

If existing groups require all of the foraging habitat they use, then they should suffer reduced productivity or survival when some of that habitat is lost due to timber harvest or other factors. The data on impact of harvesting on territory size are equivocal. Jerry Jackson and Stephen Parris examined a population on Fort Polk in Louisiana. Prior to harvesting undertaken to develop a military training area, the average territory size was 135 hectares. Average territory size expanded to 253 hectares following extensive clearing to develop the training area. However, cavity trees were also impacted and group density declined, so the possibility exists that factors other than foraging sufficiency were involved. Jackson and Parris did, however, detect declines in bird weights, suggesting foraging impacts. The declines were not significant, perhaps due to the small sample size.

The large size of Red-cockaded Woodpecker territories, even those in the densest populations in the best habitats, has been the subject of much speculation. One factor is obviously the presence of helpers and the long period of dependency of juveniles. The territory must be large enough to support the breeding pair, any helpers that may be present, and young of the year for 6 months or more. Still, Red-cockaded Woodpecker territories are 10 times larger than expected from body size (Figure 7.3); groups are not large enough to completely account for this discrepancy. An additional factor may be prey density in the fire-maintained pine forests. It has been suggested that compared to other habitats, these forests may support a low biomass of potential prey species. The patchy distribution of abundant prey sources—i.e., dying trees—could also increase the value of large territories. Consequently, large territories may be required to supply adequate prey.

On the other hand, the large size of territories may not even be a result of prey availability and foraging patterns. It is possible that large territories are a mechanism to ensure adequate numbers of potential cavity trees. This would be especially critical in the present forests, where older trees are very rare. It will require additional research to determine whether foraging requirements, cavity tree availability, or social interactions driven by other factors ultimately determine territory size.

FORAGING PREFERENCES

The association of Red-cockaded Woodpeckers with open pine forests with sparse hardwood midstory has been frequently noted, generally in the context of the habitat in the vicinity of cavity tree clusters. The characterization of foraging habitat has received less attention. Joseph Skorupa, working in South Carolina, first reported a preference for larger trees as foraging sites. Data gathered by Hooper and Lennartz, also in South Carolina, demonstrated a strong preference for trees greater than 25 cm DBH (diameter at breast height) over smaller trees. In this and other studies, preference is measured by the difference between use and availability of the resource. Hooper and Lennartz saw no indication in the data of any preference for the very largest trees among those greater than 25 cm DBH. However, Engstrom and Sanders, working on the Wade Tract, reported some intriguing results. The Wade Tract is one of the few surviving virgin tracts of longleaf pine, habitat in which the woodpeckers have access to a full range of sizes and ages of foraging trees (Plate 1). In this location the birds prefer the largest trees available for foraging. For approximately 50% of all foraging observations, the birds were on trees 50 cm DBH or larger, even though trees of this size comprised only 15% of the total.

There generally is a correlation between the size and age of pine trees. Therefore, the preferences reported in most studies, in which only size has been measured, may reflect a response to either size or age (or both) of the foraging trees. Zwicker and Walters examined foraging tree selection at Camp Lejeune Marine Base in coastal North Carolina and measured both age and size. In this area the oldest longleaf pines, which are mostly trees rejected by timber cutters years ago, tend to be smaller than the most mature second-growth trees, even though the latter are 30–100+ years younger. Age and size are correlated among the second-growth trees, but not when comparing second-growth trees 60–80 years old to remnant trees 100–300 years old. At Camp Lejeune, all species of pines selected by the birds for foraging averaged larger and older than the pool of trees avail-

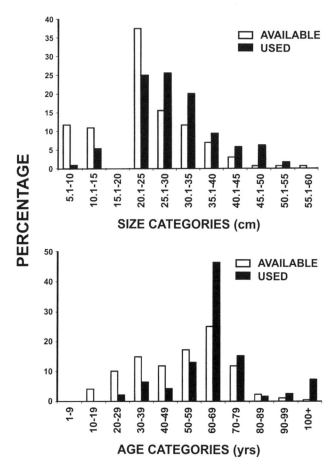

FIGURE 7.4 The relationship between use and availability of pines for foraging, in relation to diameter at breast height (DBH) and pine age (yrs). Modified from Zwicker and Walters 1999.

able, as in other studies (Figure 7.4). However, the old-growth trees were highly preferred despite their not being the largest trees available. Trees younger than age 50, which were also the smallest trees, were selected against. These results suggest that Red-cockaded Woodpeckers respond to tree age, independent of tree size, in selecting trees for foraging, although they respond to size as well. Jeff Hardesty and colleagues from the University of Florida also documented a strong preference for old-growth trees on Eglin Air Force Base, further supporting the thesis that the birds prefer the oldest trees available.

Prior to the studies by Zwicker and Walters and by Hardesty and colleagues, age preferences had been measured only at the level of the stand rather than the level of the individual tree. Particularly influential had been the studies by Hooper (with Lennartz, and with Richard Harlow) on the Francis Marion National Forest, which indicated that stands age 60 or greater were preferred and that stands less than 30 years old were avoided. This led to the notion that 30-year-old trees represented adequate forag- ing and 60-year-old trees preferred foraging. Many stands are not uniform in age, however, so it is not clear from such studies whether stand selec- tion was dependent only on the typical trees within stands or might have been affected by the selection of individual atypical trees, notably very old ones. In none of these studies were there old-growth stands that the birds could select. Interestingly, both Zwicker and Walters' and Hardesty and colleagues' studies of individual tree selection showed that trees age 60 years or more were indeed used equal to their availability and thus repre- sented adequate foraging. However, the use of younger trees was less than expected from studies of stand selection; in the Zwicker and Walters study, for example, trees younger than age 50 years were selected against.

We can conclude from these various studies that Red-cockaded Wood- peckers prefer older, larger trees for foraging. Although Red-cockaded Woodpeckers regularly forage on pines that are much too young to be used for cavity trees, they select against the youngest, smallest trees, using trees in the youngest age classes little if at all. The ages of the trees used appear to depend on what is available, suggesting that birds are forced in some areas to use trees they would avoid in better habitat elsewhere. Thus in one area birds used trees age 30 and younger infrequently, but used 40-, 50-, and 60- year-old trees extensively, whereas in another area with more older trees available, they used 60-year-old trees but avoided those age 40 and 50. Of course, soil conditions and other site factors likely account for some differ- ences, younger trees being used more in better sites, presumably because they are larger. Pine species is also a factor. For example, loblolly pines ap- pear to be used at younger ages in North Carolina than longleaf pines.

Most data on individual tree selection come from longleaf pine habitat. For longleaf pine, all results to date are consistent with a preference for the oldest trees available, with the use of younger trees, especially those less than 60 years old, dependent on the availability of older ones. That is, the more old trees available, the greater the minimum age at which trees begin to be used. If there is an age beyond which there is no selection, it must be greater than 100–150 years. In no study have age classes beyond this age been separated in order to examine selection. Similarly, there is as yet no evidence of an optimum age beyond which trees are selected against. Much the same can be said of size: as yet there is no evidence of an optimum size or a size beyond which there is no selection, and the data are consistent with preference for the largest trees available. And again, it is not possible with existing data to determine if demonstrated preferences are based on age alone, size alone, or both age and size.

Preference for large trees may simply reflect the greater surface area for foraging that large trees provide. Preference for old trees is more puzzling. The bark of old trees may differ from that of younger trees, resulting in some difference in arthropod density, availability, or community composition. Hooper, however, found no relationship between arthropod density and age of longleaf pines in South Carolina in winter. Another possibility is that because Red-cockaded Woodpeckers formerly lived in mixed-age forests, in which only a small portion of the trees were young, they are somehow better adapted for foraging in old trees. It may also be that old trees contain more of the foraging substrates heavily used by breeding males, such as self-pruning branches.

Preferences among pine species have also been evaluated primarily at the level of the stand. At this level, a preference for longleaf pine over other pines is often evident (see above). However, at the level of the individual tree there is little evidence of selection by species. Zwicker, for example, found that longleaf, loblolly, and pond pines were all used in proportion to their availability in coastal North Carolina.

Besides the pines themselves, the structure of the surrounding vegetation also influences foraging patterns of Red-cockaded Woodpeckers. Hardwood midstory vegetation is frequently abundant in Red-cockaded Woodpecker habitat, due to the reduced impact of fire in many forests. Hardwood midstory has been demonstrated to have a detrimental impact on the nesting habitat of Red-cockaded Woodpeckers, as detailed in Chapters 6, 9, and 10. The impact of hardwood midstory vegetation in the foraging habitat has been less intensively studied. In Texas most loblolly and shortleaf pine habitats have an abundant hardwood midstory: most forests can be characterized as having a pine canopy and a dense layer of hardwood midstory foliage ranging from 5–20 m in height. At the time of this writing, most cavity tree clusters have experienced recent mechanical midstory removal as one aspect of intensified management in response to declining populations and a court order (see Chapter 10). The forest landscape thus consists of a mosaic of forest with dense midstory, cavity tree clusters with midstory removed, seed-tree cuts with a sparse canopy and no hardwood midstory, and clear-cuts and young plantations.

The foraging patterns of the birds on this landscape mosaic are striking.

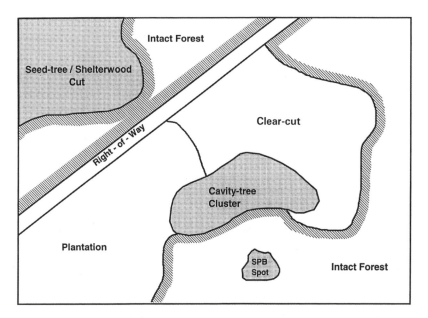

FIGURE 7.5 Diagram of areas of concentrated foraging by Red-cockaded Woodpeck-ers in a typical loblolly or shortleaf pine landscape in eastern Texas. Stands with con-centrated use (shaded) are cavity tree clusters and seed-tree/shelterwood stands with limited hardwood midstory, and southern pine beetle (SPB) spots with very abundant prey. Intact forest with abundant hardwood midstory is used to a lesser extent, and clear-cuts and young plantations are unsuitable for foraging. Intact forest (hatched) adjacent to preferred stands and open habitats also receive concentrated use.

Most of the foraging activity occurs within the cavity tree clusters, within seed-tree cuts with scattered mature trees, and within 100 m or less of large openings (i.e., clear-cuts, young plantations, etc.). Very little foraging occurs within the expanses of mature forest with well-developed hardwood mid-story, even though this is the dominant forest type (Figure 7.5). The major exception within the mature forest is the extensive use of occasional dying pines infested by bark beetles, a very rich source of prey (see Chapter 8). We interpret these patterns as an attempt by Red-cockaded Woodpeckers to avoid foraging in locations with dense hardwood midstory. The birds use areas devoid of dense midstory (cavity tree clusters and seed-tree cuts), as well as sites adjacent to openings where midstory was absent in at least one direction.

This remarkable pattern of spatial use of pine forests in the presence of hardwood midstory contrasts sharply with the situation in longleaf pine forests. A major difference between longleaf pine forests and those dominated by other species of pines in Texas is the minimal amount of hardwood midstory present. In longleaf pine forests, foraging activities occur throughout the forested habitat.

A study by DeLotelle and coworkers in Florida and Georgia supports these conclusions about the effects of hardwood midstory and provides additional insights about the factors affecting the use of habitat for foraging. They examined the relative use of different stands for foraging and found that older stands and stands with greater foraging substrate were used more than expected. They also found that the presence of hardwood midstory reduced stand use to below that expected based on other factors. Finally, they found that stand use increased with decreasing distance to the cavity tree cluster. Walters and coworkers obtained similar results in the North Carolina Sandhills: stand use was positively related to stand area and the presence of cavity trees, positively related to pine age and size, and negatively affected by hardwood midstory.

These studies indicate that the value of a tree of a given age and size is altered by additional factors, especially proximity to the center of the group's activities, the cavity tree cluster, which increases value, and by prevalence of hardwood midstory in the immediate vicinity, which decreases value.

FORAGING NEEDS

Ideas about what Red-cockaded Woodpeckers require in the way of foraging habitat are based on observations about what the birds use on current landscapes. This is acceptable for identifying the habitat types in which they are able to live, that is, for determining that they should be provided with longleaf pine savannas instead of hardwood forests. It is a poor basis for determining what is required within suitable habitat, however. That is, it is a poor basis for determining how much area they require, and what ages, sizes, and number of trees they require. Foraging requirements should be linked to foraging success and ultimately to bird fitness. In northern Florida, Fran James and coworkers found that group size and productivity were positively related to the percent of herbaceous vegetation in the ground cover and a compound variable made up of the difference between the densities of large and small trees. Walters and coworkers found group size in North Carolina to be related to density of large pines, presence of old pines, and presence of hardwood and pine midstory. These studies suggest that at least in longleaf pine forests, habitat with a particular structure provides the best foraging, namely, open parklike stands of older pines with a rich herbaceous layer and sparse hardwood or pine midstory. This corresponds to the type of habitat the birds select (see above). In this context it is interesting that the birds studied by Engstrom and Sanders on the Wade Tract, whose foraging trees were much older and much larger than those available to birds in other populations studied, appeared to have had very high fitness. We are only beginning to determine the precise foraging requirements of Red-cockaded Woodpeckers.

Relationships with Other Insectivorous Birds

Foraging by Red-cockaded Woodpeckers does not occur in a vacuum. The forests inhabited by Red-cockaded Woodpeckers support a diverse array of other bird species that are primarily insectivorous. The ecological literature is replete with examples of species limited or otherwise impacted by competition. In many cases behavioral and morphological differences have evolved that minimize the impact of competitive interactions. As a result, competition with most other species presumably is minimal. Other woodpeckers, however, are potential competitors for available prey. It has been hypothesized that Red-cockaded Woodpeckers are subject to reduced competition from other woodpeckers because they inhabit a forest with few hardwoods and limited numbers of dead trees due to frequent fires, as well as a relatively low abundance of arthropods. Their specialized foraging techniques, large territory size, and presence of helpers may all be adaptations conferring a selective advantage to Red-cockaded Woodpeckers, compared to other woodpeckers, in fire-maintained pine forests.

MIXED-SPECIES FORAGING FLOCKS

A number of avian species form mixed-species foraging flocks that include Red-cockaded Woodpeckers. Flock formation occurs during much of the year, waning only during the nesting period. The formation of mixed-species foraging flocks is a common phenomenon in many bird communities, and the selective forces leading to the evolution of foraging in mixed flocks have been the subject of lively debate. Hypotheses include more efficient location of food, rapid detection of potential predators, and detection of prey disturbed by other species. Several selective forces are probably involved.

In any event, being present at a Red-cockaded Woodpecker cavity tree cluster at dawn sets the stage for observing a spectacular example of a mixed-species foraging flock. Red-cockaded Woodpeckers are relatively late risers compared to other insectivorous species in southern pine ecosystems. Consequently, many other birds begin calling and foraging half an hour or more before Red-cockaded Woodpeckers emerge from their roost cavities. Typically, a large mixed-species foraging flock begins to assemble in the vicinity of the woodpecker cavity trees prior to the emergence of the Red-cockaded Woodpeckers. These flocks may number 50 or more individuals of several species. Once the woodpeckers emerge and begin foraging, they join the assemblage as it moves through the forest as a well-defined mixed-species foraging flock.

Brown-headed Nuthatches, Pine Warblers, and Carolina Chickadees (*Poecile carolinensis*) are the most common resident species within these mixed flocks in Louisiana and eastern Texas. Red-bellied Woodpeckers, Downy Woodpeckers, Tufted Titmice, and a number of other species frequently join them. A considerable number of migrant species also join these flocks during winter, including Yellow-rumped Warblers (*Dendroica coronata*), Ruby-crowned Kinglets (*Regulus calendula*), and Brown Creepers (*Certhia americana*). In North Carolina, Golden-crowned Kinglets (*Regulus satrapa*) and Eastern Bluebirds also are regular flock members. As many as seven woodpecker species have been observed in a single flock in North Carolina. Interestingly, summer residents that are potential members of these flocks (vireos, additional warblers) are seldom encountered. Summer Tanagers are a notable exception.

The assembly of the flock prior to the emergence of the Red-cockaded Woodpeckers from their roost cavities suggests that the woodpeckers are important constituents of the mixed flock, even though they comprise a small minority of the individuals. Perhaps their extreme predictability in time and space provides a convenient assembly point for the other flock members.

The mixed flocks show a strong tendency to remain together for several hours, generally until the Red-cockaded Woodpeckers decrease their foraging activity in late morning. The movements of the flock appear to be influenced at times by the foraging pattern of the Red-cockaded Woodpeckers. Two observations support this view. In Texas, if the Red-cockaded Woodpeckers encounter a particularly rich food source, generally a tree recently killed by southern pine beetles, they may spend considerable time foraging on that individual tree. In these situations the mixed flock tends to remain in the vicinity until the woodpeckers resume foraging movements. In situations where the Red-cockaded Woodpeckers undertake extensive movements (several hundred meters) between foraging sites, generally when crossing less suitable foraging habitat, the flock tends to follow.

Some or all flock members may drop out during these movements, presumably because the woodpeckers have moved beyond the home range of individual flock members. At other times, the Red-cockaded Woodpeckers lag behind and appear to follow the flock. In North Carolina, sometimes a woodpecker group will leave a flock and fly several hundred meters to join another flock. Often two or three flocks form within the territory of a single Red-cockaded Woodpecker group, indicating that their presence is not necessary for flock formation.

Red-cockaded Woodpeckers appear to pay little attention to other species in these foraging flocks, other than to coordinate their movements with those of the flock. An exception is when flock members give warning calls at the approach of potential predators. However, some species frequently pay detailed attention to the activities of the Red-cockaded Woodpeckers. Individuals of several species have been observed following individual Red-cockaded Woodpeckers very closely, often remaining within a meter or less of the woodpecker for extended periods. Attending individuals frequently trail the woodpeckers closely when they change their foraging position to another tree. Presumably these individuals are primarily interested in obtaining prey disturbed by the woodpeckers. Insects are often observed flying from the woodpecker's foraging site, and these are frequently pursued and captured by attending individuals. Brown-headed Nuthatches and Pine Warblers are the species most frequently involved in these interactions. Perhaps significantly, these two species, like the Red-cockaded Woodpeckers, are permanent residents, and both species, especially the nuthatch, spend considerable time foraging on the boles of pines independent of woodpeckers. Occasionally a larger species, usually a Red-bellied Woodpecker, will displace a Red-cockaded Woodpecker from a foraging site. Direct interactions such as these rarely occur within the flock, however, as each species concentrates on its favored foraging locations and techniques.

CHAPTER EIGHT

Red-cockaded Woodpeckers and Bark Beetles

A Love-Hate Relationship

Forest Landscape Interactions

Southern pine beetles and other bark beetles are found throughout the range of the Red-cockaded Woodpecker. Bark beetles, woodpeckers, and pine trees have coevolved for thousands of years, producing some interesting interactions. Historically, bark beetle infestations (beetle "spots") have served as an important process by which loblolly and shortleaf pine forests have been regenerated. Large and small beetle infestations with resulting timber losses occur in the loblolly and shortleaf pine forests throughout the southeastern United States, from Texas to Florida and from Oklahoma and Missouri to Delaware. Infestations by southern pine beetles are relatively rare in longleaf pine; longleaf pines are more resistant to bark beetle attack than loblolly and shortleaf pines, and thus infestations in this pine type are usually infrequent and of less severity. They are mostly restricted to more western areas, and occur when massive beetle irruptions occur in loblolly pine and spread out of that pine's cover type into longleaf areas. Generally, beetle infestation frequency and size is greater in the western half of the range of southern pines than in the eastern half. In the North Carolina Sandhills, small infestations may occur after a hot crown fire, but normally attacks are restricted to single trees. Such is not the case in the western portion of the range, where major infestations (epidemics), which seem to recur every 7 to 10 years, may kill thousands of hectares of pines within weeks.

In addition to tornadoes and hurricanes, discussed in Chapter 2, beetle infestations significantly influence the stand composition and mosaic of age classes that compose the forest landscape, particularly in loblolly and shortleaf pine habitat. Over the years, small beetle infestations, which are far more frequent than major epidemics, can create a very patchy, uneven-aged forest mosaic. Large epidemics produce extensive tracts of even-aged forests.

Although fire was probably the major force, bark beetles also played an important role in shaping loblolly and shortleaf pine forests. In addition to (and aided by) storms and lightning, bark beetles serve as a natural means to regenerate pine ecosystems. The size and shape of forest areas killed by bark beetles determine the composition and distribution of stands of different ages in the forest landscape. Patches killed by beetles are cleared by fire, permitting seeds from surrounding pines or existing young stems to replace what was lost.

It has been suggested that old-growth pine forests cannot be maintained for Red-cockaded Woodpeckers because bark beetles and other natural forces will destroy the forest without the presence of humans to harvest and regenerate it. This opinion is nonsensical in light of the evolution and coexistence of bark beetles, woodpeckers, fire, and pines in pre-Columbian times. On a landscape scale, the forces that shaped the southern pine ecosystems must have created and maintained a dynamic ecosystem with numerous old pines. Certainly, disturbances that killed pines on large and small scales occurred. But Europeans would not have found Red-cockaded Woodpeckers or vast areas of mature pine forests had such a dynamic balance not predominated. The existence of thousands of hectares of 175- to 320-year-old shortleaf pines in the McCurtain County Wilderness Area in southeastern Oklahoma is strong evidence of the persistence of old pines in a dynamic ecosystem. The instances of die-offs and excessive bark-beetle-caused mortality of loblolly pines in modern times that give rise to the "nonsustainable forest" view are often the result of loblolly pines being planted off-site and in extremely overstocked densities in areas formerly containing longleaf or shortleaf pine because the faster-growing loblolly pine is commercially favored.

Landscape-level interactions affect the age and distribution of pine habitat used by Red-cockaded Woodpeckers, but more direct relationships exist among the woodpeckers, bark beetles, and pines. The interaction between bark beetles and pine trees over the millennia may be the primary reason that pines produce large quantities of sticky oleoresins, or pine gum. The pine's production of resin and transport of gum to wounds has evolved as the pine's primary defense against bark beetles. As discussed in Chapter 5, the Red-cockaded Woodpecker makes use of this wound response and the resulting pine gum flow to serve as a barrier against rat snakes.

Although bark beetles—primarily the southern pine beetle—are a major cause of Red-cockaded Woodpecker cavity tree death in some areas, beetle-infested pines in these regions also serve as excellent food sources for the birds, especially when beetle infestations are small and scattered throughout the forest landscape. Large beetle epidemics provide surges in food availability over large areas, but can kill many cavity trees and result in a deficiency in foraging habitat for woodpecker groups after insects leave the infested dead pines. Thus, a classic love-hate relationship: the woodpeckers love to forage on beetle-infested pines but hate to lose cavity trees.

When beetles are at endemic (nonepidemic) population levels, clusters of cavity trees are not threatened by expanding, multiple-tree infestations, but beetles may infest and kill an individual cavity tree. Active cavity trees seem to be preferred over inactive ones by southern pine beetles. These single, infested active cavity trees typically do not trigger a growing infestation. On the Angelina National Forest, more than one-fourth of the active cavity trees in loblolly and shortleaf pine habitats were killed each year by single-tree infestations during 1992, 1993, and 1994. The vast majority of single cavity-tree infestations occur within active woodpecker clusters. Forest management activities to help the woodpecker, such as midstory removal, thinning of pines, and prescribed burning, are typically focused on active woodpecker clusters. High rates of bark beetle infestation of active cavity trees within active clusters may be associated with intensified management efforts to help the woodpecker or with the woodpecker's pecking of resin wells.

Dynamics of a Beetle Infestation

THE STAGE IS SET

A disturbance, most often caused by lightning (Figure 8.1), injures a lone pine or group of pines and sets the stage for a beetle infestation. The power associated with lightning strikes is variable. Most lightning (flow of negative electrons) strikes from the sky toward the earth, creating a gain in negative charge on the earth relative to the sky. Lightning can also flow from the earth to the sky, creating a net increase in positive charge on the earth. "Negative" lightning does far less damage and is far more frequent in occurrence than "positive" lightning. Damage to pine trees from positive lightning is far more extensive because of the greater power of the strike.

When lightning strikes a tree, various types of damage may occur. As electricity flows down the outside and just below the surface of the pine's bole, it superheats the water in the tissues. This appears to result in a rapid

FIGURE 8.1 Damage produced when lightning strikes a pine tree often includes root damage in addition to the visible blown-apart areas on the pine's bole. Lightning-struck pines are often attacked and infested by southern pine beetles, which in turn can trigger the subsequent expansion into a large, multiple-tree infestation, sometimes killing hundreds of pines. Photo by R. N. Conner.

formation of steam within the tree, usually just under the bark or within the sapwood close to the bark. The pressure created by the steam causes the wood tissue to explode outward, typically following the spiral path of the lightning's electrons down the bole, leaving a characteristic scar. Living sapwood is exposed and injury occurs laterally to the torn plant tissues. As lightning passes into the ground at the base of the pine, similar high-pressure explosions occur in the root system and ground around the roots.

Damage below ground often has the greatest impact on the tree because it destroys the pine's feeder roots, which are its main source of nutrients. If lightning does not kill the pine outright through structural damage, it greatly weakens the pine, interrupting its ability to transport water and to produce and transport pine gum.

As mentioned previously, pine gum is the pine's first line of defense against bark beetles. The structural damage caused by lightning or any other injury releases terpenes, components of oleoresin, as the pine oozes its sticky gum. However, if lightning does substantial damage, it can reduce or eliminate the pine's ability to mobilize the full capabilities of its resin defense system. Terpenes, primarily *alpha*-pinene, which are released during any injury (and are used in the manufacture of turpentine), serve as primary attractants to some species of bark beetles, particularly bark engraver beetles and black turpentine beetles. Whether *alpha*-pinene by itself serves as a primary attractant to southern pine beetles remains controversial. At present, most data suggest that it does not. What attracts southern pine beetles to susceptible pines is not yet fully understood.

Species of pines have different abilities to produce oleoresins, and resins of pines have different properties. The ability of pines to produce resin affects their potential value to Red-cockaded Woodpeckers as cavity trees (see Chapter 5). Longleaf and slash pines normally produce more resin than loblolly and shortleaf pines. Research done by John Hodges and others at Mississippi State University has shown that slash pine resin is very viscous, crystallizes slowly, and is produced in moderate amounts. Longleaf pines produce resin with moderately high viscosity in extremely high yields and at high flow rates. Loblolly and shortleaf pines generally produce resin flows of short duration and moderate to low yields. Their resin has low viscosity and a rapid crystallization rate. Thus, of these four pine species, longleaf pine would be the preferred species for Red-cockaded Woodpeckers, based on this pine's ability to produce a sustained supply of high-quality resin.

Extreme drought that induces moisture stress can also increase susceptibility of pines to bark beetles by decreasing resin production. Surprisingly, plentiful water in late spring and early summer can keep the pine in a physiological state that favors growth rather than resin production, also increasing susceptibility to bark beetles. In general, any pines that are in a weakened condition caused by disease, moisture stress, overcrowding, or injuries are more susceptible, and can serve as centers of an initial beetle infestation.

Southern pine beetles that emerge and disperse during spring are probably the pioneers that initially start the majority of beetle infestation spots. Beetles emerging later are attracted to existing infestations. Southern pine beetles emerging in spring and fall have higher body fat content than beetles emerging during other seasons. The greater the fat content, the farther a beetle can travel to find a suitable host pine to initiate an attack. Females generally have a higher fat content than males.

As mentioned earlier, it is not clear what initially attracts southern pine beetles to a pine that they attack, or if a primary attractant even exists. One line of current scientific thinking is that southern pine beetles fly through a forest randomly but are visually attracted to dark, vertically oriented objects. This line of thought also holds that the beetles attack all pines at random. Once on a host tree, female southern pine beetles bite the bark in response to chemical stimuli they perceive. If a female determines the tree is a suitable host, she begins boring.

THE ATTACK BEGINS

Regardless of how beetles locate trees, the combination of arriving beetles and a susceptible pine results in a bark beetle attack and infestation. Most infestations appear to start in one to several pines. The three species of bark engraver beetles (*Ips grandicollis, I. calligraphus,* and *I. avulsus*) typically attack a pine in the upper region of the bole, in the branches of the crown. Southern pine beetles normally attack the pine in the middle regions of the bole, whereas black turpentine beetles attack near the base of the bole, usually in the lowest 3 m of the trunk. Although infestations can consist of only one species, many attacks involve several if not all five species of pine bark beetles.

Signs of beetle attack are not always obvious at first. If a pine is not producing much resin because of season or weakness, the only signs of attack will be bits of chewed bark dust that may fall onto cobwebs around the base of the pine. The 2- to 4-mm-long adult southern pine beetles make tiny holes in the bark (about 1 mm in diameter) when they bore into the pine, and only the most experienced observers using binoculars to scan the bole of a newly attacked pine can spot these holes. Soon after this initial attack, the pine usually begins to ooze resin at the wound sites in an attempt to "pitch out" the adult beetles (Figure 8.2). Sometimes the volume of resin is sufficient to expel the beetles directly, but often a crystallization process is necessary for a successful defense. Resin transported to wound areas via resin canals begins to crystallize after arriving at the injury site. Crystallized resin takes on the appearance of yellowish white popcorn as it flows

FIGURE 8.2 When southern pine beetles attack a pine, they chew through the bark into the living cambium, causing pine resin to flow out of the wound site. The crystallized pine resin at an attack site produces popcorn-like formations on the pine's bole. Photo by R. F. Billings, Texas Forest Service.

out the small hole made by a beetle and hardens around the wound opening (Figure 8.3).

Resin in young pines begins to harden rapidly and can plug the attack hole made by the beetle. Resin crystallizes more slowly in older pines, possibly reducing their ability to block the open wound. Normally, an infestation must kill the pine in order to allow successful reproduction by southern pine beetles. A sufficient number of adult beetles (generally thousands) must attack a pine and cause enough damage to substantially block transport of sap and nutrients in the phloem tissue, killing the tree and shutting off its defenses; otherwise, the attack is repelled. However, bark beetles are often aided by allies in their attack on pine trees.

Female southern pine beetles as well as some other bark beetle species have specialized structures called thoracic mycangia located on the back of their chitinous exoskeleton. Stan Barras, an entomologist with the U.S. Forest Service Southern Forest Experiment Station in the 1970s, and Kier

Klepzig, a mycologist/entomologist with the U.S. Forest Service Southern Research Station in Pineville, Louisiana, in the 1990s, studied the interaction among three species of fungi and southern pine beetles. The mycangia carry two species of mycangial fungi, *Ceratocystiopsis ranaculosus* and *Entomocorticium* sp. A (a fungus discovered by Stan Barras: SJB-122), into the pine as the beetles gnaw their way through the bark and into the cambium. In other studies, beetle egg galleries were twice as long and had 18 times more beetle progeny when mycangial fungi grew in the galleries. Research by Matt Ayres, with Dartmouth College, and coworkers suggests that the mycangial fungus *Entomocorticium* sp. A may supplement the southern pine beetle's larval diet with nitrogen, which is scarce in the phloem tissue of pines.

Bluestain fungus (*Ophiostoma minus*) can also be carried into the pine on the surface of the southern pine beetle's body. Passage of adult female beetles into the pine serves to inoculate the pine with fungi. Large quantities of resin can impede or block the fungal infection, as well as adult beetles, by a process called resinosis, but the fungus spreads rapidly if it

FIGURE 8.3 Multiple attacks on a Red-cockaded Woodpecker's cavity tree by numerous southern pine beetles appear to cover the pine's bole with popcorn. A massive attack of a pine resulting in the tree's death is essential for the life cycle of southern pine beetles to be completed and a new generation produced. Photo by R. F. Billings, Texas Forest Service.

does not encounter copious resin. During the infestation of a pine, blue-stain fungus often grows into the xylem tissue of the sapwood and can interfere with the pine's ability to transport water.

Although the presence of bluestain fungus may help southern pine beetles kill a pine, its presence may slow the beetle's invasion, because the beetles have to contend with the bluestain fungal mycelia in their tunnels. Of 32 infestations examined throughout the South, 6 did not have blue-stain fungus associated with the southern pine beetles. In all 6 of these exceptional situations, beetle densities were very high, suggesting that the beetles can produce more brood in the absence of bluestain.

Because pine resin can block the growth of fungi, as well as deter the entry of bark beetles, low numbers of attacking beetles in a vigorous pine are probably not sufficient to successfully inoculate fungi or destroy the cambium of the pine. Thus, most trees are able to survive low-intensity attacks. Low-intensity attacks probably occur most often when southern pine beetle populations in an area are low. When southern pine beetles are scarce, they are often found in pines and branches infested by bark engraver beetles.

The infestation of Red-cockaded Woodpecker cavity trees and other pines is largely controlled by chemical signals between beetles. An under-standing of the beetle's chemical communication during the initial infes-tation of a pine is important for the development of methods to prevent beetle-caused mortality of cavity trees. As adult females begin to chew through the bark, they release a pheromone (also called a semiochemical, a chemical signal with a communicatory function). The pheromone released during early stages of an attack on a pine by female southern pine beetles is frontalin. Frontalin is produced in the hindgut of the female beetle and is released as she chews her way into the cambium and phloem tissues and defecates. The combination of frontalin and terpenes (primarily *alpha*-pinene) in the pine resin serves as a strong attractant to other females and males, which are drawn to the attacked pine by following a scent trail of these compounds as the chemicals drift through the air. *Alpha*-pinene is re-ported to be the single most important host tree odor, but during field tests *alpha*-pinene alone did not serve as an attractant to southern pine beetles. Together, frontalin and *alpha*-pinene cause female beetles to aggregate and concentrate a massive attack on the pine, further attracting male southern pine beetles and resulting in the pine's death. The pine must die for the successful emergence of a new generation of southern pine beetles.

Female southern pine beetles also produce the pheromone *trans*-verbenol. The function of this pheromone is not entirely clear, but it is believed to augment the attractiveness of frontalin, particularly when the production of the host pine's resin (terpenes) ceases during a successful attack. *Trans*-verbenol may also serve as an arrestant that causes southern pine beetles to stop dispersal when they detect the pheromone.

An infested pine could in theory fill up with too many bark beetles. Male southern pine beetles address this problem by producing a different pheromone, verbenone, in their hindgut, which is released as they join in the infestation to mate with females. Male southern pine beetles arrive at the infested pine only after the females have selected and successfully attacked the host pine tree and high concentrations of frontalin and *alpha*-pinene are being produced. The male pheromone is also an attractant chemical, but only at the low concentrations associated with the initial arrival of males at an infested pine. At this time, verbenone may play a role in males' location of the entrance holes of female beetles. At the high concentrations that occur when pines fill up with southern pine beetles, verbenone becomes an attack inhibitor and causes male and female southern pine beetles near the tree to avoid the pine already under mass attack.

THE ATTACK SUCCEEDS

Successful infestations permit adult male and female bark beetles to mate and construct galleries and egg chambers (also called niches) in the cambium and phloem where the females lay (oviposit) their eggs (Figure 8.4). Red-cockaded Woodpeckers regularly exploit the larvae and pupae of developing beetles within the chambers constructed in the phloem and bark of infested pines. When egg laying is complete, adult beetles can reemerge to fly and infest other pines. Southern pine beetle eggs are opaque and pearly white, and are about 1.5 mm long and 1 mm wide (Figure 8.4). Eggs can hatch in as few as 3–11 days at temperatures of 30–15 °C, or take as long as 34 days at temperatures as low as 10 °C. Eggs hatch into larvae that eat the cambium and phloem tissues, forming tunnels as they progress. Upon emergence from the egg, larvae are about 2 mm long and yellowish white. The head and last abdominal segment have long white bristles, and the head has well-developed mouth parts. When mature, the larvae are about 5–7 mm in length with reddish heads and reddish black mandibles (Figure 8.4). Larvae go through four stages, termed instars, prior to pupation and metamorphosis into adults. Overall, the larval stage can last 15–40 days at temperatures of 25–15 °C.

When mature, the larvae pupate. Pupae are fragile, yellowish white, and range from 3 to 4 mm in length (Figure 8.4). The pupal stage generally

A B

C D

E F

FIGURE 8.4 Life stages of the southern pine beetle, most of which occur under the bark of a pine tree. Female southern pine beetles chew *(A)* galleries just under the bark and lay *(B)* eggs in egg niches within the beetle galleries. Eggs hatch into *(C)* larvae, which eat cambial tissue and eventually metamorphose into *(D)* pupae. Pupae metamorphose into *(E)* callow adults, which in time darken and emerge from the pine as *(F)* adults. Photos by R. F. Billings, Texas Forest Service (galleries and adult beetle), and from U.S. Forest Service.

lasts 5–17 days at temperatures of 30–15° C. Pupae go through metamorphosis in the outer bark layer, and the newly formed adult beetles, called callow adults, are yellowish white in color (Figure 8.4). They turn reddish brown and finally dark brown in 6–14 days, typically in about a week. The adult southern pine beetle is brown to black (Figure 8.4). When mature, the beetles chew their way toward the exterior of the bark and emerge to fly off and infest other pines.

The beetle completes its life cycle in a shorter period of time in warmer months than in cooler months. Under ideal summer conditions, as few as 26 days are required between egg laying and brood emergence, whereas 54 days may be necessary in winter. Southern pine beetles may have as few as three generations per year in the northern portion of their range, whereas seven overlapping generations can occur in the southern United States. The number of southern pine beetles may increase by as much as 10 times in a single generation. The longer the eggs, larvae, pupae, and callow adults are inside the pine tree, the more time predators like woodpeckers have to excavate through the bark and feed on the insects. Thus, rapid population growth is more likely in the warmer months than during winter, when the extended life cycle provides more time for predators to reduce the size of the beetle population within a tree.

ENEMIES ARRIVE

Checkered clerid beetles (*Thanasimus dubius*) soon follow attacking southern pine beetles and feed extensively on the adult southern pine beetles on the surface of bark (Figure 8.5). Clerid beetles are the major insect predator of southern pine beetles, and healthy clerid populations can reduce the successful infestation rate of southern pine beetles on Red-cockaded Woodpecker cavity trees. Not only do adults eat adult southern pine beetles, but the larvae of clerid beetles prey heavily on the larval stages of southern pine

FIGURE 8.5 Clerid beetles are a major predator of southern pine beetles and are often seen climbing on the bole of an infested pine, where they attack and eat adult southern pine beetles. Photo by R. N. Conner.

FIGURE 8.6 Long-horned sawyer beetles (Cerambicidae) arrive on pines infested by bark beetles after the attack is well under way. Feeding activity of their larvae can disrupt developing southern pine beetles. The large larvae of the sawyer beetles make excellent prey for foraging woodpeckers. Photo by R. F. Billings, Texas Forest Service.

beetles. Tenebrionid beetles (*Corticeus* spp.) are also often very abundant around pines infested by southern pine beetles. Tenebrionids often enter pines through entrance holes made by southern pine beetles and appear to prey on some life stages of the southern pine beetle.

During a successful attack on a pine, numerous other species of beetles are drawn to the dying pine. Although not predators, many of these beetles compete with southern pine beetles for the same food supply. Ambrosia beetles (Scolytidae, *Platypus flavicornis*), long-horned sawyer beetles (Cerambycidae, *Monochamus* spp., and *Neacanthocinus obsoletus*), flat-headed sawyer or metallic borer beetles (Buprestidae, *Chalcopha* sp.), and click beetles (Elatoridae, *Alaus* spp.) are among the more common species that come to oviposit in the pine. These beetles consume some of the cambium and phloem tissues that larval southern pine beetles eat. In large numbers, some species of these beetles appear to exert a negative competitive impact on southern pine beetles. The southern pine sawyer beetle (*Monochamus titillator*) can significantly reduce southern pine beetle survival within a pine host by consuming large amounts of pine tissue, reducing the amount left for developing southern pine beetle larvae (Figure 8.6). Collectively, sawyer beetles can consume up to 20% of the cambium and phloem tissue and cause a southern pine beetle mortality rate of up to 14%.

Parasitoids (primarily Hymenoptera and Diptera) can also affect southern pine beetle populations. An abundance of parasitoids can hamper the ability of southern pine beetles to successfully infest pine trees. In general, parasitoids develop from egg to adulthood on a single host insect, feeding on the host without killing it until the parasitoid reaches maturity. There are about 35 known parasitoids of southern pine beetles, many of which also attack other species of bark beetles, such as bark engraver beetles. Parasitoids typically respond to the odors produced by southern pine beetles and other beetles, as well as the host odors of the infested pine. Parasitoids normally remain on the outer bark of the pine and sting their host through the bark of the pine prior to egg laying, preferring third or fourth instar larvae as their victims. Methods used by female parasitoids to locate southern pine beetles within pines are not fully understood. Some evidence suggests that they orient toward the sounds made by larvae and adults under the bark. Some parasitoids, particularly *Roptrocerus xylophagorum,* enter the bark through southern pine beetle attack and air holes and lay their eggs in the southern pine beetle egg galleries. Most parasitoids affecting southern pine beetles appear at an infested pine only after large numbers of the beetle hosts are present. Bark thickness has a major effect on the success of parasitoids; the thicker the bark, the fewer the parasitoids. Shortleaf pines typically have very thin bark compared to that of loblolly pines and thus provide a much better environment for parasitoids.

Southern pine beetles are also affected by a variety of tarsonemid mites. Some species of mites prey on southern pine beetles, others can affect their ability to fly and move, whereas an additional group of mites benefits southern pine beetles by preying on parasitic nematodes. About 15 species of phoretic mites are commonly associated with southern pine beetles. Phoretic mites typically are nonparasitic and use southern pine beetles as a means of transportation. They can successfully feed and reproduce on blue-stain fungus and the mycangial fungus *Ceratocystiopsis ranaculosus* that are carried by southern pine beetles, but not on the mycangial fungus *Entomocorticium* sp. A. About one-third of southern pine beetles carry these and other mites. Beetles with mites generally attack lower portions of the pine's bole, whereas individuals without mites attack higher regions of the bole. Flying southern pine beetles can carry up to about 20% of their body weight in mites, indicating that mites may have a substantial impact on the dispersing ability of the beetles during some seasons. Mites are typically most abundant during summer and fall.

BEETLES AS WOODPECKER FOOD

As discussed in Chapter 7, the aggregation of insects within an infested pine can be a food bonanza for Red-cockaded Woodpeckers throughout the year, but particularly during the breeding season when adults are trying to feed hungry nestlings. Beetle-infested pines can be a valuable food source for many species of woodpeckers. Various species of woodpeckers prey heavily on adult bark beetles and associated insects by gleaning them off the surface of the pines, as well as excavating into the cambium and phloem tissues to feed on developing larvae and pupae. Cerambycid and buprestid larvae are prize items because of their large size. Checkered clerid beetles, predators of bark beetles, are preyed upon by the woodpeckers.

Woodpeckers are attracted to beetle spots to exploit the abundant food source found in recently infested pines. Most woodpecker foraging activity is concentrated on the midbole, where the majority of the beetles congregate. Sixty-four percent of the beetles taken by woodpeckers are brood adults, whereas eggs are only 4% of items taken. Woodpeckers appear to forage more on beetle-infested shortleaf pine than on loblolly pine, apparently because the bark on shortleaf pine is thinner, making access to developing beetle broods easier.

Many woodpecker species are known to concentrate their foraging efforts on beetle-infested pines. Pileated, Downy, Hairy, Red-bellied, and Red-cockaded woodpeckers are all major predators of southern pine beetles and other insect associates in infested pines. The density of woodpeckers in infested pine stands can be 3 to 58 times greater than the density in non-infested stands. The beetle mortality rate caused by woodpeckers typically exceeds what the birds actually eat, since bark knocked loose during woodpecker foraging activities falls to the ground and dries in summer or chills in winter, killing the larvae and pupae within it. The value of infested pines as a food supply dwindles rapidly after young beetles emerge. The highest number of woodpeckers are usually found in beetle spots during summer and fall, while lower numbers appear during winter. Unfortunately, woodpecker predation on beetles is lowest in spring, when southern pine beetle spots become active and proliferate.

THE ATTACK SPREADS

Many factors affect whether the initial beetle attack will spread to other trees. In longleaf pine forests, such as those in the Carolinas, attacks almost never spread beyond the initial tree. Beetle populations remain at low levels, launching attacks on individual pines that have become stressed or damaged. In other areas, such as many loblolly and shortleaf pine forests, outbreaks in which attacks spread and beetle populations grow occur regularly, but even here some attacks are limited to one or a few pines. As pines become infested and beetles emerge to attack other pines, the collection of infested and dead pines is called a beetle spot (Plate 15).

Beetle spots may grow as pines continue to fill up with the increasing numbers of bark beetles, dwindle because there is an insufficient number of beetles to kill infested pines, or die out because of disruption in the attractant chemical trail. Many factors affect whether spots grow or die out. The age and vigor of the pines, their diameter and spacing, soil moisture and air temperature, the number of infested pines, and the abundance of insect predators and parasitoids all influence beetle spot dynamics. A greater amount of brood can be produced in larger pines. Dense timber stands that have older or larger pines, or both, have a greater probability of being successfully infested by southern pine beetles than stands of younger pines that are widely spaced. Pine spacing that maintains at least 7 to 8 m between pine stems can reduce the hazard of beetle infestation for a stand and is the minimum spacing of mature pines recommended throughout Red-cockaded Woodpecker nesting and foraging habitat in loblolly and shortleaf pine.

The total basal area of loblolly and shortleaf pines within woodpecker foraging habitat and cluster areas also affects the likelihood of infestation. Loblolly and shortleaf pine stands with basal areas that exceed 22 m^2/hectare have a higher hazard of beetle infestation and are in urgent need of thinning. To reduce the potential for beetle spot growth within Red-cockaded Woodpecker cluster areas, basal areas less than 16 m^2/hectare are recommended. The soil quality or site index of an area affects the likelihood of infestation by southern pine beetles in loblolly stands. Thus, pines on high-quality soils (high site index) are more likely to be infested by beetles than pines growing on poor soils. Tree basal area and density as well as soil quality affect diameter (radial) growth of pines. Pines with slow radial growth are known to be more susceptible to beetle infestation than faster-growing pines. Collectively, the above characteristics permit forest stands to be rated to determine the hazard potential, or likelihood of beetle infestation. Stands exhibiting high hazard potential can be identified by these ratings and appropriate management taken to reduce the hazard.

Beetle Epidemics

Southern pine beetle population irruptions, or epidemics, occur about every 5 to 7 years. Entomologists do not know for certain which environmental and biological factors trigger beetle epidemics. When they occur, more than 1,500 hectares of pine forest can be killed during one infestation, as happened on the Sam Houston National Forest in Texas during 1983 and 1984. A southern pine beetle spot acts and spreads in a fashion similar to a forest fire, with pheromones regulating the attraction of beetles from smaller spots around the forest. During epidemics, spots may contain massive numbers of southern pine beetles and spread rapidly; spots on the Sam Houston National Forest spread 15.2 m per day along a 5.6 km front!

EFFECTS ON RED-COCKADED WOODPECKERS

Beetle attacks on individual noncavity pines have little negative impact on Red-cockaded Woodpeckers. Loss of one foraging tree in a 100-hectare territory is trivial. Loss of a cavity tree is more critical, although in many instances the less vigorous resin flow that enables the attack greatly reduces the value of cavities in the tree anyway. Beetle spots are a more serious matter, as they may consume vigorous, valuable cavity trees. If located in the center of a cluster, even a small spot can eliminate all cavity trees within a territory, thus reducing the woodpecker population (see Chapter 9). If Red-cockaded Woodpecker cavity tree clusters are in the path of a spreading epidemic, all cavity trees in multiple clusters can be killed, resulting in serious damage to the population. During the Sam Houston epidemic in the mid-1980s, more than 50 woodpecker clusters lost all cavity trees. More than 300 cavity trees were killed in the forest between 1982 and 1984.

During epidemics, not only cavity trees but also other mature pines around cavity tree clusters are often killed. This has a severe impact on a woodpecker group if they have no potential replacements for cavity trees near their dead cavity tree cluster. Habitat used for foraging by the woodpeckers can also be killed during epidemics, or by large beetle spots. As discussed in Chapter 7, this is not a problem at first, because infested pines are food bonanzas for woodpecker groups. Only later is the impact of insufficient foraging habitat felt if too many pines have been lost.

A major infestation of loblolly and longleaf pine occurred on the Kisatchie National Forest and in the Kisatchie Hills Wilderness Area in Louisiana during 1985 and 1986. Longleaf pine is not normally infested by southern pine beetles because of its ability to produce copious amounts of pine resin, which serves to pitch the beetles out. However, if a massive population of southern pine beetles is able to grow in loblolly or shortleaf pines, the beetles can mount an attack sufficient to successfully infest and kill neighboring longleaf pines. Spots in longleaf pine typically die out, but much timber can be lost in the process. This is exactly what happened in the Kisatchie area. By March 1986, 13,294 hectares of pine forest on the Kisatchie National Forest had been infested by southern pine beetles. Cut-and-leave treatment of 3,971 hectares and cut-and-salvage treatment of 7,705 hectares were employed in an attempt to control the epidemic. In addition to the total of 11,676 hectares cut, 635 hectares of pine snags were left standing because it had been determined that beetles had already emerged from these trees. The impact on the Kisatchie Hills Wilderness Area was considerable. The total area affected was 1,571 hectares, 45% of the wilderness area.

Twenty Red-cockaded Woodpecker clusters of unknown status were present on Kisatchie Hills Wilderness Area prior to the epidemic. A survey of the clusters in 1986 revealed 12 active, 7 inactive, and 1 destroyed by the epidemic. The continued existence of many of the remaining 12 active clusters is uncertain because of the loss of foraging habitat. Red-cockaded Woodpeckers used white oaks (*Quercus alba*) extensively as foraging sites following the epidemic. White oaks are not normally a major foraging substrate for Red-cockaded Woodpeckers (see Chapter 7). An intense fire followed the infestation and consumed many of the dead pines left by the bark beetles, as well as the logging slash left behind in cut-and-leave treatments.

MANAGING BEETLE SPOTS

In view of the damage that southern pine beetles can inflict on Red-cockaded Woodpecker cavity trees and foraging habitat, it is important that epidemics and beetle spots in general be prevented, or at least reduced in magnitude. Various methods of treating southern pine beetle spots have been proposed and used over the past 20 years. An environmental impact statement (EIS) developed by the U.S. Forest Service in the mid-1980s and accepted in 1987 described alternative means to attempt control of southern pine beetle spots, synthesizing new knowledge to develop effective methods of control. Because southern pine beetles depend on pheromones to mount a mass attack on an infested pine, treatment to control beetle spots has focused on ways to disrupt the pheromone trail.

The best approach for the prevention of beetle damage in woodpecker habitat is an integrated pest management approach. This approach includes using all technology available for the prevention and treatment of beetle spots. It encourages thinning of pine stands to maintain proper spacing of pine trees, practices that promote the health of beetle predators such as woodpeckers and checkered clerid beetles, early spot detection, and prompt application of scientifically valid treatments to control beetle spots before they achieve massive proportions.

The first step is prevention, which is promoted by maintaining the proper pine species for a given site and maintaining stands in suitable condition, as described above.

The second step is detection. Aerial surveys are often used to scan a forest for bark beetle activity, because the crowns of infested pines take on a reddish color as they begin to die (Plate 15). Unfortunately, recently attacked pines cannot be differentiated from normal pines from the air, especially if soil moisture levels are high. Four to 6 weeks usually need to pass during spring before the tops of infested pines begin to discolor, by which time an attack may be well developed. While flying aerial surveys, entomologists mark beetle spots on maps for subsequent ground verification. Entomologists then visit each beetle spot, measure the number of pines infested and other environmental parameters, and determine the stage of beetle brood development by removing a small portion of the infested pine's bark.

Evaluation is the third step. On a computer, data collected at each spot are entered into a spot growth model to mathematically project the probable development of the spot through a simulation of ecological conditions. According to entomologists, all beetle spots will pass through at least three phases if left untreated: initiation, expansion, and decline to inactivity. However, the amount of timber lost during infestations can be high if spots are allowed to expand. Therefore, spots that the model predicts will grow are recommended for treatment, in an attempt to control the spot and protect cavity trees and foraging habitat. Some spots may require further watching and field checks to collect data for additional model simulations.

The last step and final resort is treatment. Four different treatment methods to attempt direct control were adopted in the Forest Service EIS. The first of these, cut-and-leave, involves cutting all pines that have live adults or developing larvae within them, as well as cutting uninfested pines to create a buffer area between the edge of the beetle spot and the forest toward which the beetle spot is moving. When this method is employed during warm, dry months many of the developing larvae die within the felled infested pines. The success of this method depends on environmental factors that cause some beetle mortality in felled pines, plus the buffer between felled and healthy pines, which disrupts the pheromone trail. The adult beetles that do emerge, it is hoped, will be unable to successfully mount a massive attack at the treated spot. However, they may be able to fly to nearby areas to join in attacks on other spots or help create new spots. Movement from a treated spot to form another spot is termed spot proliferation, the possibility of which is one of the arguments often used against cut-and-leave treatment, since this treatment does not directly kill the beetle population associated with the treated spot. Southern pine beetles are also known to infest healthy pines felled in the buffer area, once the pines are on the ground. Attack holes bored by female beetles are typically on the bottom side of the felled pine in such instances. Although Red-cockaded Woodpeckers prefer to feed on insects in standing live and dead pines, they will also forage on the logging slash (boles and branches) left during cut-and-leave operations.

The second treatment, cut-and-remove (salvage), also involves cutting pines infested with beetles and sometimes creating a buffer area by cutting healthy pines. Felled trees are removed from the area and usually hauled to paper pulp mills, instead of being left on the ground. Some of the costs of treatment are recovered in this way, and a portion of the beetle population is removed from the area.

The third treatment, cut-and-spray (chemical treatment), involves felling infested pines, cutting the boles into smaller sections, and spraying all sides of the bole sections with an insecticide like lindane or Dursban. The active life of the insecticides is relatively long after it has been absorbed by the bark. Emerging adult beetles must chew their way through the insecticide-impregnated bark to exit, and in doing so they consume a lethal dose of poison. Care must be taken during this treatment to ensure that all sides of the bole are sprayed. A potential problem with cut-and-spray is possible consumption of the insecticide by Red-cockaded Woodpeckers when they forage on logging slash that has been sprayed.

The fourth treatment is cut, pile, and burn. Logs and branches from felled infested pines are pushed into a pile and ignited with a fuel. If the burn is complete, all developing young and unemerged adults are killed.

Selection of the best method to control southern pine beetles around woodpecker cluster and foraging areas depends on the season and location of the woodpecker cluster. The ideal treatment would remove the least amount of forest habitat, preserving the older pines particularly. Treatment that involves cutting large buffer areas to help break the pheromone trail removes more forest than treatments designed to remove or kill beetles within trees. This is important when beetle spots are in or near woodpecker cluster areas.

Treatment is especially difficult when beetle spots are around cavity tree clusters in wilderness areas because motorized equipment and wheeled vehicles are not permitted in wilderness areas, which limits the methods available for the protection of Red-cockaded Woodpecker habitat. Under the 1987 southern pine beetle EIS, treatment to control southern pine beetle outbreaks affecting Red-cockaded Woodpeckers in wilderness areas is permitted only when spot growth models predict that cavity trees or critical foraging habitat will be lost if control is not effected. As to methods of treatment, cut-and-remove is not usually an option unless the beetle spot is close to the edge of the wilderness area and cables can be used to extract felled pines. Cut, pile, and burn is not allowed, making cut-and-leave and cut-and-spray the only real options. In special situations, the U.S. Forest Service's regional forester can approve the use of wheeled vehicles and chain saws in a wilderness area, but such approval is extremely rare and requires an emergency situation. Some wilderness user groups disapprove of the treatment of beetle spots in wilderness areas in general, as well as the emergency use of equipment in wilderness areas to treat spots. They feel that such actions violate the intent of the wilderness designation, and that the actions could be abused in order to gain access to timber in wilderness areas.

LEGAL CONSIDERATIONS

Because of these objections from wilderness users and other groups, court litigation has been a constant companion of efforts to control beetle spots around Red-cockaded Woodpecker habitat in and out of wilderness areas. An initial court case in federal court in Washington, D.C., filed by Sierra Club and The Wilderness Society in 1986–1987 (*Sierra Club et al. v. Froehlke*), failed in its attempts to challenge the beetle spot treatment methods proposed by the U.S. Forest Service and the effectiveness of control efforts in wilderness areas. In this case, the federal district court ruled that southern pine beetle control could be conducted in wilderness areas to prevent timber losses to private landowners adjacent to wilderness areas.

In April 1985 the Texas Committee on Natural Resources (TCONR) and Sierra Club Legal Defense Fund (representing the Sierra Club and The Wilderness Society) filed suit in federal district court in Tyler, Texas (*Sierra Club et al. v. Lyng et al.*), alleging that treatment cutting to attempt control of southern pine beetle spots was not effective. Additional briefs filed with the court alleged that beetle treatment and clear-cutting in general were causing population declines in Red-cockaded Woodpeckers and thus violating Section 9 of the Endangered Species Act by causing harm to the woodpecker. During pretrial hearings in late 1987 and early 1988, Federal Judge Robert Parker ruled that the issue of effectiveness of treatments in the control of southern pine beetles had already been decided in federal district court in Washington, D.C., and denied a hearing of that portion of the case.

Treatment of beetle spots to protect Red-cockaded Woodpecker cavity trees and foraging habitat in Little Lake Creek Wilderness on the Raven Ranger District of the Sam Houston National Forest was challenged twice in federal court in the spring and early summer of 1990. The Texas Committee on Natural Resources (TCONR) filed suit in federal district court in Tyler, Texas (*Sierra Club et al. v. Lyng et al.*), to obtain an injunction to stop treatment efforts. TCONR claimed to have significant new information about the initiation of beetle spots and asserted that the U.S. Forest Service's method to check for brood development in pines that had already been infested augmented the infestation. TCONR alleged that the *alpha*-pinene in pines was released when infested pines were felled and hacked to examine brood development, and that the terpenes in combination with the pheromone frontalin became an attractant that caused beetle spots to grow.

Scientific studies previously completed by Ronald Billings, an entomologist with the Texas Forest Service, and by Thomas Payne, an entomologist at Virginia Tech, and coworkers had already demonstrated that the terpene *alpha*-pinene by itself was not a primary attractant to southern pine beetles, but that a combination of frontalin and terpenes was a strong primary attractant in the initiation of a beetle spot. TCONR's claim that their information was new was not valid, since the original Forest Service environmental impact statement recognized the pheromone-terpene relationship. Nevertheless, a federal magistrate granted an initial injunction. The U.S. Forest Service appealed the injunction, asking for an emergency hearing with the Fifth Circuit Court of Appeals. Within 3 days, the appellate court overruled the injunction and allowed treatment of infestations to continue. Eventually, the Tyler court ruled in favor of the Forest Service's approved treatments and permanently withdrew the overruled injunction. Thus the injunction lasted only 3 days. Unfortunately, beetle spots near cavity trees grew rapidly during the brief injunction, permitting brood from infested pines to successfully emerge and attack other pines and cavity trees within some cavity tree clusters.

RECENT ADVANCES

The question of efficacy of treatment methods to control southern pine beetle infestations during endemic (nonepidemic) population levels remains controversial. There is so much variation in beetle spot size, soil conditions, pine species, timber spacing, and tree size that entomologists maintain it would be too expensive, as well as logistically impossible, to experimentally examine whether treatment efforts are effective. Spots to be treated and spots to be left as experimental controls would have to be randomly selected, and then either treated and monitored or just monitored for final outcome. The sources of variation just mentioned likely would produce sufficient statistical variance to prevent detection of significant successful treatment. As an alternative, Billings and others have used beetle spot growth models to test treatment efficacy. The models are used to predict whether a spot will continue to grow or die out. If the model predicts the spot will grow, treatment is applied. A statistical comparison is then made between the actual effects of treatment and the model's predicted outcome. Clearly, accurate, mechanistic spot growth models are absolutely essential; otherwise, comparisons between predicted and actual outcomes are scientifically useless. Several models and variations thereof are available; among them are those developed at Texas A&M University (TAMBEETLE), the University of Arkansas (HOG Model), and Clemson University (CLEMBEETLE).

During the past decade, Payne and Billings have examined the use of pheromones to inhibit or stop the growth of southern pine beetle spots. Artificially manufactured frontalin, the female beetle's primary attractant pheromone, has been mixed with *alpha*-pinene to produce a southern pine beetle chemical lure (Frontalure). Frontalure can be painted on dead pines from which adult beetles have already emerged. Southern pine beetles massing an attack that would normally be directed at living pines are instead attracted to the already dead pines. Thus, beetles waste energy reserves and egg-laying efforts if treatment is successful.

Conversely, both Billings and Wayne Berisford, with the University of Georgia, have recently applied high concentrations of verbenone, the pheromone produced by male southern pine beetles, to live uninfested pines that lay in the path of the expanding beetle spot. On many occasions, the high concentrations of verbenone served as an attack inhibitor and helped deter infestation of the live pines on the leading edge of the beetle spot. Both of these pheromones can be applied together or in conjunction with other traditional treatments, such as cut-and-leave or cut-and-salvage.

In Texas, southern pine beetles account for more than 50% of cavity tree mortality in years when epidemics are not occurring (see Chapter 5). Particularly vulnerable are active loblolly and shortleaf pine cavity trees that have been used for a year or two, as well as those that have been nest trees in the previous breeding season. Some protection may result from direct application of high concentrations of verbenone to these vulnerable trees in September and October, when they are extremely susceptible to infestation and become overwinter sites for southern pine beetle brood.

Chemicals produced by pines (host compounds) also appear to be potential aids in the protection of vulnerable cavity trees. In addition to the terpenes discussed above, pines also produce the chemical estragole (4-allylanisole). Recent research by Jane Hayes, Brian Strom, Lary Roton (with the Southern Research Station's lab in Pineville, Louisiana), and Leonard Ingram (with Mississippi State University) suggests that estragole may act as a repellent against southern pine beetles. Estragole levels in injured and stressed pines appear to drop below those of healthy pines. Loblolly and shortleaf pine cavity trees are often stressed by woodpecker activity at resin wells and may have below-normal estragole levels, possibly leaving them open to beetle attack. Artificial use of estragole on such trees in the late summer and early fall may reduce the number infested by southern pine beetles for winter brood trees. Final results of extensive field trials of estragole's ability to protect single pines are still pending.

The use of pheromones and/or host chemicals to control southern pine beetle spots around Red-cockaded Woodpecker clusters in wilderness areas would be ideal. Many of the largest beetle spots in recent times have been in locations where treatments that involved cutting were not allowed or were delayed, notably in wilderness areas. These new methods capture the spirit of the "minimum tool" rule that is appropriate for wilderness management; they avoid cutting, or reduce the number of pines that would need to be cut, thus minimizing disturbance to wilderness areas. Unfortunately, development, experimentation, and testing of such techniques have not advanced to the point of implementation as accepted treatment methods.

CHAPTER NINE

The Causes of Population Declines

Our goal in this chapter is to relate the decline of the Red-cockaded Wood-pecker to specific features of its ecology and population dynamics. We have described the changes in the species' environment that accompanied its decline. What changes in the woodpecker's population dynamics did these alterations in the environment produce, and why did these changes lead to such a severe reduction in numbers compared to coexisting species?

It is useful to think about animal populations in terms of carrying capacity and vital rates. Carrying capacity refers to the amount of suitable space available in the environment. For Red-cockaded Woodpeckers, we can define the carrying capacity as the number of potential territories available, which is a function of the quantity and quality of appropriate habitat. Vital rates are birth rates and death rates of individuals within the population. A population that exceeds the carrying capacity cannot be sustained, but a population may be less than the carrying capacity if deaths exceed births over some period of time. In fact, populations can be extirpated from suitable habitat if deaths exceed births for an extended period.

Viewed in these terms, populations decline for two reasons. First, the carrying capacity may be reduced, typically by a loss or alteration of habitat. With fewer places for individuals to live, the population becomes smaller. For example, some Hawaiian honeycreepers (Drepanididae) are thought to have become rare or extinct because the plants on whose nectar they fed were greatly reduced in number or eliminated. In this case, the extent of the nectar resource defined the carrying capacity of the environment. Second, changes in birth or death rates may cause deaths to exceed births, so that the population becomes smaller even though unoccupied space is available. Osprey (*Pandion haliaetus*) and Brown Pelicans (*Pelecanus occidentalis*) exposed to the pesticide DDT produced thin-shelled eggs that often cracked during incubation. The resulting reduction in birth rates caused populations to plummet for several decades, even though habitat was readily available. When DDT exposure was reduced and birth rates returned to normal levels, populations rapidly increased to fill underutilized habitat.

Of course, population dynamics are not actually so simple. Carrying capacity is a product of levels of several essential resources, and levels may change over time in response to variation in climate, interactions with competitors, and other factors. Hence carrying capacity is not a constant, and this introduces another element of variation in population numbers. Death rates and birth rates vary in response to many factors, causing numbers to dip below carrying capacity regularly. Indeed, populations of some species may seldom approach carrying capacity. The demographic and environmental fluctuations that cause this variation in birth and death rates are discussed in relation to population viability in Chapter 6. Also, population dynamics vary among habitats used by a particular species. Habitats typically vary in their quality or suitability for a particular species; that is, carrying capacity is lower on a per unit basis in some habitats than in others, and the balance between birth and death rates is less often favorable in some habitats than in others.

Although it may be difficult to interpret slight population fluctuations within this framework, the framework is useful in analyzing longer-term, major changes in population size associated with the decline of a species. Changes in carrying capacity and changes in vital rates are very different problems, requiring different kinds of solutions. The solution to a carrying capacity problem is to provide more habitat; in contrast, providing more habitat or protecting existing habitat accomplishes little if the problem is changes in vital rates. Instead, the problem that caused the change in vital rates must be identified and fixed.

The Decline of the Red-cockaded Woodpecker as a Carrying Capacity Problem

The decline of the Red-cockaded Woodpecker is due almost entirely to change in carrying capacity. Reductions in southern pine forests, especially longleaf pine, since the arrival of Europeans (see Chapters 2 and 4) greatly reduced the carrying capacity by eliminating habitat. This affected all species using southern pine forest ecosystems. The Red-cockaded Woodpeckers suffered greater losses than other species because alterations in most of the remaining habitat eliminated resources critical to them, rendering the habitat unsuitable for them.

The most critical resource for Red-cockaded Woodpeckers is the cavity tree. The unusual population dynamics resulting from the species' cooperative breeding system are such that birds will only accept territories with existing cavity trees (see Chapter 6). Thus, in the short term, carrying capacity is equivalent to the number of territories containing cavity trees. Over the long term, areas that contain trees suitable for cavity excavation but have no cavity trees represent potential territories; therefore, the potential carrying capacity may be higher than the current carrying capacity. Over the short term, population dynamics are dominated by changes in occupancy of current territories, with only occasional additions from occupancy of potential territories.

The birds require old trees for cavity excavation (see Chapter 5). The loss of old-growth pine in the remaining habitat therefore greatly diminished the woodpecker's carrying capacity in the Southeast far beyond the reduction in carrying capacity resulting from elimination of pine habitat alone. For example, most private timber plantations have no old trees, and therefore the carrying capacity for Red-cockaded Woodpeckers on these plantations, both current and potential, is zero. In nearly all areas where populations still exist, they are persisting by using remnant old-growth trees spared from past timber harvests. As these trees die from lightning strikes, disease, windstorms, and other natural causes (see Chapter 5), the pool of old-growth trees shrinks, because second-growth trees are not yet old enough to serve as potential cavity trees. When all the cavity trees are lost, the territory is lost; and the carrying capacity of the landscape is reduced another notch.

The great reductions in population size beginning with the arrival of Europeans and continuing through the massive timber cutting of the early 1900s (see Chapter 4) were no doubt due to loss of habitat and elimination of old growth in the remaining habitat. The better documented declines of the past few decades also resulted from reduced carrying capacity, but the forces involved in the reduction were somewhat different. Outright loss of habitat, although still a factor, was not the dominant force that it was previously. Elimination of old growth in otherwise suitable habitat, on the other hand, did continue to play a dominant role. The invention of the chain saw in the 1940s altered the way that timber was harvested, favoring more complete clearing of land. The few cull trees that were left behind during the timber harvests of the early 1900s represent the remnant old-growth trees on which the species now depends. The seemingly devastated landscapes of the early 1900s, with vast areas of only two to four trees per hectare, in many cases were more suitable habitat than modern landscapes containing

FIGURE 9.1 Aerial photograph of the northern Angelina National Forest showing habitat fragmentation and cutting patterns around active Red-cockaded Woodpecker clusters. In addition to temporary removal of mature forest habitat by harvesting, conversion of forest habitat to agricultural uses, pastures, roads, and lakes has a more permanent effect on forest fragmentation than does modern silviculture. Photo by J. A. Neal, U.S. Fish and Wildlife Service.

many patches of fairly dense forest and occasional clear-cuts (Figure 9.1). The latter contain many more pines, but in most cases fewer old pines, and thus have a lower potential carrying capacity. The private timberlands mentioned above are a good example.

The other significant difference between previous declines and those of the past few decades is the emergence of forces that do not cause cavity trees to be lost but prevent cavities from being used. Some of these have become primary agents of carrying capacity reduction. Many studies have shown that when hardwood midstory encroaches on cavities, they are abandoned (Figure 9.2). The reason that Red-cockaded Woodpeckers are reluctant to use such cavities is unclear. They may be bothered by obstructions to their aerial approach to the cavity, or their behavior may be an adaptation to avoid predation. Encroaching hardwoods may provide predators, particularly snakes, with a pathway around the resin wells on the bole of the cavity tree, and thus greatly improved access to the cavity. It is not clear that the risk to nests or roosting adults imposed by encroaching hardwoods is actually very large, but the birds may perceive cavity trees surrounded by hardwoods—or perhaps any habitat with a substantial midstory—as unsuitable.

A

B

FIGURE 9.2 *(A)* Hardwood encroachment within Red-cockaded Woodpecker cavity tree clusters and foraging habitat is most prevalent in irregularly burned or unburned loblolly and shortleaf pine habitat. It can also occur in longleaf pine habitat if fire is excluded from the landscape. Hardwood encroachment causes Red-cockaded Woodpeckers to abandon cavity tree clusters and will reduce their use of pine forest as foraging habitat. Mechanical means such as tree mulching and bush-hogging are sometimes necessary to return clusters to *(B)* the desired open parklike condition required by Red-cockaded Woodpeckers. Photos by R. N. Conner.

This must have been an excellent adaptation in the forests before the arrival of Europeans, when birds had many unobstructed cavities they could switch to on those rare occasions when hardwoods encroached on a cavity, and when using cavities that were even slightly less safe would have been selected against. Under current conditions, however, the behavior renders the woodpeckers vulnerable to another major alteration in southern pine ecosystems: fire suppression (see Chapters 2 and 4). Under conditions of fire suppression, the encroachment of hardwood midstory on cavities is widespread, and avoidance of such cavities can leave territories with no suitable cavities at all, causing them to be abandoned and thereby reducing carrying capacity. This may represent an overreaction under current conditions, but the birds' psychology has been shaped by generations of evolution in an environment quite different from the one in which they find themselves today.

It is abundantly clear that territory loss due to hardwood encroachment has been a major factor in recent declines of the Red-cockaded Woodpecker. Populations in areas or habitats such as loblolly forests, where hardwood development has been most extensive, generally have suffered—and continue to suffer—the greatest declines. The decline of the population inhabiting the McCurtain County Wilderness Area in Oklahoma is an instructive example. This population inhabits old-growth forest, and therefore has experienced no reduction in carrying capacity due to loss of older trees. Nevertheless, the population declined by 62% between 1977 and 1989, coinciding with the development of a tall, dense hardwood midstory resulting from the long-term exclusion of fire. The territories that remain occupied are those in which the hardwood midstory is least developed. In this case, carrying capacity was reduced by hardwood encroachment and not loss of old growth, whereas in most recent declines in other regions both factors are involved.

Why Red-cockaded Woodpeckers have suffered from the changes that have occurred in southern pine forests is thus quite clear. Not only have they suffered great reductions in habitat, like all species inhabiting these forests, but the two primary alterations of the remaining habitat, elimination of old-growth trees and fire suppression, directly reduced their carrying capacity.

As cavities have become an ever more scarce, ever more critical resource, threats to cavities that previously were insignificant problems have emerged as significant factors in recent population declines. First, some of the other species that use cavities excavated by Red-cockaded Woodpeckers render them unsuitable for the endangered woodpecker by enlarging them. The species of most concern is the Pileated Woodpecker, since it renders cavities totally unsuitable (see Figure 5.8) and often damages several cavities within the same cluster in a short period of time (see Chapter 5). An occasional loss of a cavity may be balanced by the excavation of new cavities, but the sudden loss of many cavities is a blow from which groups may not be able to recover. This phenomenon can cause a previously viable territory to suddenly be abandoned. In recent times, cavity damage due to Pileated Woodpeckers may have reduced carrying capacity more than cavity tree death or hardwood encroachment in some areas.

Species such as Red-bellied and Red-headed woodpeckers, which usurp active cavities and enlarge them slightly, may have a similar but much smaller effect. Often Red-cockaded Woodpeckers will still use a cavity after it has been enlarged by these species (see Chapter 5). Furthermore, even when Red-bellied or Red-headed woodpeckers render a cavity unsuitable, this is usually an isolated event affecting a single cavity. It is unlikely that the loss of cavities to any species other than Pileated Woodpeckers has been an important factor in the population declines of Red-cockaded Woodpeckers.

The second factor that can cause cavities to be abandoned is related to resin flow. Birds apparently will abandon cavities if gum flow from resin wells is too low, but the dynamics of this relationship are not as well known as those of abandonment due to Pileated Woodpecker damage, tree death, or hardwood encroachment. This factor may be revealed in differences between tree species, in terms of the length of time cavities of different species are used (see Chapter 5). Population declines in certain habitat types, particularly loblolly forests, in the absence of loss of old growth or hardwood encroachment may be due to this factor. The extinction of the population in old-growth loblolly at Tall Timbers in Florida as reported by Wilson Baker in 1983 is a possible example. Generally, the dynamics of excavation and use of cavities excavated in loblolly are less favorable than the dynamics of cavities excavated in longleaf. Less vigorous resin flow in loblolly pine is one of several factors that may be responsible for this difference (see Chapter 5). The result is that territories in loblolly habitat are more likely to become deficient in cavities, and thus unsuitable, than territories in longleaf. The overall increase in the abundance of loblolly pine relative to other pine species has led to increased reliance of Red-cockaded Woodpeckers on loblolly pine. This too has contributed to the decline of the bird.

Finally, an increasing threat to cavities results from renewed emphasis on prescribed burning, especially growing-season burning. Because high fuel loads have accumulated during decades of fire suppression, initial fires are often extremely hot; they occasionally kill cavity trees or ignite the gum around the cavity and burn the cavity out. Cavities today are lower than they were before the arrival of Europeans, which increases their vulnerability to fire (see Chapter 5). Still, losses to fire have been only a minor factor in recent declines, and this threat will diminish as natural fire regimes are reestablished. They do demonstrate, however, that any threat to cavity trees, no matter how insignificant historically, can have a significant impact on Red-cockaded Woodpecker populations under current conditions. Effects of southern pine beetles, discussed in Chapter 8, are another example.

Thus declines during the past few decades, like previous ones, have been due to reductions in carrying capacity resulting from losses of cavities and cavity trees. Losses to hardwood encroachment on cavities have played a much larger role in recent declines than in previous ones, due to vigorous fire suppression policies beginning in the mid-1900s. Also, with old trees becoming increasingly scarce, a variety of threats to individual trees and cavities in recent times have become more important and outright habitat destruction less important.

VITAL RATES

That population declines of Red-cockaded Woodpeckers are a function of territory loss explains one of the surprising results of demographic studies of the species: rates of reproduction and mortality seem to be independent of population trend. That is, reproduction and survival rates reported from declining populations tend to be just as high as those reported from stable or increasing populations. Changes in vital rates associated with the decline of the Red-cockaded Woodpecker are subtle, and they do not drive changes in population size.

The social system plays a large role in producing this pattern. If reproduction is unusually high in a particular year, the result is not more territories occupied or higher mortality because more individuals are competing for a limited number of territories, but an increase in the size of the helper class. The helper class grows and shrinks with changes in birth and death rates, while the number of breeders remains relatively constant. It appears to be rare that territories are unoccupied because there are too few individuals to fill them; that is, most populations remain close to or at carrying capacity, even declining ones. When populations decline because territories are being lost, productivity on the remaining territories and survival of breeders and helpers are unchanged. What is reduced is the rates of transition to breeder status of helpers and juveniles, because they have fewer places to go.

There are many more male helpers than female helpers, so the number of female breeders is not buffered by the presence of a large nonbreeder class, and therefore is more sensitive to birth and death rates than is the number of male breeders. Unpaired males occupy territories, whereas unpaired females do not, so the number of territories, and thus carrying capacity, is equivalent to the number of territorial males. The female floater class, as well as the female helper class, provides a small buffer between the number of breeding females and vital rates. Still, it appears that some males are unpaired in some years due to a shortage of females. Whereas changes in birth and death rates of males result in fluctuations in the male helper class, changes in birth and death rates of females result in fluctuations in the helper female, floater female, and solitary male classes.

These fluctuations, although a significant component of population dynamics, are not an important factor in population declines. A shortage of females may cause some territories to be unproductive in some years, but it does not cause them to be abandoned and therefore does not lead to any change in carrying capacity. A preponderance of solitary males is characteristic of declining populations, but close examination reveals that these males often are on poor territories, typically ones with severe hardwood encroachment on cavities. Males appear to be reluctant to abandon their territories once they acquire breeding status, even when territory quality is greatly reduced, but females will not accept these territories. Therefore males on territories of declining quality often remain single until they die. At that point the territory becomes abandoned, because new males will not accept it either because of its poor quality. Stable populations with solitary males reflect a temporary shortage of females, but in declining populations some males remain solitary even when nonbreeding floater and helper females are present in the area.

Solitary males in declining populations are not always indicative of poor habitat quality. Another frequent scenario is that of solitary males representing isolated territories. In many instances, formerly contiguous populations have been reduced to a collection of scattered remnant groups by the loss and fragmentation of habitat. Under these conditions, dispersal processes are disrupted (see below and Chapter 6), and males who lose their mate remain solitary because females cannot find them. When the male dies, the territory is abandoned because other males are unable to locate it before it deteriorates. In this scenario, it is again the transition from nonbreeder to breeder status, not survival or fecundity, that is altered as the population declines.

An implication of the dynamics of population declines that we have described is that factors affecting only vital rates, and not carrying capacity, have played no role in either the declines of the past few decades or earlier ones following the arrival of Europeans. Predators, which either reduce birth rates by preying on eggs and nestlings (rat snakes) or increase death rates by preying on adults (rat snakes, *Accipiter* hawks), are an example. Reduced quality of foraging habitat is another example, one to which we will return later. Species such as flying squirrels that use Red-cockaded Woodpecker cavities but do not enlarge them, and thus reduce reproduction by destroying nests or increase death rates by forcing adults to roost in the open, are a final example.

Other cavity users especially have attracted the attention of those concerned with Red-cockaded Woodpeckers because levels of interaction can be quite high. Susan Loeb found that other species occupied 18–25% of Red-cockaded Woodpecker cavities over 4 years in a Georgia population, and John Kappes reported similar rates from a Florida population. In Georgia, flying squirrels (Plate 13) occupied the most cavities (10–21% compared to 4–9% for all other species combined), whereas in Florida Red-bellied Woodpeckers did (17–29% compared to 4–9% by flying squirrels). But groups whose nests are lost to other species can renest, and adults who lose their roost cavities can roost in the open. As was discussed in detail in Chapter 5, other cavity users seem to have little or no impact on mortality and reproductive rates in most populations, and even if significant effects occur, they will produce fluctuations in the size of the helper, floater, and solitary male classes, not reductions in carrying capacity. Thus the actions of species other than Pileated Woodpeckers, like those of predators, have never posed a significant threat to Red-cockaded Woodpecker populations. Although these enemies certainly impact the lives of individual birds, they do not appear to impact populations.

Similarly, research activities have limited potential to impact Red-cockaded Woodpecker populations. The extinction of the Tall Timbers population initially stimulated speculation about adverse effects of research activities. However, the species has since proved, perhaps ironically, to be unusually amenable to such activities as capture, banding, and frequent disturbance at cavity trees. Mortality and injury during handling are much less frequent than with most birds. In the North Carolina Sandhills population, capture, banding, and even drawing blood from adults do not result in a detectable increase in mortality rates. Injury of nestlings at typical banding age is also rare. However, capturing nestlings beyond the age when their feathers begin to erupt, roughly age 12 days, does increase their injury and mortality rate. The effect becomes more pronounced with increasing nestling age (Figure 9.3). David Richardson and coworkers, with the U.S. Fish and Wildlife Service in Mississippi, developed a metal grabbing device as an alternative to the standard noosing technique to extract nestlings. This device has the potential to produce a large effect on nestling mortality, be-

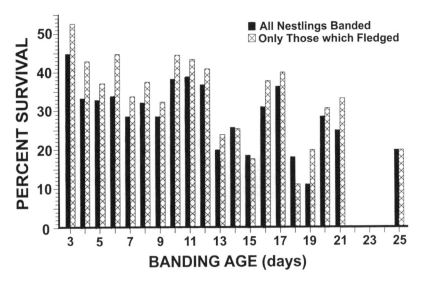

FIGURE 9.3 Survival of nestlings to age 1 year as a function of the age at which they were captured and banded. Data from the North Carolina Sandhills, courtesy of Melinda LaBranche.

cause it is known to fracture bones and is employed on nestlings beyond 12 days of age. We feel its use should be prohibited.

An unusual impact of banding was reported by John Hagan and Michael Reed, then graduate students working with Walters at North Carolina State University. They found that in the North Carolina Sandhills males fitted with red bands had lower reproductive success than males receiving bands of other colors. They speculated that because the color red is an important social signal among these woodpeckers (recall the red cockade of adult males and the red crown patch of fledgling males), the presence of extra redness could have subtle, adverse effects on social interactions. This effect, although interesting, has not been confirmed in any other population, suggesting that it may be an anomaly or artifact.

Another procedure that is problematic is attaching radio transmitters to adults or fledglings in order to track them using telemetry. During the past two decades, several research groups have attempted to use radio transmitters to track Red-cockaded Woodpeckers in order to study bird movements and home range requirements. Such studies have consistently proved problematic, as woodpeckers fitted with transmitters suffered increased predation rates by *Accipiter* hawks. Also, transmitter antennas created problems for woodpeckers as they entered and tried to use their cavities, and the breakage rate of antennas was high. Developing technology may eliminate these problems in the future. In any case, this procedure is too rarely used to have a significant impact on populations. Even if this and other effects of research activity exist, their impact would be on vital rates, not carrying capacity. Therefore there is little possibility that Red-cockaded Woodpeckers are being studied to death.

CLUSTER ISOLATION

The Red-cockaded Woodpecker remains vulnerable to the same problems —namely loss of old growth and threats to cavities such as hardwood encroachment—that have reduced its carrying capacity and caused tremendous population declines over the last few centuries. These same factors, if unchecked, could cause further declines and even extinction. The state of the current population is much different than that of the original population at the outset of decline, however, and the change makes the species susceptible to additional problems that presumably were only a trivial threat formerly. Current populations and those existing before the arrival of Europeans differ greatly in their distribution. Currently, most populations are small and isolated, whereas in earlier times they were large and continuous. Populations today face new threats such as inbreeding depression and demographic stochasticity that were not important earlier. Problems of small, isolated populations are discussed in Chapter 6.

More insidious is the spatial distribution of territories within populations. The forests of today are much more fragmented than before Europeans arrived. Patches of suitable habitat may be embedded in a mosaic of young pine plantations, developed land, and areas that have succeeded to hardwood forest. Many groups have few or no immediate neighbors, a circumstance that must have been rare in earlier days. Fragmentation of habitat is a dominant feature of most current landscapes worldwide. It is noteworthy in the case of Red-cockaded Woodpeckers, because this species is unusually sensitive to fragmentation.

The reason for the woodpecker's heightened sensitivity to fragmentation is the unusual dynamics of dispersal discussed in Chapter 6. Because many individuals become helpers and therefore are restricted to short-distance dispersal, the Red-cockaded Woodpecker is notoriously inefficient at long-distance dispersal. Isolation eliminates the possibility of immigration via the short-distance dispersal mechanisms characteristic of the species (see Figure 6.11). Therefore breeding vacancies in isolated groups are less likely to be detected than is the case for most other species. Data from the North Carolina Sandhills and Texas indicate that if territories remain vacant for 4 or 5 years, cavities become unsuitable and the territory is permanently lost.

The likelihood that a breeding vacancy in an isolated territory will remain undetected by dispersing woodpeckers depends on the size of the population and the location of the territory. It is less likely in larger populations and in centrally located territories. Long-distance dispersers may come from all directions to a central territory, but can come from only a limited set of directions to a peripheral territory (see Figure 6.11). Peripheral territories in small populations lead a tenuous existence and face a crisis with every breeder death.

In theory, then, habitat fragmentation is expected to have an adverse impact on Red-cockaded Woodpeckers, because isolated and peripheral territories are lost when breeding vacancies go unfilled. Empirical data bear out theory on this point. Even in large populations such as the North Carolina Sandhills there is a tendency for territory abandonment to be most frequent on the periphery of the population. The strongest evidence of the adverse effects of cluster isolation and habitat fragmentation, however, comes from Texas. In the Texas populations, isolated clusters are more likely to be abandoned or contain solitary males than clusters that are not isolated. The status of clusters is affected by the amount of timber cleared in the surrounding landscape, due not to loss of foraging habitat but to the extent of habitat fragmentation. This is evidenced by the fact that effects are seen in small, but not large, populations. Populations are equally vulnerable to loss of foraging habitat regardless of size, but small populations are more sensitive to the effects of habitat fragmentation.

The discontinuous nature of many small remnant populations today is a serious threat to their continued existence. Habitat fragmentation is one of the few things that can cause suitable territories to be unoccupied, and hence cause populations to be below carrying capacity. However, once the territory is abandoned for several years and cavities become unsuitable, the carrying capacity itself is affected. Thus, isolation causes temporary vacancies on suitable territories and eventually permanent reduction of carrying capacity. It has been a factor in declines over the past few decades, nibbling away at the edges of populations declining mostly for other reasons. As new management reduces the threat posed by historical problems such as cavity loss and hardwood encroachment, the adverse impacts of cluster isolation likely will loom larger and larger as a continuing threat to many populations.

A SYNOPSIS OF POPULATION DECLINE

We can summarize the dynamics of the decline of Red-cockaded Wood-peckers as follows. The species presumably was once distributed in large, continuous populations in the many areas where growing-season fire was frequent in old-growth pine forest. Once the Europeans arrived, habitat began to be cleared for human use. Loss of old growth to logging radically changed the habitat that remained, converting many previously good territories to poor ones. Populations declined in response to reductions in carrying capacity, due to both the elimination and alteration of habitat. These historical declines were extensive, but their dynamics are poorly described because they occurred prior to intensive study of Red-cockaded Woodpeckers.

Recent declines have involved continuing loss of remnant old growth, as well as losses of cavities to hardwood encroachment, Pileated Wood-pecker excavation, southern pine beetle attacks, and other threats, and loss of territories through the effects of isolation. Through all of these processes, the carrying capacity continued to erode. Where fire suppression led to hardwood encroachment on cavities, the decline sometimes was sharp and sudden, since many territories were impacted simultaneously as the hardwoods grew to a critical height over a broad area. On whatever good territories remained, the birds continued to do well, exhibiting rates of reproduction and survival that were probably similar to what they had been in better times. Remnant population size became a function of the number of suitable cavity tree clusters remaining.

Recent declines have been well described in many instances. At the Savannah River Site in South Carolina the population declined from 18 groups in 1974 to 1 in 1985, and on the Noxubee National Wildlife Refuge in Mississippi the population declined from 32 active clusters to 12 between 1970 and 1985. In Texas, on the Angelina National Forest half the active clusters were lost between 1983 and 1988, and 41% were lost on the Davy Crockett National Forest from 1983 to 1987. Development of hardwood midstory and cluster isolation were factors in all of these declines. Populations living in loblolly forest have suffered most, those in longleaf least.

Data from the North Carolina Sandhills population provide details on the dynamics of these recent declines. This population underwent a major decline in the 1970s and early 1980s, following several decades of fire suppression. Although nearly all cavity tree clusters were occupied in the

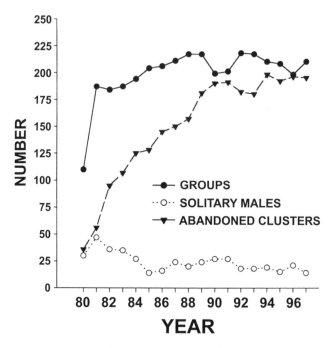

FIGURE 9.4 Population trends in the North Carolina Sandhills, 1980–1997. The number of clusters monitored increased through 1987, was constant during 1987–1994, was reduced by 13 clusters in 1995, and again remained constant during 1995–1997.

1970s, by the time intense population monitoring began in 1981, 18% of the territories were abandoned. Solitary males occupied another 15%, indicating that many territories were still in the process of being lost in 1981. By 1984 the period of decline was over: 34% of the territories monitored were abandoned, and solitary males occupied only 7% (Figure 9.4).

The presence of many solitary males on territories that subsequently are abandoned is an almost universal characteristic of recently declining populations. Data from the Sandhills confirm that these solitary males are a manifestation of declining territory quality, as described above. A typical sequence when a territory is lost is that, first, young no longer remain as helpers. Then, when the breeding female dies or moves, the territory is occupied only by the breeding male. When he dies, the territory is finally abandoned. Because of these complicated dynamics, there is a significant delay between habitat alteration and population response. Jackson suggests that it may take 4–7 years for a territory to be lost following the environmental change that renders it unsuitable.

Just as the behavior of the Sandhills population in the early 1980s demonstrates the dynamics of many recent declines, the subsequent behavior of this population illustrates how stable populations can be, despite fluctuations in mortality and reproduction, if territories are not being lost (Figure 9.4). This stability reflects the dynamics of the cooperative breeding system, in terms of both efficient filling of breeding vacancies by helpers and the lack of formation of new groups. Contrast the minute fluctuations in the number of breeding groups from 1981 to 1997 (Figure 9.4) to population fluctuations in a typical songbird (Figure 9.5).

These stable population dynamics replaced the dynamics of declining populations in better-managed, good habitat in the late 1980s. Where appropriate management was not practiced or too many clusters were isolated, the dynamics of decline continued. Interestingly, a single population was reported to increase during the early 1980s, when most others were still declining. This was the population on the Francis Marion National Forest. Robert Hooper and colleagues have related this difference to the presence of significant numbers of old-growth trees, as well as the lack of hardwood encroachment on cavities due to a long history of prescribed burning. Again, the evidence points to carrying capacity being the key to population dynamics, with cavity trees as the critical element. The evidence that the Francis Marion population was increasing was largely indirect. Unfortunately, the population suffered catastrophic losses due to Hurricane Hugo (see Chapter 11) before the details of the population dynamics could be determined or the population trend verified. No other populations were reported to be increasing in the late 1980s, although several were stable.

It is important to realize that the dependence of the Red-cockaded Woodpecker on remnant old growth from previous timber harvests will continue for several more decades. It will take 10–30 years, depending on location, for second-growth trees to mature to an age where they become suitable for cavity excavation. Until then, populations will remain dependent on the continually declining pool of remnant old-growth trees, and therefore will remain vulnerable to further decline.

The Role of Foraging Habitat

The available foraging base may affect vital rates on suitable territories, and it may also determine whether a territory is suitable, and thus affect carrying capacity. Although the relationship between foraging habitat and vital rates is not completely understood (see Chapter 7), there is growing evidence that better foraging can increase productivity and survival.

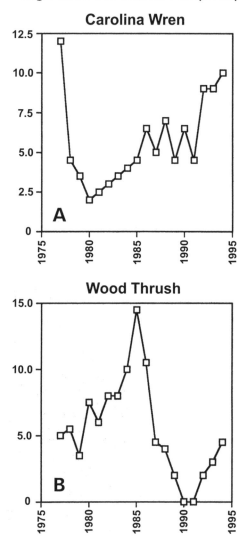

Mason Farm Biological Reserve
Big Oak Woods Census Plot (13 ha)

FIGURE 9.5 Population trends for two noncooperative breeders at Mason Farm Biological Reserve in the North Carolina Piedmont, 1977–1994. *(A)* Carolina Wren *(Thryothorus ludovicianus)*, a permanent resident, and *(B)* Wood Thrush, a Neotropical migrant. Courtesy of Haven Wiley, University of North Carolina, Chapel Hill.

Michael Lennartz and colleagues have shown that groups on the Francis Marion National Forest that have a larger foraging base are more productive. In Texas Schaefer has shown that foraging habitat affects rates of provisioning of nestlings, female weight, and mortality. In Florida, Fran James and coworkers have shown that quality of foraging habitat affects productivity of Red-cockaded Woodpecker groups. In North Carolina, Walters and coworkers have shown that quality of foraging habitat affects group size, which is correlated with reproductive success (see Chapter 7). Of course, effects of foraging habitat on reproduction and survival, like other effects on vital rates, have been minor factors in the decline of the Red-cockaded Woodpecker. The interesting question concerns the extent to which the massive loss of foraging habitat has reduced carrying capacity. Are there enough cavity trees to support more groups than currently exist in some populations, but the lack of sufficient foraging habitat precludes the existence of these groups?

One can deduce from the relationship between territory size and foraging habitat quality, as well as from the large size of territories (see Chapter 7), that there exists some minimum amount of foraging habitat required to support a group through all seasons of the year. The issue is whether the space available to groups, as determined from the distribution of remaining cavity trees, is near this minimum limit or well above it. In many populations, unused foraging habitat exists, and territorial budding sometimes occurs on relatively small territories, suggesting that surplus foraging exists. On the other hand, population densities elsewhere do not approach that observed in old-growth longleaf habitat on the Wade Tract, a fact that is hard to explain by the availability of cavities alone.

One imagines that the massive habitat loss associated with population declines following the arrival of Europeans reduced the foraging base to levels that required territory sizes to increase, thus reducing carrying capacity beyond the level attributable to the elimination of cavity trees alone. But without a better understanding of the relationship between foraging habitat quality and potential population density, we cannot determine whether lack of foraging habitat played a major or minor role in the historical population decline. We simply do not know how sharply the number of possible territories was reduced by loss of foraging habitat.

Loss of foraging habitat has not seemed to play an important role in the declines observed in the last three decades. Even in the rapid decline in Texas, one of the few declines that occurred in a site where there were major losses of foraging habitat, the evidence indicates that the population decline was unrelated to loss of foraging. The loss of particular territories was associated with the degree of encroachment of vegetation on cavities and degree of isolation, not with the amount of foraging habitat lost. It is possible, however, that the association between hardwood midstory and territory loss is due to the adverse effects of hardwood encroachment on the suitability of foraging habitat rather than on the suitability of cavities.

Territories from which a significant portion of the foraging base has been removed appear to remain viable, although their productivity presumably will decline. Hooper and Lennartz observed no loss of territories within a portion of the large population on the Francis Marion National Forest, following removal of up to 43% of the population's foraging base. Gene Wood, of Clemson University, and coworkers observed none in a smaller population in which some groups had lost 37% of their foraging base (see also Chapter 7). These results suggest that the densities of recent populations have not approached the limits imposed by foraging requirements, and thus that foraging habitat has been largely irrelevant to the dynamics of current populations.

Possibly reductions in carrying capacity due to loss of foraging habitat have not been detected because their time course is prolonged. Certainly the usual 4–7-year delay between habitat change and territory abandonment described above would prevail, and there may be an additional delay in the case of foraging habitat loss. The territory might be unattractive to dispersing birds, and thus would be abandoned when the initial residents died, as with cavity loss, but the process might be more complicated in the case of foraging habitat loss. Loss of a territory might require that neighbors resist reoccupancy, or that a series of occupants be so unsuccessful that the territory becomes unoccupied. The groups studied by Wood and coworkers persisted over the short term and continued reproducing, but they eventually disappeared. However, whether additional events eventually occurred that can also cause territory abandonment, such as loss of cavities, is unknown in this case.

In summary, foraging habitat appears to have little to do with declines of the Red-cockaded Woodpecker during the past few decades. Its role in earlier declines is unknown. Groups were lost where habitat was cleared or degraded, but was it the absence of nesting or foraging habitat that was critical? Nesting habitat is much more sensitive than foraging habitat to the major habitat alterations associated with decline: fire suppression and loss of old growth. This suggests that cavities, not foraging habitat, have been the key to the decline of the Red-cockaded Woodpecker.

Regardless of the past, foraging habitat may become critical in the future. Red-cockaded Woodpeckers require large territories (see Chapter 7).

Given that the remaining land that can support the species is limited, as recovery proceeds populations may eventually reach the limits imposed by foraging requirements. Fran James suggests that portions of the Apalachicola National Forest have already reached this point. Although current densities may reflect mostly the availability of cavity trees, future densities, if management is highly successful, may reflect maxima imposed by availability of foraging. The area required for foraging, after all, greatly exceeds the area required for nesting. One could provide all the cavity trees required for nesting and roosting for decades on 5 hectares, but even in old growth 50 hectares of foraging would be required, judging from Todd Engstrom and Felicia Sander's work on the Wade Tract (see Table 4.1). If foraging habitat is kept young by logging, even more area per group will likely be required.

It is because of the difference in area required for nesting and foraging that management policy has focused so much on foraging requirements. Foraging requirements, as well as nesting requirements, limit how much timber can be taken from a forest in which Red-cockaded Woodpeckers reside. The irony is that arguments about foraging habitat are critical to the future of the species, but are probably irrelevant to its current condition.

CHAPTER TEN

Extinction, Legal Status, and History of Management

Extinction

Evolution of life on the planet Earth and subsequent speciation has produced many millions of different life forms, each with its own distinct genetic code. The passage of evolving species through a time continuum in which they experienced ecological, physiological, and genetic bottlenecks produced many biological extinctions while molding the species that currently inhabit our planet. Extinctions over geological time were undoubtedly the rule rather than the exception. The mass extinctions of the dinosaurs and the evolution of the taxonomic class Aves from reptilian ancestors are well known, as are the more recent extinctions of many large vertebrates discussed in Chapter 2. It is argued by some that extinctions in modern times should be viewed as natural processes and therefore should not be resisted. Many species of birds have become extinct within the past few centuries: the Dodo (*Raphus cucullatus*), Reunion and Rodrigues Solitaires (*R. solitarius* and *Pezophaps solitaria*), Giant Moa (*Megalapteryx didinus*), Great Auk (*Pinguinus impennis*), Carolina Parakeet, and Passenger Pigeon. Others like the California Condor, Ivory-billed Woodpecker, and Puerto Rican Parrot (*Amazona vittata*) are perched on the edge of oblivion. All of these extinctions and severe declines are the direct result of human activities.

Indeed, human activity likely has been a direct cause of an enormous number of extinctions over the past several thousand years. David Steadman, with the New York State Museum, uncovered fossil evidence suggesting that prehistoric humans most likely caused the extinctions of more than 2,000 bird species in the Pacific islands during the past three millennia. Human colonists arrived on islands in the Pacific Ocean between 1,000 and 3,500 years ago, clearing forests for agriculture and raising domestic animals. These activities significantly impacted the biodiversity of Pacific island ecosystems.

A more precise view of extinction is obtained by examining extinction rates through time as summarized by Norman Myers, who serves with the International Union for Conservation of Nature and Natural Resources in Gland, Switzerland. The best scientific data indicate that life evolved at least 3.6 billion years ago. Fossil evidence from the past 600 million years suggests that extinction has occurred at a rate of no more than one species per year, except during periodic episodes of mass extinctions, such as that which eliminated the dinosaurs. Estimates of present-day extinction rates range from one to several thousand species per year. If global destruction of habitat and pollution continue, it is estimated that extinction rates could rise to an average of 100 species per day, more than 35,000 per year. Thus, we are beyond the realm of normal extinction processes, and sit instead on the verge of the sixth mass extinction episode in the history of Earth, one that has focused on the largest species and spanned 10,000 years (see Chapter 2).

There is controversy about the agents of mass extinctions during previous episodes. A leading theory proposes that collisions with huge meteors were involved in some episodes. There is no doubt about the cause of the current episode: it is we, through our direct destruction of species (the cause of numerous losses early in human history) and our current destruction, alteration, and fragmentation of habitat (that can escalate the episode to mass extinction levels). John C. Briggs, with the University of South Florida, stated that "the catastrophic extinction under way today as the result of human destruction of the biosphere is far faster and greater than the events of 65 million years ago," when one of the greatest mass extinctions recorded in fossil history occurred.

Early Roots of Conservation Efforts

The human species is apart from any that have preceded it primarily because of our ability to reason. Unlike other species, we have the ability to make unprecedented alterations to our environment that cause impacts of global magnitude. The human species is the ultimate habitat modifier but also has the ability to understand its collective actions as cause and effect relationships. These abilities enable us to trigger mass extinction but also give us the capacity to prevent it from happening. This has led to an ongoing philosophical, moral, and political struggle among exploitationists, conservationists, and preservationists over the past two centuries in the United States.

George Perkins Marsh expressed early conservation concepts in 1864. Marsh described the destructive effects of excessive clearing and burning of forests and tried to inspire educated citizens to effect changes in public policy. His efforts, however, were largely defeated by political and economic unrest in the aftermath of the Civil War. In the late 1800s preservationists like John Muir proposed that places of great natural beauty like Yosemite be preserved in their natural state. Counter to this concept was the desire by some to have most of the natural areas that remained in the United States distributed among the public so people would have a chance to share in the economic benefit gained in the exploitation of the nation's natural resources.

The U.S. Department of Interior's General Land Office deeded large tracts of land to private interest groups in the 1800s, which resulted in vocal complaints by conservationists about the destruction of forests and water supplies. The conservation movement resulted in the establishment of parks, refuges, and most important to the Red-cockaded Woodpecker, the national forests. In 1891 the General Land Law Revision Act permitted Presidents Benjamin Harrison and Grover Cleveland to place 18 million acres (7.3 million hectares) of western public land into reserves. Individuals interested in exploiting resources opposed the newly created forest reserves. As a compromise, the Forest Management Act was passed in 1897, which permitted the government to regulate the occupancy and use of the reserves. The act permitted grazing, commercial logging, and development of hydroelectric power. By 1905 Gifford Pinchot, through his advice to President Theodore Roosevelt, had transformed the forest reserves into a managed national forest system and gave the direction for its expansion. Pinchot claimed to have been the first to use the word "conservation" in an environmental sense. "Conservation," he wrote,

> means the wise use of the earth and its resources for the lasting good of men. Conservation is the foresighted utilization, preservation, and/or renewal of forests, waters, lands, and minerals, for the greatest good of the greatest number for the longest time.

By the 1940s, concepts embracing principles of conservation and land ethics were fluently expressed and amplified by the renowned conservationist Aldo Leopold in his *Sand County Almanac* and other writings. Leopold visualized a land ethic supported by three legs, similar to a tripod. The legs, integrity, stability, and beauty, support the biotic community. His writings aided the development of an environmental conscience for humanity in North America. Timely was his idea of keeping all the parts of the "ecological clock" intact while we altered ecosystems. Concerns for all components of ecological systems, as well as the realization that their presence may be necessary for the proper functioning of ecosystems on which humans depend, led to the biopolitical desire to preserve species that were threatened with extinction.

Designation of Endangered Status and Initial Guidelines

In the late 1960s and early 1970s, the federal government began to develop a list of threatened and endangered species. The U.S. Bureau of Sport Fisheries and Wildlife (now U.S. Fish and Wildlife Service) identified many species, including the Red-cockaded Woodpecker, as endangered species in 1968.

THE FIRST U.S. FOREST SERVICE GUIDELINES

In 1969 the U.S. Forest Service in Texas issued one of the first Forest Service manual supplements that directed protection of all Red-cockaded Woodpecker cavity trees and recognized that replacement pines would be needed for new cavity trees as old ones died. The supplement also indicated that 15 to 20 older pines should be left in a cluster around cavity trees for the protection of "colonies" that had only a single cavity tree. It also directed that an aggregation of mature and nearly mature pines be left intermingled with clusters of cavity trees and that buffer strips of uncut pines be retained around this cluster of cavity and noncavity trees.

In 1970, the Red-cockaded Woodpecker was officially listed as an endangered species (35 Fed. Register 16047, 13 Oct. 1970). Daniel Lay, a biologist with the Texas Parks and Wildlife Department, was instrumental in the initial efforts to list the woodpecker as a result of local population losses and declines he had observed in eastern Texas. Throughout the South, the Red-cockaded Woodpecker had been extirpated from much of the land area that it had originally occupied, and old-growth pine habitat known to be needed by the woodpecker for nesting appeared to be dwindling, as is detailed in Chapters 4 and 9. A thorough and quantitative analysis of the Red-cockaded Woodpecker's distribution in the early 1970s by Jerry Jackson indicated that the species had contracted within its former range and that fragmentation of habitat and isolation of remaining populations had increased.

THE FIRST RED-COCKADED WOODPECKER SYMPOSIUM

A 1971 symposium held at the Okefenokee National Wildlife Refuge in Folkston, Georgia, stimulated further research on the woodpecker. It provided the first rangewide population estimates (see Chapter 4) and summarized what was currently known about the woodpecker's life history and ecology.

LEGAL PROTECTION ACHIEVED

In 1973 Congress passed the Endangered Species Act, which mandated the development of plans and actions to recover populations of listed threatened and endangered species. With the 1973 act, the Red-cockaded Woodpecker gained legal protection. Section 9 of the Endangered Species Act made it illegal to "do harm" to the woodpecker. Harm is generally defined as killing (taking), as well as activities that prevent a species from performing behaviors necessary for it to complete its life requirements such as breeding and foraging. Section 7 of the act made it necessary for federal agencies to consult with the U.S. Fish and Wildlife Service on management activities or habitat manipulations that might adversely affect the species. This consultation process had to conclude with an opinion of nonjeopardy from the Service before the agency could proceed with an activity or habitat alteration. Otherwise an agency had to alter planned management so that it did not adversely affect the species. If an agency's implemented management plan for a protected species resulted in a population decline, or if the agency was unable to fully implement an approved plan, the agency was required by law to enter into immediate consultation with the U.S. Fish and Wildlife Service to resolve the problem.

U.S. FOREST SERVICE MANAGEMENT GUIDELINES — THE NEXT GENERATION

As a result of the Endangered Species Act and information presented in the first Red-cockaded Woodpecker symposium, the U.S. Forest Service developed regional guidelines for the management of Red-cockaded Woodpeckers and their habitat and amended Chapter 400, Section 420, of its *Wildlife Habitat Management Handbook* accordingly in July 1975. With this action began a management paradigm that focused on two life requirements: nesting habitat (cavity tree clusters) and foraging habitat. The 1975 amendment indicated that Red-cockaded Woodpeckers would be managed as a featured species and discontinued the practice of leaving isolated clumps of cavity trees completely surrounded by a regenerating pine plantation. Preexisting timber sale contracts were grandfathered and thus exempt from management guideline changes. Timber sold under grandfathered contracts sometimes was not cut until 5 years after the original sale, delaying the beneficial effects of management changes.

The 1975 amendment also required leaving 0.2 to 2.0 hectares for the cluster of cavity trees plus a 61-m buffer of uncut forest around the aggregate of cavity trees. The function of the 61-m buffer was to prevent wind damage to cavity trees. The cluster area (3–6 hectares) was designed to provide pines for nesting, roosting, and replacement trees. A 16-hectare even-aged support stand of pines 20+ years old was also provided adjacent to cluster areas to provide for foraging and formation of new groups of woodpeckers. Timber harvesting within cluster areas in forest compartments where the woodpeckers were featured would be done selectively and was limited to August through February, when the birds were not breeding. The amendment also encouraged prescribed burning, removal of trees that blocked cavity entrances, and treatment of southern pine beetle spots within cluster areas.

The 1975 guidelines were well intended, but proved grossly inadequate for the recovery of the woodpecker, largely because insufficient scientific information was available to understand the ecological problems faced by the woodpecker.

A recovery team for the Red-cockaded Woodpecker with Jackson as team leader was appointed in 1975. The team was composed of biologists from academia, government, and industry and other personnel. From the beginning, heated debate and disagreement was the rule rather than the exception during team meetings. The desire to maintain substantial timber yields from the South's forests, as well as a lack of scientific information to demonstrate why such yields would be harmful, prevented the development of a recovery plan that would halt woodpecker population declines and recover the species. In 1979 the U.S. Fish and Wildlife Service accepted the compromise recovery plan the team produced. In the plan, the team recognized four major goals as necessary for the recovery of the woodpecker: (1) determination of an accurate inventory of populations and periodic resurveying of populations to establish the effects of management efforts; (2) protection of existing cavity trees and cluster areas and management with appropriate silvicultural techniques to avoid loss of groups or a reduction of their reproductive potential; (3) reintroduction of Red-cockaded Woodpeckers into parts of their former range from which they had been extirpated; and (4) linkage of existing populations by corridors of suitable habitat. The second goal was elaborated to include avoidance of short timber rotations (100 years was suggested for longleaf pine and 80 years for other pines), connection of foraging habitat to cavity tree clusters, retention of other mature pines in cluster areas to provide foraging habitat and replacements for cavity trees (with at least a 61-m buffer retained around cavity tree cluster areas), retention of snags in cluster areas for other cavity nesters, thinning to maintain open stands, prescribed fire to produce low ground cover and open forests, removal of hardwoods that blocked entrances to Red-cockaded Woodpecker cavities, and research to study the benefits or harm caused by removal of beetle-infested pines in woodpecker cavity tree cluster areas. In 1982 the recovery team for the Red-cockaded Woodpecker was dissolved.

FURTHER REVISIONS OF U.S. FOREST SERVICE MANAGEMENT GUIDELINES

In 1979, 1980, and 1981 the U.S. Forest Service amended Chapter 400, Section 420, of its *Wildlife Habitat Management Handbook* as a result of the recovery plan. The revisions removed the provision of support stands and directed the designation of recruitment stands to provide nesting habitat for new groups. These stands were to be at least 10 hectares in size and over 80 years old for loblolly and other pines and over 100 years old for longleaf pines. Recruitment stands were to be at least 400 m and not more than 1,200 m from established cluster areas. If the general forest rotation was 80 years for loblolly and other pines and 100 years for longleaf pine, recruitment stands were not necessary. A theoretical goal of five active clusters per 405 hectares was set, with an average home range size of 81 hectares. All cavity trees were to be marked with paint or tags and cluster perimeters delineated and marked. Cluster areas were to be connected to foraging habitat by stands at least 20 years old. Not more than one-third of any forest compartment with woodpeckers was to be in the zero to 20-year age classes. Only beetle-infested pines were to be cut in cluster areas, and any pine with an active nest was not to be cut. The exchange of land containing Red-cockaded Woodpeckers or cutting of hazardous cavity trees required consultation with the U.S. Fish and Wildlife Service. Although a step forward, the 1979 recovery plan and resulting Forest Service guidelines were still inadequate for the recovery of the woodpecker.

Expansion of Scientific Information and Refinement
of Management Guidelines

THE SECOND RED-COCKADED WOODPECKER SYMPOSIUM

In January 1983 the second Red-cockaded Woodpecker symposium was held at Panama City, Florida. Similar to the first symposium, the second assembled and synthesized knowledge on the biology of the woodpecker. The second symposium produced revised estimates of the status of the woodpecker and provided new information on its biology and management. Michael Lennartz and others estimated that more than 3,000 groups probably existed on federal lands and indicated that habitat fragmentation and population isolation seriously threatened the genetic variability and viability of most of the remaining populations. Only 3 of 37 areas with woodpeckers that had been censused had populations that exceeded 250 groups. In his overview of the second symposium in 1983, Richard Thompson discussed the diversity of conflicting management approaches for the Red-cockaded Woodpecker and concern about the lack of progress toward stopping population declines and degradation of habitat:

Some of those conflicts have been resolved as a result of the tremendous growth in our knowledge during the past 12 years, but other conflicts still remain and many questions remain to be answered. Severe inroads are still being made on populations and on habitat quality, and as a result the bird's continued existence is still very much in doubt.

Later on in the same paper, Thompson related:

There are divergent points of view about management of the Red-cockaded Woodpecker, and sometimes a lack of, or inadequate, communication. This has frequently created a controversial yet stimulating atmosphere over the past 12 years. The glossary of terms submitted at the first symposium provided a communication bridge that allowed us to identify and relate the requirements of the Red-cockaded Woodpecker to the nomenclature of other forest and land use management needs. I hope this second symposium will allow us to further enhance the management expertise and to maintain a cooperative atmosphere among diverse interests.

THE SECOND RED-COCKADED WOODPECKER RECOVERY PLAN

During 1983 the U.S. Fish and Wildlife Service contracted with the U.S. Forest Service to have Lennartz write a revision of the Red-cockaded Woodpecker Recovery Plan. The revised plan, authored by Lennartz and Gary Henry, a biologist with the U.S. Fish and Wildlife Service, was accepted by the U.S. Fish and Wildlife Service in 1985. The revised plan sought to perpetuate viable populations in the major physiographic provinces and forest types where the Red-cockaded Woodpecker currently existed. The 1985 recovery plan identified 15 populations that needed to achieve viability before the woodpecker could be considered recovered as a species. A minimum population size of 250 family groups was established as necessary for viability. Managers of areas with an insufficient land base to achieve 250 groups were encouraged to manage for as many groups as possible and to try to obtain lands or cooperative arrangements so that a population of 250 groups could be reached. The plan also stated that a population of more than 250 groups that had an adequate management plan should be judged recovered. Populations between 125 and 250 groups, as well as any population with over 250 groups that was without a management plan, should be considered threatened. Populations with less than 125 groups, as well as populations with between 125 and 250 groups with diminishing habitat, should be considered endangered.

The 1985 recovery plan listed recovery activities that included (1) surveys and monitoring to assess the status and trends of populations, (2) implementation of protection and management of nesting and foraging habitat on federal lands, (3) encouragement of protection and management on private lands, and (4) research on habitat needs and management, population dynamics, and genetic variation. Protection of Red-cockaded Woodpecker cluster areas was increased by managing cluster areas as stands and not separating them from adjacent forest cover and foraging habitat. Reduction of hardwood vegetation to a basal area less than 4.5 m^2/hectare and total removal of hardwoods within 15 m of cavity trees in cluster areas was deemed necessary to prevent woodpecker abandonment of cluster areas. The 1985 recovery plan added the requirement of maintaining a space of 6 to 7.5 m between pines within cluster areas to minimize the probability of bark beetle infestation and spread. The plan suggested that potential cavity trees could be provided by lengthening rotations, by leaving old-growth remnant pines throughout younger stands or in patches, or by any combination of these methods; but subsequent U.S. Forest Service guidelines only lengthened rotation ages minimally, to 70 years for loblolly and 80 years for shortleaf and longleaf pines.

The new plan added the requirement that 50 hectares of acceptable foraging habitat should be contiguously provided within 800 m of all active cluster areas. The 50 hectares needed to be well-stocked pine stands greater than 30 years old, and 40% of the 50 hectares needed to be over 60 years old. An equivalent amount of foraging habitat needed by a family group would be 21,250 pine stems with a total basal area of 1,930 m^2 and 6,350 pine stems equal to or larger than 25 cm diameter at breast height. Unfortunately, none of the new requirements of the 1985 recovery plan addressed Red-cockaded Woodpecker population demographics and the dispersal dysfunctions that were occurring within small woodpecker populations.

FOREST SERVICE GUIDELINE REVISIONS

During 1985 the U.S. Forest Service concurrently revised its *Wildlife Habitat Management Handbook* section on Red-cockaded Woodpeckers to be in accord with the revised recovery plan. The revised Red-cockaded Woodpecker handbook chapter paralleled the recovery plan's recommendations for foraging habitat needs. If only younger stands and sparsely stocked stands were present, more habitat needed to be reserved as foraging habitat for family groups. Replacement stands were to provide additional nesting habitat for existing groups. These were to be at least 60 years old and 4 hectares in size, and located near existing active cluster sites. Inactive cluster sites could be used as replacement or recruitment stands. As before, recruitment stands were to be located between 400 and 1,200 m from existing active clusters. Unfortunately, the second recovery plan and resulting Forest Service guidelines proved inadequate to recover the endangered woodpecker. The plans did not provide sufficient nesting habitat or solve the demographic problems faced by the majority of the woodpecker populations.

Management and protection of Red-cockaded Woodpecker nesting and foraging habitat was a major issue in the final environmental impact statement (EIS) for the suppression of southern pine beetles that was approved by the U.S. Forest Service in 1987. The record of decision for the environmental impact statement gave direction to the U.S. Forest Service on how to treat southern pine beetle infestations within and around Red-cockaded Woodpecker cluster areas and foraging habitat (see Chapter 8 for methods of treatment).

As described in Chapters 4 and 9, Red-cockaded Woodpecker populations continued to decline rapidly despite the adoption of these management guidelines. Although it is now clear that the decline was large and pervasive, at the time many disputed its occurrence, particularly those with a vested interest in timber harvest. The management guidelines called for less harvesting of timber over the short term than might otherwise have occurred. Since the guidelines did not focus on cavities as the key to population dynamics specifically, and were not derived from the view of population dynamics presented in Chapter 9, it is not surprising that their effectiveness was erratic. Some measures necessary to halt declines were included; others were not. No measures that might have stimulated population expansion were included, whereas many predicted to have little, if any, impact were.

THE AOU COMMITTEE REPORT

In late 1983 Thomas Howell, then president of the American Ornithologists' Union (AOU), at the request of Warren King of the United States section of the International Council for Bird Preservation formed the AOU Committee for the Conservation of the Red-cockaded Woodpecker. The AOU is the preeminent association of scientists studying avian biology in the United States, and five such scientists were appointed to the committee: Peter Stacey, Richard Conner, Carl Bock, Curtis Adkisson, and David Ligon as chair. The committee's charge was to review the status of the Red-cockaded Woodpecker, evaluate its conservation and management, and provide scientific advice and suggestions to managers. Over the next two years, the committee reviewed research results and management documents and visited many woodpecker populations, managers, and researchers. In 1986 they published their findings in the October issue of *The Auk*, the scientific journal of the AOU.

The committee concluded that the Red-cockaded Woodpecker unquestionably was declining, and indeed that there was no evidence that any populations were increasing. They also concluded, as have we, that the decline was without doubt due to habitat loss and habitat alteration, specifically the loss of old growth pine and fire suppression. The committee reviewed the 1985 recovery plan and found it deficient in several respects. They were especially critical of foraging habitat recommendations, which they found to be poorly justified and perhaps deficient (see Chapter 7), and of the recovery criteria. They questioned whether the number of recovered populations required for down-listing was sufficient, and pointed out that the recovery plan was in error in equating 500 birds of breeding status to an effective population size of 500 (see Chapter 6).

The report included a number of recommendations. First, programs designed to increase Red-cockaded Woodpecker populations needed to be developed, as no increase under the proposed management had ever been demonstrated. Second, greater accountability was needed, as managers were failing to implement even the minimal guidelines of their own agencies. Third, habitat management guidelines needed improvement. Forests needed to be older, cavity trees needed to be managed differently, and more old trees needed to be provided for foraging than the provisions of the recovery plan dictated. Most important, recruitment stands should be abandoned as a means to promote population expansion. The recruitment stand strategy depended on rates of pioneering that were unrealistic and indeed had already proved to be a failure. Fourth, problems of cluster isolation and habitat fragmentation, ignored in the recovery plan, needed to be addressed. Finally, more public involvement in the conservation of Red-cockaded Woodpeckers was needed.

Basically, the AOU committee concluded that the brand-new recovery plan would not lead to recovery or even bring an end to ongoing population declines. They discerned the same inadequacies in the approach to population dynamics that we discuss in Chapters 9 and 11. Coming so soon after the release of the recovery plan, the impact of such criticism from a distinguished group of scientists, backed by a highly respected society such as the AOU, was great. Clearly the recovery plan did not represent a scientific consensus. There was a feeling that the recovery plan was too much a concession to timber interests, too little an attempt to increase populations, and too optimistic in its assessment of the current situation. Thus in the late 1980s the Red-cockaded Woodpecker was doing poorly and management was in disarray. The situation was bleak.

Litigation

Other events that ultimately affected the management of Red-cockaded Woodpeckers began in the spring of 1985. The poor performance of populations managed under the second recovery plan and guidelines derived from it left agencies vulnerable to legal action. The Sierra Club, The Wilderness Society, and the Texas Committee on Natural Resources filed a lawsuit against the U.S. Forest Service in Texas, alleging that treatment of southern pine beetle spots was destroying wilderness areas and did not control beetle populations (see Chapter 8). A requested injunction to stop treatment efforts while the case was pending was denied following a hearing in April 1985 in federal district court at Tyler, Texas. The plaintiffs took no further action for over 2 years.

In July 1987 studies by Conner and Rudolph demonstrated severe Red-cockaded Woodpecker population declines on the national forests in eastern Texas. Their data indicated that the woodpecker population on the Angelina National Forest had dropped by at least 42% between 1983 and 1987. Data from the Davy Crockett and Sabine National Forests also indicated declines of a similar magnitude. A draft manuscript released for review and a subsequent publication indicated that an abundance of hardwood midstory in cluster areas, isolation of clusters, and forest fragmentation were associated with the population declines.

This information prompted Sierra Club Legal Defense Fund and Texas Committee on Natural Resources attorneys to amend their original allegations to include new charges that management by the National Forests and Grasslands in Texas (part of the U.S. Forest Service) had caused, and would continue to cause, harm to the Red-cockaded Woodpecker. The plaintiffs obtained a court discovery order that permitted them to bring a photocopy machine to the Forest Service Laboratory in Nacogdoches, Texas, and copy all requested data, computer analyses, draft papers, and correspondence pertaining to the observed declines on the three northern national forests in Texas. They were also permitted to request and copy records located at the forest supervisor's office in Lufkin, Texas.

A trial on the merits, with Judge Robert Parker presiding, was held in Tyler, Texas, from 29 February through 3 March 1988. During the trial, the plaintiffs charged that clear-cutting, southern pine beetle treatment, and failure to correctly manage habitat for the woodpecker had caused harm to the species and violated Section 9 of the Endangered Species Act by causing "take." The plaintiffs also charged that the Forest Service had failed to consult in a timely manner with the U.S. Fish and Wildlife Service (violation of Section 7 of the Endangered Species Act) when they became aware of severe population declines.

In June 1988, Judge Parker ruled in favor of the plaintiffs in the case (*Sierra Club et al. v. Lyng et al.*), deciding that the National Forests and Grasslands in Texas had violated both Section 7 and Section 9 of the Endangered Species Act:

> The Court hereby finds and holds that the conduct of the Forest Service has detrimentally impacted upon the Red-cockaded Woodpecker in violation of the Endangered Species Act and the applicable regulations, and

thereby constitutes a 'taking' of the species within the meaning of Section 9 of the Endangered Species Act. The Court further finds that actions of the Forest Service have jeopardized the woodpecker within the meaning of Section 7 of the Endangered Species Act.

The court also found that even-aged silviculture was detrimental to the woodpecker and ordered the U.S. Forest Service to take actions that included applying single-tree selection during harvesting (uneven-aged silviculture) to protect the species. The existing management was based on even-aged silviculture, primarily involving clear-cutting of relatively large patches that regenerated as stands of uniform age to create a mosaic of timber stand ages across the landscape. The court further directed that a management plan be developed that would incorporate specific management practices within 1,200 m of all woodpecker clusters on national forest lands in eastern Texas. The initial plan developed by the U.S. Forest Service during the summer of 1988 included modified shelterwood harvesting, rather than single-tree selection, and was rejected by the court after a hearing on 28 September 1988.

Shelterwood harvesting typically removes most trees, while retaining a moderate density of mature trees (residuals) to provide seeds and shelter for the regeneration of the new pine forest. Normally, residual pines are eventually harvested after young pines are fully established. Modified shelterwood harvesting would retain most residual pines so they could serve as sites for woodpecker cavities and foraging. In contrast, under uneven-aged management employing single-tree selection, pines of all ages (sizes) are harvested, leaving a much denser stand than that of a shelterwood system while retaining a constant pine midstory. In single-tree selection, harvests are usually conducted every 5 to 7 years. (See Chapter 11 for a more detailed description of even- and uneven-aged timber management.)

In October 1988, Judge Parker issued a final order that accepted parts of the initial plan but rejected aspects that directed the use of even-aged silviculture. A second plan, submitted to the court in December 1988, included single-tree selection and other specific management requirements ordered by the court in the October 1988 final order. The second Red-cockaded Woodpecker management plan directed the use of modified single-tree selection harvesting, and specified that older pines should not be harvested within 1,200 m of any woodpecker clusters. This second, court-ordered plan also directed that pine stands within the 1,200-m zones be thinned to a basal area of 14 m^2/hectare to reduce the potential for infestation by southern pine beetles, that prescribed fire be vigorously applied to control hardwood midstory encroachment in woodpecker habitat, and that vehicular traffic in cluster areas be reduced to the minimum essential for the management of the woodpecker. The court also approved the use of cavity restrictors (described below) and translocation of young woodpeckers to augment clusters that contained only a single woodpecker (see below).

As of this writing the court has made no further official comment on the second woodpecker management plan. Shortly after implementing the second court-ordered management plan in December 1988, the U.S. Forest Service appealed the district court's decision to the U.S. Fifth Circuit Court of Appeals and entered into formal Section 7 consultation with the U.S. Fish and Wildlife Service over the second plan. In September 1989, the U.S. Fish and Wildlife Service decided that the second court-ordered plan would likely jeopardize the continued existence of the Red-cockaded Woodpecker.

In March 1991 the appellate court issued a decision that affirmed the original decision of the federal district court in Tyler, Texas. The Fifth Circuit Court agreed that the National Forests and Grasslands in Texas had violated both Section 7 and Section 9 of the Endangered Species Act by failing to make timely consultation with the U.S. Fish and Wildlife Service, and by causing harm to the woodpecker with its management. However, the appellate court also ruled that although the federal district court had the authority to order the formulation of a management plan for the Red-cockaded Woodpecker, it had to allow a due consultation process with the U.S. Fish and Wildlife Service and could not dictate specific technical aspects of a plan. National Forests and Grasslands in Texas was directed to submit a new plan to the federal district court, which the court could only approve or disapprove based on an arbitrary and capricious legal standard.

Judge Parker's initial decision and the appellate court's sustainment of that decision made legal precedent that would affect the woodpecker's management throughout the South and that of other threatened and endangered species. Population declines resulting from a failure to manage endangered species habitat on federal lands were now considered "take" under the Endangered Species Act.

This interpretation of the Endangered Species Act and the legal precedent set by it and similar rulings regarding logging in the Pacific Northwest and the Spotted Owl, as well as management of critical habitat for the Palila (*Loxioides bailleui*) in Hawaii (*Palila v. Hawaii*), were challenged by the timber industry during 1994 in the U.S. Supreme Court (*Babbitt v. Sweet Home*). In a six-to-three vote in June 1995, the Supreme Court agreed with the government's interpretation of the Endangered Species Act and ruled that destruction of an endangered or threatened species' habitat, which significantly impaired essential behavior patterns, constituted legal "harm" to the species and violated Section 9 of the act. The Supreme Court's ruling applied to both public and private lands.

FOREST SERVICE RESPONSE: THE INTERIM GUIDELINES

While litigation was progressing in Texas, other events were unfolding within the Forest Service elsewhere in the South. A report released during 1989 by the Regional Office of the National Forest System in Atlanta, Georgia, indicated that most Red-cockaded Woodpecker populations on national forest lands in the South were declining. As described in Chapter 4, Ralph Costa and Ron Escaño had analyzed data collected in the mid-1980s to determine population status and trends on all southern national forests. Their report indicated that declines were associated with hardwood midstory encroachment problems in cluster areas, a lack of prescribed fire, isolation of active clusters, and an inadequate number of sufficiently aged pines for cavity trees. Considerable concern about the plight of the Red-cockaded Woodpecker began to sweep through the regional office of the National Forest System in Atlanta. All was not going as well with the woodpecker as many of the individual national forests had led the regional office to believe.

In addition, each national forest with Red-cockaded Woodpecker populations received a letter from the Sierra Club Legal Defense Fund attorneys notifying the U.S. Forest Service of intent to sue under the Endangered Species Act. The notice of intent to sue and the realization that most woodpecker populations were plummeting led to a March 1989 meeting between the Sierra Club and the Forest Service in Dallas, Texas. Scientists representing the U.S. Forest Service were Richard Conner and Michael Lennartz, while Jerry Jackson and Robert McFarlane, now a private consulting ornithologist in Houston, were scientists for the Sierra Club Legal Defense Fund. Legal counsel for the U.S. Forest Service included Jean Williams and Vicki Bremen, while Douglas Honnold represented the Sierra Club. Marvin Meier, deputy regional forester, represented the Forest Service administration. The goal of the meeting was to determine and agree on what interim management steps needed to be taken while an environmental impact statement was developed to revise material on Red-cockaded Woodpeckers in the *Wildlife Habitat Management Handbook* (Ch. 400, Sec. 420). Immediately following the March meeting, regional forester Jack Alcock issued a policy on timber cutting within 1,200 m of woodpecker clusters that permitted modification of existing timber sales as necessary to protect Red-cockaded Woodpecker habitat. The policy restricted timber harvesting within 1,200 m of woodpecker clusters and was designed to protect the species while problems faced by the woodpecker were explored and an EIS was drafted.

Concepts discussed at this March meeting, and later solidified in June 1989, became known as the Interim Red-cockaded Woodpecker Management Guidelines. A notice of decision and environmental assessment were released during May 1990 putting most of the South under the interim guidelines. Key elements of the interim guidelines permitted only seed-tree and shelterwood harvesting between 400 and 1,200 m of woodpecker clusters with the stipulation that residual pines left as seed and shelterwood trees would not be removed during the interim period. Remnant (relict) pines and other potential cavity trees would be left during seed-tree and shelterwood harvesting. Further, no regeneration timber harvesting would occur within 400 m of woodpecker cluster areas, and stands within this zone were to be thinned to a basal area of $14-16$ m^2/hectare to reduce probability of southern pine beetle infestation initiation and spread. A vigorous prescribed fire program was urged and reduction of midstory in cluster areas mandated. Marking and tagging of cavity trees was directed to aid in their protection. All cavity trees used by a group of woodpeckers were to be put under one protective 61-m buffer. Establishment of recruitment and replacement stands, as outlined in previous guidelines, was also directed. Liberal use of newly developed management techniques like cavity restrictors, augmentation, and drilled and inserted cavities (described below) was encouraged.

The Kisatchie and Apalachicola National Forests were excluded from the interim guidelines in the May 1990 notice of decision because their populations were larger than 250 groups. Texas national forests were not included under the interim guidelines because management during 1989, 1990, and 1991 was determined by the court-ordered plan requiring single-tree selection silviculture. The Francis Marion National Forest was included under the interim guidelines because Hurricane Hugo had reduced the population by about 50% on 22 September 1989 (see below).

By January 1991 data became available indicating that portions of the Red-cockaded Woodpecker populations on both the Kisatchie and Apalachicola National Forests were declining. Fran James collected data indicating declines on parts of the Apalachicola. The Apalachicola National Forest provided habitat for the largest extant Red-cockaded Woodpecker population, and indications of problems there were alarming. In response, the U.S. Forest Service issued a supplemental environmental assessment to include both the Apalachicola and the Kisatchie National Forests under the interim guidelines. A notice of decision including both forests under the interim guidelines was issued in May 1991.

As a result of the Fifth Circuit Court's ruling in the Texas case, the Regional Office of the U.S. Forest Service released an additional environmental assessment supplement during June 1991 in an attempt to bring national forests in Texas under the interim guidelines that directed woodpecker management in the rest of the South. During 1993 Forest Service attorneys requested that Judge Parker permit the Texas national forests to be managed under the interim guidelines. In March 1994, Judge Parker issued another court order (Civil Action No. L-85-69-CA) that denied the Forest Service's request and added some very pointed comments:

> The defendant's currently-proffered Plan is simply the latest in a (seemingly never-ending) pattern of proffered Plans which again and again reflects the agency's dogmatic adherence to the sorts of timber practices that have pushed the Red-cockaded Woodpecker to the brink of extinction in its native home in the Texas national forests—and which have unambiguously been found violative of the clear dictates of the Endangered Species Act. Perhaps the defendants' touted, "coming soon," newest long-term Plan will demonstrate a less deadly devotion for illegal timber management practices typifying the defendants' past efforts at Red-cockaded Woodpecker management Plan-making. Hope springs eternal.

The Forest Service appealed Judge Parker's ruling on the interim guidelines in May 1994 and presented its arguments at a hearing with the Fifth Circuit Court of Appeals in July 1995. The appellate court denied the Forest Service's appeal, sustaining Judge Parker's ruling on the interim guidelines.

Red-cockaded Woodpeckers on Military Bases

The other agency besides the U.S. Forest Service whose lands house a substantial number of Red-cockaded Woodpecker populations is the Department of Defense. Management of Red-cockaded Woodpeckers on military bases also underwent changes in the late 1980s and into the 1990s. Because military bases are federal lands, they fall under the jurisdiction of the Endangered Species Act. Consequently, improved management was initially driven by rulings under this act. During 1987 the U.S. Fish and Wildlife Service issued a jeopardy opinion on management of Red-cockaded Woodpeckers on Camp Lejeune, a U.S. Marine Corps base in North Carolina. Since then, management for woodpeckers on Camp Lejeune has become exemplary. Adequate foraging habitat is being provided for family groups, midstory in cluster areas has been cleared, and intensive winter and growing-season burning programs have been established with great success.

More contentious incidents occurred on U.S. Army bases. The U.S. Fish and Wildlife Service pursued criminal action against Fort Benning in Georgia, alleging that some woodpecker clusters had been cut, that necessary habitat around other clusters had been cut, and that forest management had isolated active clusters from each other. Three civilian managers were indicted and arraigned during 1992 on seven counts of violating the Endangered Species Act on Fort Benning. Charges included a conspiracy to conceal locations of Red-cockaded Woodpecker cavity tree clusters and giving falsified maps to U.S. Fish and Wildlife Service personnel in order to harvest forest habitat on the base. The case was later settled out of court. Again, the result of this process was greatly altered and improved management on the base.

During February 1990 the U.S. Fish and Wildlife Service issued a jeopardy opinion of management of the woodpecker on Fort Bragg in North Carolina. The opinion indicated that the military needed to survey their Red-cockaded Woodpecker population, control midstory in woodpecker cluster areas, and make a composite map of forest stands in order to manage the military base on a forest-level scale to the benefit of the woodpecker. Again, changes in management followed, including hardwood control, through manual clearing and growing-season burning, and drilling of artificial cavities.

While managers were motivated by legal action to effect changes on some bases, managers on others were more proactive. The other installation, besides Camp Lejeune, that became a model of progressive management was Eglin Air Force Base in the western panhandle of Florida. This base housed a large population of Red-cockaded Woodpeckers, over 200 groups. In the early 1990s natural resource personnel on this installation, led by Rick McWhite and Carl Petrick, initiated radical changes in the base's management program. They began to burn tens of thousands of hectares annually, much of it in the growing season, adopting ecosystem management as a philosophy. They also instituted a woodpecker management program designed to increase the population that incorporated all the recent developments in research. Finally, they cultivated a diverse research program on the base involving scientists from numerous universities and nongovernmental organizations in order to better understand the ecosystem they were attempting to manage and restore.

Although there were a few bright spots such as Lejeune and Eglin, generally through the 1980s the Red-cockaded Woodpecker fared no better on military land than it did on U.S. Forest Service land. Military bases had their own timber programs, some of which were even more aggressive than those of the Forest Service, and the military, too, practiced fire suppression. In fact, there has been little difference over the past several decades between natural resources management on military bases and that on the national forests, except that in the former allowances must be made for military missions.

Military training, however, appears to have had little or no impact on woodpecker populations historically. Populations on military bases, as elsewhere, have declined because of loss of old growth and hardwood encroachment, not because of military activity. With only a few exceptions, such as Camp Lejeune, the management policies of the 1980s failed to halt the decline of the woodpecker on military lands, as on U.S. Forest Service lands. Again, a few populations were stable but most continued to decrease. Increasing requirements for space for training with modern weapons, the general deterioration of habitat under previous management policies, and the precarious position of most woodpecker populations have now combined to bring woodpecker management into conflict with military training on many bases. It is ironic that poor natural resources management has created a conflict where none existed before.

Red-cockaded Woodpeckers and Private Lands

U.S. Fish and Wildlife management policy during the 1980s for private lands was limited and generally directed landowners to avoid management that would cause "take." This guidance basically prevented killing woodpeckers or cutting down cavity trees, and required provision of sufficient foraging habitat contiguous to cluster areas similar to that on federal lands. Currently, Ralph Costa, now with the U.S. Fish and Wildlife Service, is developing a new set of guidelines for private lands.

Because the obligations of private landowners under the Endangered Species Act are not well established legally, the strategy of the U.S. Fish and Wildlife Service on private land has been to attempt to convince landowners to manage their land in a way that will allow the woodpeckers to persist. Legal action has been avoided except when active cavity trees are knowingly cut or birds are deliberately killed. The most notorious case involved the premeditated shooting of birds in Florida, a crime for which the offending individual paid a several hundred thousand dollar fine.

Management of Red-cockaded Woodpeckers on private lands poses many difficulties, which the new guidelines being developed by Costa are designed to address. Only two large populations occur partly or largely on private lands. About 15% (50+ groups) of the North Carolina Sandhills population resides on private land. Here woodpecker groups are found on horse farms, along golf courses, and in residential neighborhoods in and around the towns of Pinehurst and Southern Pines. In the Red Hills of South Georgia, a population of nearly 200 groups resides on private quail-hunting plantations in a relatively undeveloped area. But these are unusual situations. In most cases, populations on private land consist of a few, scattered remnant groups, barely clinging to existence. Generally such populations are isolated, and they have declined precipitously over the last few decades. In county after county across the region, the small number of remaining groups has slowly but surely declined. Some groups manage to hang on for years, even decades, but they eventually succumb, and there are no new groups to replace them. These losses represent the end of a once continuous distribution, which is being replaced by a distribution characterized by isolated populations on large units of public land.

Management in such situations has been highly uneven. In some areas efforts have been made to protect adequate foraging habitat, as well as cavity trees, whereas in others the land around cavity trees has been cleared, and in some locations cavity trees have even been cut.

Timber companies are major owners of private forested land in the Southeast, but the woodpeckers have fared no better on their large tracts of pine forest than on the small farms and residential lots of the small landowners. Forestry practices on most timber company lands involve very short rotations, so that old growth suitable for cavity excavation is almost nonexistent. Thus on these lands too, populations have declined to very low levels or become extinct. Only recently have timber companies made concerted efforts to preserve the few woodpecker groups remaining on their lands (see Chapter 11).

The Dawn of a New Era

It should be clear by now that through the 1980s the prognosis for the Red-cockaded Woodpecker was bleak. Management derived from the 1985 recovery plan failed to stem the tide of continuing population decline. New research indicated that views of population dynamics, and therefore management policy, needed to change if the species was to be saved. Scientists and managers developed significant new technologies that had a major impact on our ability to stabilize and increase woodpecker populations. The late 1980s and early 1990s were a volatile period during which agencies independently changed their management policies to incorporate these new technologies. We will first describe the new research and technologies, and then the changes in management policy based on them.

ARTIFICIAL CAVITIES

An insufficiency of cavities for nesting and roosting has been a recurrent problem for woodpecker populations, particularly following the harvest of the South's original pine forests (see Chapters 2, 9). In 1990 Carole Copeyon described a drilling technique to artificially create cavities in living pines (see Chapter 6). The technique requires there to be sufficient heartwood in the bole of pines for the cavity chamber to be excavated with a drill bit without penetrating living sapwood, to avoid a flow of pine resin into the cavity chamber. This technique involves drilling two holes in the pine, one through which the cavity chamber is excavated, and a second slightly lower, to serve as the actual cavity entrance. The upper hole is filled and sealed upon completion of the procedure (Figure 10.1). Eddie Taylor, working with Hooper at the Southeastern Forest Experiment Station, subsequently developed a modified version of the drilling technique.

David Allen developed a cavity insert technique in 1991. This technique involved cutting an 11-cm wide × 26-cm high × 16-cm deep section of wood out of a living pine tree with a chain saw and a wood chisel. A preformed cavity with entrance tube in a similarly sized wooden block of western redcedar (*Thuja plicata*) was then inserted into the opening and sealed with wood putty (Figure 10.1). Woodpeckers readily accepted both types of artificial cavities.

TRANSLOCATIONS

In 1987 John DeFazio, with the U.S. Forest Service's Southeastern Forest Experiment Station, and others presented a method they used to translocate young female Red-cockaded Woodpeckers to single male woodpeckers that were the sole occupants of a cavity tree cluster. The technique involves noosing nestling woodpeckers during their first week of life and

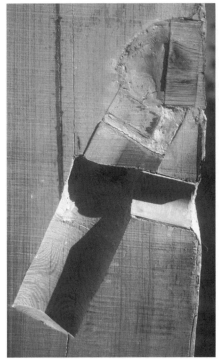

A   B

FIGURE 10.1 The use of artificial cavities, *(A)* drilled, developed by Carole Copeyon, and *(B)* inserted, developed by David Allen, is essential to the recovery of the Red-cockaded Woodpecker. Artificial cavities provide an easy way for managers to provide the most critical habitat resource this endangered species needs: cavities for nesting and roosting. Photos by D. C. Rudolph.

placing color bands on their tarsi (legs), so they can be identified when later captured or observed through binoculars and spotting scopes (Figure 10.2). During the subsequent fall and winter, the first-year females are recaptured in their roost cavities and transported to an unused cavity within a single male's cluster during the night. The female is placed in the cavity and kept inside with the aid of a wire screen that is loosely tacked to the tree. A cord is attached to the screen so that it can be pulled loose by the biologist from the ground the next morning. The female woodpecker is released when the male flies from his roost cavity in the morning with hopes that she will pair with the male and produce a clutch of eggs in the following spring. Thus, the technique can be used to establish a breeding pair of woodpeckers where only a single male was present previously. Since the first translocations of female woodpeckers to single males, Rudolph and colleagues also have translocated first-year males to single females to successfully produce breeding pairs.

Research efforts in eastern Texas have expanded this technology to enable reintroduction of Red-cockaded Woodpeckers into areas where they no longer occur. The reintroduction technique also permits managers to add breeding pairs to small populations that are composed mainly of isolated woodpecker clusters. The technique involves the simultaneous morning release of a first-year male and a first-year female that had been placed inside artificial cavities the previous night. Thus, a new woodpecker group and a new cavity tree cluster, created with artificial cavity technology and translocations, can literally be created in a day. Conner dubbed the new groups and cavity trees created by the reintroduction technique as "shake and bake" clusters.

CAVITY RESTRICTORS

Enlargement and destruction of Red-cockaded Woodpecker cavity entrances by Pileated Woodpeckers can create a significant problem if cavi-

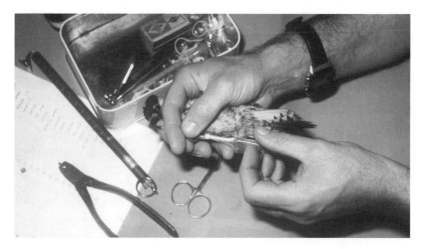

FIGURE 10.2 Banding with USFWS metal and color bands is essential for programs that reintroduce Red-cockaded Woodpeckers to sites where they have been previously extirpated, to augment single woodpeckers to replace lost breeders and to study woodpecker population demographics. Rapid replacement of lost breeders is especially important in populations that suffer from demographic isolation. Photo by D. C. Rudolph.

FIGURE 10.3 Cavity restrictors developed by Jay Carter are extremely effective in preventing damage to Red-cockaded Woodpecker cavities by Pileated Woodpeckers and restoring cavities that have already been slightly enlarged. Photo by D. C. Rudolph.

ties are in short supply and Pileated Woodpeckers are abundant (see Chapter 9). In 1989 Jay Carter and others at North Carolina State University developed a technique to repair or prevent enlargement damage to Red-cockaded Woodpecker cavity entrances. They constructed a stainless steel plate with a deep notch cut out of one side and attached the plate to the tree with screws to decrease the size of an enlarged cavity entrance or as a preventative measure on cavities that had not yet been enlarged (Figure 10.3). Cavities in which enlargement extends beyond the entrance tunnel into the cavity chamber typically cannot be salvaged with the steel cavity restrictor alone. In the early 1990s Steven Best, with the National Forests and Grasslands in Texas, developed a method to repair such cavities that uses metal hardware cloth and putty to rebuild the entrance tubes prior to the installation of restrictor plates.

HURRICANE HUGO AND USE OF NEW TECHNOLOGIES

As mentioned in Chapter 5, Hurricane Hugo swept across the Frances Marion National Forest on the coastal plain of South Carolina in September 1989, destroying 87% of the woodpecker cavity trees. A population of approximately 470 woodpecker groups was reduced to 249 groups by the following (1990) breeding season. The storm hit only months after the first successful use of artificial cavities and the revision of views of population dynamics that resulted (see Chapter 6). Thus the technology to make artificial cavities by drilling or with cavity inserts appeared in a very timely fashion relative to this disaster. Hugo provided an opportunity to put the technology to a test. Hundreds of artificial cavities were installed over the next 4 years in areas where there had been clusters and where pines of sufficient diameter to house an artificial cavity were still present. In 1990, 83 clusters contained a single woodpecker, but by 1994 only 14 clusters had single birds. Over that time span, the number of groups with two or more woodpeckers increased from 249 to 353, and the total population increased from an estimated 607 woodpeckers to 945.

Without artificial cavities, the population would have likely continued to decline following the storm, because so many territories were rendered unsuitable by loss of cavities. Indeed, one might have expected the population to decline to a level equivalent to the number of remaining cavities, that is, to around 13% of the original population. Instead, artificial cavities held the birds on their territories, so that the reduction in number of groups was only 47%, which was even less than the loss of adults (estimated at 60%). By 1994 the number of groups was 75% of the pre-Hugo level. Even with the highest observed natural levels of new group formation, such a recovery would have required more than 15 years. If the population had fallen to 13% of its original level, recovery to the 75% level would have required nearly 200 years. The value of this technology has been firmly established as a result of the devastating hurricane and the perseverance of managers and scientists like Craig Watson, Bob Hooper, and their coworkers who were determined to salvage the population.

Development of New Woodpecker Management Guidelines

New views of population dynamics and the success of new technologies based on those views caused a flurry of activity among those concerned with the Red-cockaded Woodpecker in the early 1990s, as the various agencies attempted to assess these developments and adjust their management strategies accordingly. The activity included efforts to determine the current consensus within the scientific community about the implications of these developments for conservation of the Red-cockaded Woodpecker.

THE RED-COCKADED WOODPECKER SUMMIT AT CAMP WEED

The information base needed to develop a new management strategy for Red-cockaded Woodpeckers began to form in early 1990. In March 1990 a Red-cockaded Woodpecker summit was held at Live Oak, Florida. The meeting was sponsored by the National Wildlife Federation and was conducted by the Southeast Negotiation Network from the Georgia Institute of Technology in Atlanta, Georgia. Experts on Red-cockaded Woodpecker biology, ecology, and management, as well as several silviculturists and entomologists, attended the summit to examine the scientific basis of management decisions affecting the woodpecker. To the extent possible, it was hoped that a scientific consensus could be obtained on the biological requirements of the Red-cockaded Woodpecker and the management needed for its recovery. Initial objectives of the summit were to (1) develop a consensus among scientists about the ecological requirements of the woodpecker, (2) develop a list of research needs, (3) develop guidelines for the management practices that would best stop population declines and assure survival of the species, and (4) examine the effects of Red-cockaded Woodpecker management on other forest resources.

The Southeast Negotiation Network produced a document that detailed and summarized the topics, ideas, discussion, and conclusions of the summit's proceedings. The primary management initiatives recommended were: (1) begin managing the woodpecker on an areawide basis rather than cluster by cluster, as in the past; (2) include recommended changes from the summit into the Red-cockaded Woodpecker Recovery Plan and provide sufficient financial and personnel resources to fully implement the recovery plan; (3) develop an emergency action plan to halt the declines of the small and sensitive populations; and (4) form a broad-based technical advisory committee to help government agencies develop management guidelines for the woodpecker.

Several conclusions of major significance also emerged from the summit. Sufficient numbers of pines with adequate red heart fungal infection should be provided for nesting habitat. Pine ages recommended for nesting habitat were 100–250+ years old for longleaf pine, 80–150 for shortleaf pine, and 80–120 for loblolly pine. These tree ages were considerably older than what had been deemed necessary for the woodpecker in prior management guidelines. A majority view suggested that 500 active clusters were needed in a population for it to be considered viable, and that 81 to 162 hectares of foraging habitat should be provided for each woodpecker group. The use of effective prescribed fire was agreed by all to be absolutely essential for the maintenance of pine ecosystems in which Red-cockaded Woodpeckers lived. Ecosystem management was endorsed, and its key elements identified as growing-season fire and lengthened timber rotation periods. It was also concluded that Red-cockaded Woodpecker managers needed to recognize that isolated populations in fragmented habitat were the norm, not the exception. New management methods, such as use of cavity restrictors, translocation, and artificial cavities, were strongly endorsed by all participants as emergency methods to stop population declines.

THE SILVICULTURAL SUMMIT AND FIELD TRIPS

Because the woodpecker biologists had held a woodpecker summit and appeared to be developing silvicultural recommendations, the foresters within the Forest Service decided that a silvicultural summit was needed to evaluate the recommendations of woodpecker biologists. A silvicultural summit of U.S. Forest Service personnel was held in Atlanta, Georgia, in October 1990 to evaluate and discuss items developed at the woodpecker summit and to determine what was currently known about alternative silvicultural methods that might produce the habitat that the biologists had determined was needed by the endangered woodpecker.

Significant points discussed included the maximum ages that southern pines and forest stands attained. Longleaf pine was recognized as having the greatest longevity, reaching ages in excess of 400 years old, whereas short-leaf could attain 350 years and loblolly about 180 years. Also significant was the realization that knowledge of uneven-aged silvicultural regeneration methods was generally lacking when compared to knowledge of even-aged methods for the southern pine timber types. Research on uneven-aged regeneration methods had been done in the recent past primarily by two individuals. Along with their coworkers, James Baker, a scientist with the Southern Forest Experiment Station of the U.S. Forest Service, had been studying the use of single-tree selection in harvesting loblolly pines, and Robert Farrar, also a Southern Forest Experiment Station scientist, had been studying group selection in harvesting longleaf pines. Silvicultural information was also lacking on the diameter of heartwood present in pines of different ages and various soil conditions, as well as frequency and extent of heartwood decay.

Two silvicultural field trips were conducted by the U.S. Forest Service to increase understanding and develop silvicultural alternatives to clear-cutting for timber harvesting and regeneration of Red-cockaded Woodpecker habitat. The first of these included a series of visits to various longleaf pine areas in Florida, Georgia, and Alabama from 28 January to 1 February 1991. Shelterwood and group-selection harvesting were examined on the Escambia Experimental Forest. During the trip, it became obvious that uneven-aged group selection with prescribed fire was an excellent management option for Red-cockaded Woodpeckers in longleaf pine habitat, one that could be successfully implemented in southern forests. It was recognized that retention of high basal areas of residual longleaf pines for extended periods of time within an irregular shelterwood system would affect the growth of young pines somewhat and would likely produce lower timber yields than group selection.

The second silvicultural field trip occurred from 4 to 7 March 1991 in Oklahoma and Arkansas and focused on the silviculture of loblolly and shortleaf pines. A visit to the McCurtain County Wilderness Area in southeastern Oklahoma revealed that Red-cockaded Woodpeckers were using shortleaf pines ranging from 175 to 320 years old. The condition of this area demonstrated how exclusion of fire could cause hardwood encroachment and eventual conversion of an old-growth shortleaf pine ecosystem to a forest dominated by young and middle-aged hardwoods with a virtual absence of pine regeneration. Functional single-tree-selection harvesting was examined at the Crossett Experimental Forest in Arkansas. It was clear that single-tree selection could produce Red-cockaded Woodpecker habitat, but only with extensive use of herbicides to kill competing hardwoods. Fire at regular intervals to control hardwoods appeared to be detrimental to loblolly regeneration, as most seedlings and young pines are killed by frequent fire.

With the knowledge from both summits, new research information from government, private, and university scientists, and what had been learned during the silvicultural field trips, the stage was set to develop new long-term management guidelines for the Red-cockaded Woodpecker. The U.S. Forest Service was the first agency to develop such guidelines, which were intended to direct management on national forests over the entire South into the foreseeable future. There are several important conceptual changes in the new guidelines, initially drafted by Ron Escaño, compared to previous guidelines. Rather than managing just cluster areas, foraging habitat, or radii around cluster areas, the new plan included the concept of habitat management areas. The minimum size for a habitat management area was 4,047 hectares, enough area to contain at least 50 active cavity tree clusters and their resident groups. If more than 8 km of unfragmented habitat or 5 km of fragmented habitat was present between active clusters, they were considered to belong to separate populations, each requiring a habitat management area for 50 active clusters. This greatly encouraged managers to fill in the gaps with active clusters, thus creating a larger demographically functional population.

Rotation ages in previous management guidelines were essentially 70 years for loblolly and slash pine and 80 years for longleaf and shortleaf pine. The new guidelines included rotations of 100 years for loblolly and slash pine and 120 years for longleaf and shortleaf pine. Rotation age for Virginia pine was 70 years. Another major change was the linkage between management intensity and population size. Use of shelterwood harvesting, with retention of residual pines, was mandatory for small populations, but clear-cutting was permitted in recovered populations, defined as more than 250 breeding groups with stable or increasing population trends. Acceptable uneven-aged timber management methods were permitted in all pine species at all population levels. If ownership patterns had caused forest removal around active clusters in small populations, no further forest removal could occur, because it would cause forest fragmentation. Intensive artificial management, such as the use of cavity restrictors, woodpecker translocations, and artificial cavities, was required for all small populations. Retention of at least 15 relict or residual pines per hectare was required on all populations with fewer than 250 groups and in decreasing populations with more than 250 groups.

The revised guidelines for Red-cockaded Woodpecker management were approved in June 1995. These guidelines incorporated a new management strategy very different from the 1985 recovery plan. Other agencies have responded similarly to the new information available. The army has produced new guidelines, the U.S. Fish and Wildlife Service has produced new draft guidelines both for private lands and for their own wildlife refuges, and timber companies have produced their own management plans. In January 1996 the U.S. Fish and Wildlife Service formed a new recovery team for the Red-cockaded Woodpecker with Ralph Costa as the team leader. The 17-member team began drafting sections for a new recovery plan with an initial anticipated completion date in 1998. The public review draft of the new recovery plan was released in September 2000. In the next chapter we look at the new management in detail and examine its implications for the future of this woodpecker.

CHAPTER ELEVEN

State-of-the-Art Management

The New Management Strategy

What we refer to as the new management strategy is described in no policy document or scientific publication. Portions of it appear in the recent management guidelines adopted by various agencies described below, and each part of it can be found, with justification based on research, somewhere in the scientific literature. It is perhaps best imagined as the logical extension of the new view of population dynamics described in Chapter 9. We have discussed each element of the new strategy, along with its rationale and effect, in detail in previous chapters. Here we place each element in the context of the total strategy, with a brief reminder of what each accomplishes.

The strategy is based on the premise that territory occupancy depends on availability of high-quality cavities. Elements of the strategy (Table 11.1) designed to avoid loss of existing groups and to stimulate formation of new groups are derived from this premise. To avoid loss of existing groups, cavity trees are of course protected from cutting. Also, in clusters where other species begin to enlarge cavities, cavities are protected by cavity restrictors. Excessive hardwood midstory is removed where it currently exists, and prevented from developing where it does not exist, so that cavities are not lost to hardwood encroachment. The preferred method for hardwood control is growing-season burning, since this is beneficial to the ecosystem generally, as well as to the Red-cockaded Woodpecker specifically. Where the hardwood layer has developed to the point that growing-season burning is risky, hardwoods may be removed mechanically or even chemically. Where old growth is sparse so that replacement cavity trees are in short supply, artificial cavities are added to clusters that are deficient in cavities. In theory, this set of actions should reduce territory abandonment to a minimum.

Table 11.1. *Elements of the New Management Strategy*

	Short-term Strategy *Method and Purpose*
Cavity tree protection	Prohibit cutting; prevents territory abandonment
Cavity restrictors	Restrict cavities attacked or threatened by Pileated Woodpeckers to prevent enlargement; prevents territory abandonment
Growing-season fire	Burn to control midstory and promote ground cover; prevents territory abandonment, increases productivity
Hardwood control	Treat well developed hardwood midstory by chemical or mechanical means, or with winter fire, to reduce to level that can be controlled with growing-season fire; prevents territory abandonment
Cavity management	Monitor number of suitable cavities, provide artificial cavities to maintain sufficient number in each territory; prevents territory abandonment
New cluster creation	Construct new clusters of artificial cavities (recruitment clusters) in abandoned territories or vacant habitat; induces new group formation to expand populations
Distribution improvement	Place recruitment clusters in locations designed to improve spatial configuration of population, specifically to create clusters of territories within helper dispersal range; improves dispersal, reduces territory abandonment
Translocation	Move individuals or pairs to unoccupied territories, especially isolated ones; increases rate of population expansion
	Long-term Strategy *Method and Purpose*
Growing-season fire	(see above)
Timber management	Manage timber to promote development of old-growth forest; provides nesting habitat for population expansion, trees for replacement cavities to reduce territory abandonment, improved foraging habitat

Construction of artificial cavities in abandoned territories and in areas not previously occupied by the birds is the method for stimulating the formation of new groups. Clusters with new artificial cavities (termed recruitment clusters, not to be confused with recruitment stands) are located appropriately relative to existing groups, not so close that existing groups simply take them over and not so far that birds dispersing from existing groups have difficulty finding them. In the new management strategy, cluster isolation is recognized as an important problem, due to the penchant of this species for short-distance dispersal. Recruitment clusters are located so as to reduce the isolation of existing groups and fragmentation of the population. Ideally, recruitment clusters are located close enough to neighboring groups that breeding vacancies may be filled by helpers dispersing from these nearby groups. Total dependence on long-distance dispersal by juveniles to fill breeding vacancies is to be avoided. The new management strategy includes translocation as a method to establish new populations, to add new groups to populations in isolated locations and to small populations, and to fill breeding vacancies in isolated groups. As new groups are established in appropriate locations, the need for translocation diminishes.

The new management strategy is a short-term, emergency strategy. The appropriate long-term strategy (Table 11.1) for managing the species once it is recovered is to maintain the southern pine ecosystems in which it occurs in healthy condition. The key elements of the long-term strategy are growing-season burning and a timber management program that results in an uneven-aged forest with large numbers of well-distributed old trees. The long-term strategy provides for abundant nesting and high-quality foraging habitat capable of supporting high densities of Red-cockaded Woodpeckers. In short, the long-term strategy is to maintain forests that resemble those that existed when Europeans first arrived in North America, and to allow the woodpeckers to fend for themselves within the ecosystem to which they are so well adapted.

The short-term strategy represents single-species management; the long-term strategy, ecosystem management. The long-term strategy is less expensive than the short-term one, and its benefits to other members of the community are more extensive, but it is inadequate in the short-term. Under current conditions, ecosystem management alone will at best maintain existing Red-cockaded Woodpecker populations, and likely will result in declines of at least some populations. This is because current forests and current populations do not resemble those of the past, and until they do the woodpecker needs special attention. The natural rate of new group formation is too slow, and too many groups face shortages of old trees on their territories. Once open stands and old trees abound, we can expect populations to slowly increase under ecosystem management alone. But while the stands are being cleared of hardwood midstory and the trees age, the woodpecker will decline under ecosystem management alone, as it has under previous management strategies. Therefore the extent to which the new short-term management strategy is employed over the next 20 or 30 years is the key to the eventual fate of this bird. Recovery and extinction both are realistic possibilities, as well as everything in between.

The new management strategy has been adopted as the management philosophy in several locations, although not all the techniques that are a part of it have been used in each case. The new strategy is not so much the entire set of techniques just described, but more a particular view of population dynamics that dictates the proper course of action in any set of circumstances. We will next describe the successes that have occurred where the new strategy has been adopted. These success stories are among those highlighted in Chapter 10, but here we relate success to employment of the new management strategy.

CAMP LEJEUNE MARINE BASE, NORTH CAROLINA

Since the mid-1980s the natural resources personnel at Camp Lejeune Marine Base in coastal North Carolina have employed a progressive management regime, characterized especially by extensive use of prescribed fire. From 1986 to 1991, the population (measured as usual by the number of potentially breeding groups) was stable under this management regime, varying between 27 and 31 groups (Figure 11.1). In 1986 there were 28 groups; in 1991, 27 groups. Beginning in 1991 the base adopted the new management strategy. Elements of the strategy included protecting existing cavities from enlargement, constructing artificial cavities in clusters deficient in natural cavities, increasing the burning program further to include growing-season fire, and constructing artificial cavities in abandoned territories. From 1991 to 1998 the population increased from 27 to 46 groups (Figure 11.1), an increase of 70% in 7 years, with a mean rate of increase of 8.4% per year. As of this writing, the population continues to increase at the rate of 10% per year, reaching 55 groups in 2000.

CROATAN NATIONAL FOREST, NORTH CAROLINA

The population on Croatan National Forest in coastal North Carolina, only a few miles distant from the Camp Lejeune population, was one of the

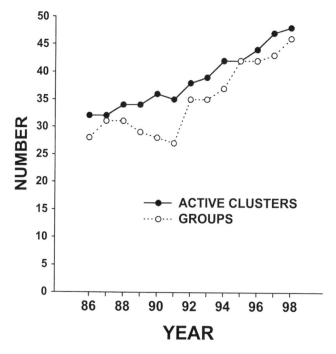

FIGURE 11.1 Population trends at Camp Lejeune, North Carolina, 1986–1998. The entire population was monitored throughout this period.

many rapidly declining populations on Forest Service land during the period when timber was king. This population continued to decline through the 1980s, reaching a low point of only 38 groups in 1990 (Figure 11.2), which probably represents less than half the population that existed in the early 1970s. In 1990 the new management strategy was adopted on Croatan National Forest. In many respects, the new management followed the tenets of the Forest Service's new long-term guidelines that were being drafted in the early 1990s (see below). Elements of the strategy included those employed at Camp Lejeune: protecting existing cavities from enlargement, constructing artificial cavities in clusters deficient in natural cavities, increasing the burning program and including growing-season fire as part of it, and constructing artificial cavities in abandoned territories. Additional elements on Croatan included constructing recruitment clusters in areas not previously used by the birds, placing recruitment clusters so as to reduce the isolation of existing groups, and translocating females to isolated territories occupied by solitary males. Under the new management strategy, the population increased from 38 groups in 1990 to 57 groups in 1995 (Figure 11.2), an increase of 50% over 5 years, with a mean rate of increase of 8.6% per year.

SAVANNAH RIVER SITE, SOUTH CAROLINA

The potential of the new management strategy to restore very small populations has been demonstrated at the Department of Energy's Savannah River Site (managed by the U.S. Forest Service) in west-central South Carolina. This population was reduced to the smallest level possible—only one breeding group—in 1985. Aggressive timber cutting virtually eliminated old growth, reducing the habitat to an extremely poor condition for Red-cockaded Woodpeckers. Although the new management strategy has not been adopted in full at Savannah River, elements of it have been the cornerstone of recovery efforts. Artificial cavity construction has been used in tandem with an enormous translocation effort to establish additional

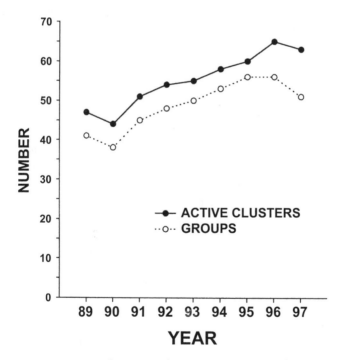

FIGURE 11.2 Population trends on Croatan National Forest, North Carolina, 1989–1997. The entire population was monitored throughout this period.

groups and thus increase the population. By 1992 the population had been increased to 7 groups in this manner, and by 1995 to 20 groups. Although the population is still quite small, it has moved considerably farther from the brink of extinction. This case also demonstrates the utility of the cavity insert technique for constructing artificial cavities in areas lacking the old trees necessary for woodpecker-excavated and -drilled cavities.

THE TEXAS NATIONAL FORESTS

In the National Forests and Grasslands in Texas, the new management strategy has been employed vigorously. Here cavities have been protected with restrictors, hardwoods have been removed mechanically and with fire, and construction of artificial cavities has been used to stimulate both reoccupancy of abandoned territories and occupancy of new areas. More-over, recruitment clusters have been placed so as to reduce fragmentation of populations, and translocation has been used to establish groups in arti-ficial clusters. In fact, the combination of cavity construction and translo-cation has been used to reestablish the species in areas from which it had been completely extirpated. The technique of translocating pairs to estab-lish new groups quickly in recruitment clusters was pioneered here.

The new strategy has been as successful in Texas as elsewhere. The three smaller populations, whose dramatic declines in the early 1980s caused the Forest Service to be sued (as described in Chapter 10), have increased sub-stantially since reaching their low point in 1989–1990. By 1993, 25 new active cavity tree clusters had been added to these populations, which had only about 55 active clusters at their low point. The populations increased at a rate of 6–7% per year during this period (Figure 11.3). Thus the site that had been the epitome of failed management became a symbol of the promise of the new management strategy. Progress was slower in the much larger population on the Sam Houston National Forest, but management activity was focused on the smaller populations, rather than the larger one.

EGLIN AIR FORCE BASE, FLORIDA

The previously described examples of use of the new management strategy all involve populations of small or modest size. The first large population to be managed with the new strategy is the population of more than 200 groups on Eglin Air Force Base in Florida. Eglin represents management on a huge scale: the base covers 187,780 hectares, and more than 81,000 are being managed for Red-cockaded Woodpeckers. Natural resource managers

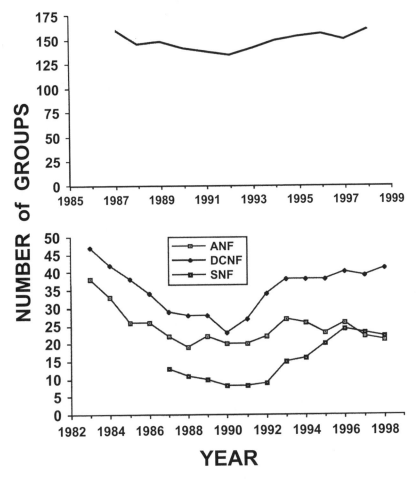

FIGURE II.3 Population trends of Red-cockaded Woodpeckers on the four national forests in eastern Texas. The top graph represents the Sam Houston National Forest. The bottom graph represents the Angelina National Forest (ANF), Davy Crockett National Forest (DCNF), and Sabine (SNF) National Forest. The number of active clusters was used to estimate the number of woodpecker groups.

Rick McWhite and Carl Petrick have led the effort at Eglin. The institution of the new management strategy began with the adoption of a new natural resources management plan in 1993. This plan was all-encompassing, covering all the natural resources on the base, but included within it were management guidelines for Red-cockaded Woodpeckers loaded with elements of the new management strategy. An aggressive hardwood-control program, which emphasized growing-season fire, was adopted, and as of

1995 16,000–20,000 hectares were being burned in the growing season each year. Military pilots were employed to assist in igniting fires from helicopters.

The ongoing management program also includes protecting cavities with restrictors where appropriate; drilling artificial cavities in those clusters in which too few cavities remain; construction of recruitment clusters to induce formation of new groups, both in vacant areas and abandoned territories; and translocation to help establish groups in recruitment clusters. Recruitment clusters are being placed so as to reduce fragmentation of the existing population. The management program was first fully implemented between the 1995 and 1996 breeding seasons, although some components began as much as 2 years earlier. All the elements are employed on only some parts of the base, because the response of the population to different levels of management intensity is being tested experimentally. Therefore a maximum response is not expected on the base as a whole.

The efforts at Eglin have already begun to produce positive results. Construction of recruitment clusters began in the winter of 1993–1994, and 92 recruitment clusters had been created by the breeding season of 1999, of which 54 were active. The population increased by 36% from 1994 to 1999. Interestingly, the early results indicate that the population on portions of the base for which at least some elements of single-species management are being employed is increasing at a rate of 10% a year or better. In contrast, the part of the population for which ecosystem management alone is employed is increasing at only 2% per year, which matches rates of growth expected through budding and pioneering.

Successful employment of the new management strategy at Eglin is a significant event, as it has not been attempted on this scale previously. If it works on this scale—and it clearly is—then recovering a population will become something that is possible rather than an elusive dream. All the elements for success are in place: an excellent, dedicated natural resources staff, commitment and support from the military, and an aggressive management plan that incorporates the new management strategy. All eyes should be on Eglin as we enter the new millennium.

A CAUTIONARY NOTE

Although the successes of the early 1990s at the five sites just described provide ample grounds for renewed optimism about recovery efforts, more recent developments remind us that significant obstacles remain. The population at Eglin continued to grow in 1996 and 1997 under effective implementation of the new management strategy. But during these same 2 years, the Savannah River population grew by only 10% and the Camp Lejeune population by only 2%, while the Croatan and Texas populations declined. With the exception of Savannah River, recent problems with these populations are political and logistical rather than biological. At Camp Lejeune there were no more abandoned clusters in which cavities could be placed to stimulate new group formation, and creation of recruitment clusters in unoccupied habitat was postponed due to delays in the adoption of a new management plan for the base. Subsequently, the plan was adopted and construction of new recruitment clusters was initiated; currently the population again is growing rapidly (see above).

On Croatan, growing-season burning has ceased. In Texas, management efforts are hindered by ongoing legal proceedings and have become a product of court decisions. Especially damaging are misguided efforts by local environmentalists to eliminate prescribed burning, hardwood reduction, and other elements of Red-cockaded Woodpecker management, due to unfounded fears about impacts on songbirds and other species (see Chapter 12). Even where success has been greatest, the contrast between what is possible and what is actually accomplished is all too apparent.

The Optimistic and Pessimistic Futures

The success stories related above are not the only ones. We have already described the spectacular recovery of the population on the Francis Marion National Forest immediately following Hurricane Hugo (Chapter 10). In addition, under the new management strategy, the number of active clusters was increased from 16 to 26 on the Noxubee National Wildlife Refuge in Mississippi by David Richardson and James Stockie, and from 33 to 44 on Fort Polk in Louisiana by Ross Carrie, Kenneth Moore, and coworkers. New groups have been established on private land as a mitigation procedure, especially by Jay Carter and Associates. These are just a few of several examples of recent successes. In fact, we conclude that the new management strategy has been successful wherever it has been employed. It appears that we have the ability to recover populations where and when we desire. Insufficient scientific knowledge and management technology are no longer impediments to recovery, and there appears to be no technical reason that managers cannot increase the population on their land to the maximum level the amount of suitable habitat will support. Given the suitable habitat that still exists in the region, there is no technical reason for not increasing many populations to viable levels, so that the species can be removed from the endangered species list. In the optimistic future, recovery is a reasonable goal, and one that can easily occur within the lifetime of most readers of this book. We know how to recover the species, and in the optimistic future we will do so.

But the issue that remains is whether we have the will as a society to achieve recovery. There is certainly no guarantee that management policy will be guided strictly by biology. The politics of alternative land uses is omnipresent, and society may well decide to place other land use priorities above Red-cockaded Woodpecker recovery, thereby precluding recovery. This can come about if agencies adopt management plans based on criteria other than biological ones, so that management efforts come to differ greatly from the new management strategy we have described. The greater the compromises made to other interests, the less effective management will be and the less likely recovery will be. There are many places where managers have elected not to employ the new strategy, despite the abundant evidence of its effectiveness, and in these places populations continue to be at best stable and at worst declining. For example, the North Carolina Sandhills population is distributed across several management units. Although some elements of the new management strategy are employed in some areas, the full strategy is employed in none, and in some areas no elements are used. Despite the large size of this population, it declined slightly during the mid-1990s (see Figure 9.4), while the smaller Camp Lejeune and Croatan Forest populations about 100 miles away were rapidly increasing. Other places where populations have suffered in recent years, despite readily available means to avoid such problems, are described in Chapter 10.

The Sandhills situation illustrates well the obstacles to recovery. Some of the population occurs on private land, where incentives to increase populations have been few, resistance has been vocal, and federal policy murky. It is not surprising that the new management strategy has not been employed on private land, but that is changing with the advent of new programs such as Safe Harbor that address the legal problems and economic costs of endangered species conservation on private lands (see below). Another part of the Sandhills population occurs on state lands, where neither the new management strategy nor ecosystem management was practiced until very recently. State lands generally have fallen into the void between policy on federal lands and policy on private lands. They have been ignored for the most part, and their woodpecker populations, like that in the Sandhills, have been neglected. The remaining portion of the Sandhills population occurs on an Army base, Fort Bragg, whose personnel have been the most vocal objectors to constraints on military training resulting from woodpecker conservation (see below). Not coincidentally, elements of the new strategy that promote the retention of existing groups have been employed, but those designed to increase the population (and with it, according to Fort Bragg, constraints on training) have not been. Thus it is not surprising that the population on Fort Bragg has been stable to slightly declining, rather than increasing.

In the Sandhills, conflicts with other land uses and differences in philosophy that affect implementation of conservation measures are well illustrated. The ability to recover the population exists, but at the local level the will to do so has not been evident. Local politics and even the values and beliefs of key individuals can determine whether the new management strategy, even if adopted, is implemented effectively. At a higher level, the extent to which the new strategy is adopted by the U.S. Fish and Wildlife Service and by the various agencies on whose lands woodpeckers occur is critical to the future of the bird as well. Their decisions, too, are subject to conflicting philosophies about land use. Finally, politics at the highest levels of government must be considered. It is possible that the Endangered Species Act will be weakened or even eliminated by Congress, and therefore that recovery of Red-cockaded Woodpecker populations will cease to be a required management objective. So there is a potential pessimistic future as well, one in which other interests prevail over conservation of woodpeckers and other plants and animals, and the Red-cockaded Woodpecker continues to be endangered, or perhaps becomes extinct.

Currently, one indicator of whether an optimistic or pessimistic future will prevail is the content of new management guidelines that have been adopted by agencies since the successes of the new management strategy have become known. All involved agencies have responded to the new developments in management during the late 1980s and early 1990s by seeking access to new information and then revising their management guidelines (see Chapter 10). The third Red-cockaded Woodpecker symposium, held in January 1993, was a manifestation of communication between researchers, managers, and policy makers about this new information. Some agencies have completed this process and adopted new guidelines, whereas others are still developing them.

The era symbolized by the 1985 recovery plan (see Chapter 10) is clearly over, and a new one has begun. New agency guidelines bear little relationship to the 1985 recovery plan, which has become obsolete, but instead are a reaction to new knowledge. The degree of resemblance between new agency guidelines and the new management strategy we have described (or lack thereof) indicates the will of an agency to recover the species—the extent to which conservation of Red-cockaded Woodpeckers is a priority within the agency compared to other interests. In the next section we review the various new guidelines and discuss what they are likely to achieve biologically, what they indicate about the priorities of the various agencies, and what they indicate about the future of the Red-cockaded Woodpecker.

New Management Policies

THE FOREST SERVICE LONG-TERM GUIDELINES

In June 1995 the U.S. Forest Service officially adopted and released new long-term management guidelines for the Red-cockaded Woodpecker, to replace the interim guidelines described in Chapter 10. The new guidelines are quite different from the interim guidelines and quite different from the 1985 recovery plan. In fact, the management strategy contained in the recovery plan was one of the alternatives considered by the Forest Service and rejected in favor of the alternative adopted. In essence, the Forest Service recognized the old management strategy as inadequate, and instead adopted a new one that incorporates new knowledge about the biology and management of the woodpecker.

The new Forest Service guidelines include the elements of the new management strategy expressed through complicated management instructions. It is clear that the authors of the document were aware of and understood the new information available. Therefore the document can be said to accurately indicate Forest Service priorities. That is, if the guidelines and their implementation fall short of the best possible management biologically, it is not because of ignorance but because of concessions to other interests.

The new Forest Service guidelines incorporate the new management strategy to a much greater extent than any previous management document (see also Chapter 10). Creating a favorable spatial distribution of Red-cockaded Woodpecker groups relative to the unusual dispersal behavior of the species is at the heart of the management approach. Managers must evaluate the distribution of the population within their management area and divide it into subpopulations based on distances between groups. The distances used are appropriate for determining whether groups will interact sufficiently to be considered part of the same population demographically. Through this process, the demographically distinct units within the area are identified and deemed subpopulations.

The guidelines also include the other key elements of the new management strategy, namely growing-season burning, artificial cavity construction, use of cavity restrictors, and translocation. However, in contrast to analysis of spatial distribution, the extent to which these techniques are used is somewhat at the discretion of the local manager. Managers are required to employ the new strategy but have considerable leeway as to how vigorously they do so. There is some incentive to implement the strategy aggressively, namely that monitoring requirements and intensity of management are tied to population size and trend. If a manager can increase the size of the population and produce a positive population trend, then the time and funds that must be devoted to woodpecker management can be reduced and other activities—notably timber cutting—can be increased. The incentive to employ the new management strategy is that it works, but a manager who is unaware of this, or does not care whether the population size and trend change, can choose not to implement it vigorously. For example, the construction of artificial cavities is required except in large, increasing populations, but putting them in recruitment clusters to induce new group formation is a lower priority than placing them in active clusters deficient in cavities. How many recruitment clusters are provisioned each year, and thus become suitable for occupancy, depends on how the local manager implements the guidelines. In this and other ways, through their interpretation and implementation of the guidelines, managers will determine whether populations remain stable, grow slowly, or grow rapidly.

One can conclude that the Forest Service has adopted the new management strategy but does not require its vigorous implementation. The likely result is uneven population behavior, with populations increasing rapidly on some forests and remaining stable or even continuing to decline on others as a function of differences among managers and perhaps availability of funds. In developing the guidelines, much attention was paid to managing the age distribution of trees, in order to provide a sustained presence of older trees through time and meet legal requirements for timber harvesting. The guidelines will result in larger numbers of old trees on the landscape than previous management, certainly a sufficient number to provide required nesting habitat. The most contentious issue has been managing foraging habitat. Since foraging habitat is not a critical element in the new, short-term management strategy so long as there is enough to preclude territory abandonment, and the amount provided under the guidelines appears sufficient for this purpose, the decisions made about timber rotations in formulating the guidelines are not critical. These decisions do, however, greatly affect compliance with the long-term management strategy. They affect the ecosystem greatly and most certainly will affect other species. Longer timber rotations (beyond 120 years), irregular shelterwood harvesting, and uneven-aged management techniques such as group selection will produce forests that more closely resemble those that existed prior to the impact of Western culture than do current forests. These practices are permitted by the guidelines in all instances but are required only where woodpecker populations are in poor condition (see below). Thus, the degree to which areas designated as foraging habitat come to resemble ancient forests will vary as a function of the condition of the woodpecker population and the will of the local manager. There is considerable latitude for managers to express varying philosophies about land use in the implementation of foraging habitat guidelines.

There is a decision about woodpecker populations implicit in the debate over foraging requirements, however. If forests were allowed to revert to their historical age distributions and fire regimes, there is good reason to believe that the density of woodpecker groups would be much higher than current densities, and therefore that management units could support populations considerably larger than objectives set under the guidelines, at least in longleaf pine. In loblolly pine there are complicating concerns about relationships between forest management, southern pine beetle outbreaks, and woodpecker population dynamics (see Chapter 8). Even in longleaf, however, the debate has not been about target densities and how these could be achieved. Instead, it has been about the schedule of timber cutting and age distribution of trees used for foraging by populations of a particular density, namely that observed in the better habitat on current landscapes. The alternatives debated may differ concerning the densities that can potentially develop through territorial budding, but still these represent a restricted range of densities compared to what is possible. Implicit in the guidelines is the decision to manage with the goal of less than maximum woodpecker densities in order to permit timber production. We point this out not to criticize, but to indicate how the guidelines reflect the will of the agency and the consequences of congressional mandates. Overall, the philosophy seems to be to maintain and possibly increase woodpecker populations of modest densities, while continuing a reasonable—albeit reduced compared to the preceding 30 years—level of timber production.

MILITARY GUIDELINES

Military bases rank with the national forests as stewards of significant populations of Red-cockaded Woodpeckers (see Chapter 10). Besides the many populations on army bases (see below), there is one large population (>200 groups) on an air force base (Eglin) and one modest population (43 groups in 1996) on a marine base (Camp Lejeune). There are also two small (<25 groups) populations on air force bombing ranges and one on another marine base. For the marines and the air force, management is based on individual base management plans, with approval through consultation from the U.S. Fish and Wildlife Service. Eglin's management is guided by a plan implemented in 1993, and Camp Lejeune adopted a new management plan in 1999. Both these bases were discussed above as shining examples of the implementation of the new management strategy. Thus their plans, and the management requirements contained therein, are models of enlightened management, and the populations on both bases already are increasing. These two branches of the military are arguably the leaders in progressive management, because they have not only incorporated the new management strategy into their guidelines but also have implemented enlightened management successfully.

Both Eglin and Camp Lejeune continue to be heavily used for military training. Although military training has not been reduced much if at all by woodpecker management, it has been significantly altered by restrictions on activities within clusters and on clearing forest for new range construction. This has had less effect on Eglin, since air force training does not involve ground maneuvers through the forest and ranges generally can be constructed, although their location and alignment may be constrained by the distribution of woodpeckers. On Camp Lejeune, however, units must deploy and maneuver so as to avoid woodpecker clusters except for transient movement through them. Establishment of fixed positions, excessive disturbance (e.g., noise simulators, smoke production), and off-road movement of tracked vehicles, among other things, are prohibited in clusters. This detracts from the realism of training and makes execution of some training exercises difficult in some areas, causing shifts in the location of training.

Woodpecker management has had a large effect on timber production on these bases. On both bases, new policies include extended rotations and emphasize restoring longleaf pine and maintaining it in a more natural state, as opposed to the rapid timber production on even-aged plantations that characterized previous management.

Restrictions on training activity are a major issue on Camp Lejeune, as they are on army bases (see below). The military command on Camp Lejeune is willing to live with the constraints imposed by existing woodpecker clusters on the base, but is wary of increased constraints resulting from population growth. The new management plan calls for the construction of recruitment clusters to increase the Red-cockaded Woodpecker population, but only half are to be subjected to training restrictions. In this way, the impacts of training activities currently prohibited within clusters can be measured. If these impacts are minimal, it may be possible to ease restrictions on training, and thus increase compatibility between military training and management for population recovery. The impacts of training most likely occur through behavioral disturbance, which might increase the rate of nest failure. Since a few more nest failures are not expected to have much impact at the population level, according to our view of population dynamics, this may well be a negative impact that can be tolerated in order to increase the incentive of the military to promote positive activities such as recruitment cluster construction.

More problematic are activities such as the digging of fixed positions and movement of tracked vehicles, which might kill cavity trees by damaging their roots. If such damage proves common, it may be necessary to retain some restrictions. Still, prohibiting activities within a limited area around cavity trees to prevent their being damaged would restrict training less than current guidelines, which prohibit activities within a cluster several hectares in size. In fact, avoiding individual trees detracts little from the execution or realism of most training exercises.

Although management by the marines and air force is exemplary, these branches of the military are responsible for only one significant population each. It is the army that is the primary steward of Red-cockaded Woodpeckers within the Department of Defense. Three bases—Fort Bragg, Fort Benning, and Fort Stewart—have populations with 150 groups or more, and another, Fort Polk, has one of more than 50. There are three additional bases with very small populations (<10 groups). The army has developed its own set of guidelines, adopted in 1994 and revised in 1996, from which the management plans of individual bases are to be derived. These are the army equivalent of the Forest Service's long-range guidelines discussed above.

The 1994 army guidelines incorporate the new management strategy to a large extent. Cavity management is required, with an objective of maintaining at least four good cavities in each active cluster by using both restrictors and artificial cavity construction. Growing-season burning, both in clusters and in foraging areas, is to be used to control hardwood development. Construction of recruitment clusters in vacant habitat and abandoned territories is to be used to increase populations to the population goal established for each base. The guidelines are vague in describing what the population goal should be, except for stating that it must be at least as large as the current population. The current population size, distribution, and habitat, as well as military mission requirements, are to be considered in setting the population goal, but precisely how these factors should interact to set the goal is not specified. The guidelines state that cluster isolation and fragmentation of populations and habitat are to be avoided. All the elements of the new management strategy are featured in the guidelines, but specifications about appropriate amounts and distributions of activities required to increase and improve spatial arrangement of populations are lacking. Therefore, the effectiveness with which the guidelines are implemented will, as usual, depend to a large extent on the will and ability of the local manager. The guidelines also include new timber management directives, such as increased rotations and increased emphasis on unevenaged management designed to improve foraging habitat and increase the amount of old growth, and restrictions on training activities within cluster boundaries.

There were immediate objections to the 1994 guidelines within the army, especially from Fort Bragg. These objections were quite serious; they resulted in Senate hearings on the subject, conducted by the Senate Committee on Environment and Public Works, and in subsequent interaction between the Interior and Defense Departments that resulted in a directive to revise the guidelines. The primary issue was the conflict between military training and population recovery just discussed in reference to Camp Lejeune. Since criteria for setting population goals were vague, there was a fear that bases would set goals that were much lower than what could easily be achieved, thus avoiding further constraints on training.

In 1996 revised guidelines were proposed and adopted. Many aspects were unchanged, including nearly everything related to the new management strategy. The exception was that a required rate of construction of new recruitment clusters was specified, a change that improves the ability of the guidelines to promote population increases. The greater emphasis on increasing populations in the revised guidelines is tied to a new method of setting population goals and reduced restrictions on military training. Each base is required to set an installation mission compatible goal (MCG), which represents the maximum population size for which restrictions on activities in clusters do not significantly impede the military mission. This goal seems to be roughly equivalent to the goal in the 1994 guidelines, but whether it is similarly constrained to be at least as large as the current population size is unclear.

Under the revised guidelines, each installation must in addition set an installation regional recovery goal (RRG), which will be higher than the

MCG and will better reflect the ultimate potential of the base to support Red-cockaded Woodpeckers. Installations will manage to increase populations to the RRG, but clusters beyond the MCG will not be subject to training restrictions. Thus, the revised guidelines promote population expansion beyond the level that military trainers are willing to accept by eliminating training restrictions beyond that level.

The new army guidelines also reduce restrictions within the clusters that are still protected, that is, included in the MCG. Certain activities previously excluded are now allowed, including transient off-road vehicle travel, hand digging of hasty individual fighting positions, various forms of noise, and use of smoke grenades and flares. In addition, restrictions no longer apply to entire clusters, but to 61-m buffers around individual cavity trees. The net result of these changes will be increased transient disturbances around cavity trees, disturbances of longer duration closer to cavity trees, and greater risk of damage to the roots of cavity trees. Damage to cavity trees potentially is a serious problem, although the guidelines require the construction of artificial cavities to replace any lost through such damage.

The effect of reduced training restrictions bears watching, and installations are required to evaluate the impacts of training under the revised guidelines. Since no clusters will be subject to previous levels of training restrictions, it will not be possible to measure the effect of reduced restrictions on clusters included in the MCG, despite language in the revised guidelines that suggests such a comparison is possible. However, it will be possible to compare clusters for which reduced restrictions apply (clusters included in the MCG) to clusters for which there are no restrictions (clusters excluded from the MCG). Also, the frequency with which critical impacts of training occur can be measured. For example, one could measure the frequency with which cavity trees are lost due to training in order to assess the potential of such activities to result in territory loss. If this frequency is high, it will be essential to reinstitute restrictions against vehicular traffic near trees, even in clusters above the MCG level.

The effects of disturbance on birds are a different matter. There are almost no data on this topic. A small-scale study Walters conducted at Camp Lejeune indicates that excessive noise and movement around nest trees may result in the temporary reduction of feeding rates, but that groups are highly variable in their response. Some seemed totally unaffected by noise and movement, whereas these factors completely inhibited others from approaching their nests. In the biological assessment supporting the guidelines revision, Tim Hayden, a biologist with the Army Corps of Engi-

neers Construction Engineering Research Laboratory, attempted to ana-
lyze the potential impact of reducing training restrictions. He used unpublished data from Jack Mobley, Jay Carter, and coworkers at North Carolina
State University, which correlated productivity of birds in clusters on Fort
Bragg with historical uses of the cluster areas. In these data, productivity
appeared lower in clusters near parachute drop zones and traditional biv-
ouac areas. Hayden used a computer model to simulate effects of reduced
training restrictions by assigning unprotected clusters (i.e., those above the
MCG level) these lower productivity values, while he assigned the remaining clusters the higher values associated with control clusters in the Fort
Bragg data. He found that population viability was not reduced until a fairly
large percentage of clusters were unprotected, and used this result to justify
reduced restrictions on training.

The revision of the guidelines has been controversial, due primarily to
these changes in training restrictions. The Environmental Defense Fund
objected formally to the revision, arguing that the evaluation of effects of
reduced training restrictions was inadequate, and generated a letter supporting its position signed by several well-known scientists. The army re-
fused to reconsider its position or to reinitiate consultation with the U.S.
Fish and Wildlife Service on this point. Whether or not the data used are
inappropriate as EDF contends, Hayden's analysis is indeed inadequate,
because the computer model used was based on old views of population
dynamics and is completely inconsistent with the new understanding of
population dynamics. Therefore his simulation results cannot be trusted.
The issue of the adequacy of the analysis, however, is a red herring. The
point of contention is the degree to which the increased disturbance of
birds might affect reproduction and survival, parameters to which population dynamics are relatively insensitive. Therefore the effect of disturbance, within the range of magnitudes conceivable, is expected to have
only a negligible effect on rate of change in population size. Birds do not
seem to abandon territories due to training activities, nor does training activity appear to have been a major factor in population declines on military
bases (see Chapter 10).

Since the positive effects of new territory creation are expected to out-
weigh the negative impacts of training, we expect that installations will be
able to increase their populations to the RRG level under these guidelines.
Perhaps group sizes might be larger and turnover lower if training restrictions applied to all clusters, but still populations will be larger and therefore
more viable than if limited to the MCG level. Overall, then, the revised
army guidelines, if properly implemented, appear to be a prescription for

employing the new management strategy to increase populations. Issues of contention, which include reduced foraging requirements for clusters above the MCG as well as removal of training restrictions, have more to do with the expected condition of the ecosystem (long-term management) than the expected behavior of Red-cockaded Woodpecker populations in the short term. There are, however, important loopholes in the guidelines that necessitate close monitoring by the U.S. Fish and Wildlife Service to ensure that they are implemented in good faith. Most important, since the army is issued an incidental take permit for clusters above the MCG, the army could eliminate all such clusters in the name of training range construction. To avoid this, MCGs should be as large as possible, and no clusters should be designated as supplemental (i.e., excluded from the MCG) until the MCG is met. The U.S. Fish and Wildlife Service could also require population growth as evidence of good faith implementation of the guidelines.

NATIONAL WILDLIFE REFUGE GUIDELINES

The U.S. Fish and Wildlife Service is directly responsible for managing some lands on which endangered species may occur, namely those lands included in the National Wildlife Refuge System. Red-cockaded Woodpeckers occur on 10 refuges in the Southeast, although only one refuge (Carolina Sandhills) houses more than 30 groups. It and three others are part of designated recovery populations. Thus there is some incentive to manage for increasing populations in a number of refuges, although refuges will not be as important to the future of the species as national forests and military bases. Since refuges do not have the conflicting land use requirements of the latter two types of land, and because the U.S. Fish and Wildlife Service may develop management guidelines completely internally in the case of refuges, one expects refuge guidelines to epitomize enlightened management. Therefore it will be especially interesting to evaluate these guidelines, though currently only in draft form, against the new management strategy.

The refuge draft guidelines do in fact contain the new management strategy in full. They require that existing clusters be managed to ensure that sufficient cavities remain available. This includes controlling hardwoods, the preferred method being growing-season burning, protecting cavities with restrictors, and constructing artificial cavities as needed so that each cluster has at least four good cavities. Also, managers are required to construct recruitment clusters to increase populations. Thus the guidelines provide for increasing the availability of suitable territories, the key element in the new management strategy.

Managers are further required to designate subpopulations as in the Forest Service guidelines (see above), and are encouraged to improve the spatial distribution of the population and provide corridors linking isolated components. The guidelines provide for standard amounts and composition of foraging habitat, and for timber management similar to that prescribed in other federal guidelines. One might expect more innovation and variety, especially where providing more old growth for foraging is concerned, in the absence of a mandate for timber harvesting. Another respect in which the guidelines can be said to be less than optimal with respect to the new management strategy is that the desired population density is lower than could be achieved. The U.S. Fish and Wildlife Service uses the typical standard of one group per 80–120 hectares of suitable habitat, which originated as a concession to other interests (see Forest Service guidelines above), not as an optimum. Currently, one group per 60 hectares could easily be achieved, and eventually even higher densities could be reached. Overall, the U.S. Fish and Wildlife Service guidelines provide a template for management policy designed to increase woodpecker populations but fall short of the ideal in setting population goals.

In some respects the refuge guidelines actually are overzealous. The guidelines include both single species and ecosystem management in the short term, to be succeeded by purely ecosystem management in the long term. This is an appropriate approach (see above), as short-term and long-term strategies are quite compatible. The elements of single-species management included in the new management strategy, such as growing-season burning and cavity construction, have a positive effect on the community as a whole as well as on the Red-cockaded Woodpecker (see Chapter 12). However, this is not true of other elements of single-species management included in the refuge guidelines but excluded from the new management strategy. The refuge guidelines call for control of both rat snakes, which prey on Red-cockaded Woodpeckers and their nests, and flying squirrels, which usurp their cavities. Control of these species is expected to have little impact on woodpecker population dynamics but highly negative effects on squirrels and snakes, and could have indirect effects on other species with strong ecological interactions with squirrels and snakes. Indeed, as Sheridan Samano and coworkers have documented in Mississippi, some devices used to control snakes occasionally kill Red-cockaded Woodpeckers! This seems to us single-species management in its worst sense, which sets a poor example for other agencies.

GUIDELINES FOR PRIVATE LANDS

Federal agencies are legally obligated to develop effective management plans for their lands under the Endangered Species Act. The preceding discussion of the current guidelines of the various agencies addressed the extent to which these guidelines embrace the new management strategy and thus provide the most biologically sound approach. Where deviations from optimum woodpecker management exist, they can usually be attributed to concessions to other land uses, for example, timber harvesting and military training. The situation on private lands is much different. Here the philosophy of the U.S. Fish and Wildlife Service has been to restrict its enforcement of the ESA to prevention of actions that result in "take" as defined in Section 9, that is, the elimination of birds (see Chapter 10). Because of the delicate issue of imposing economic hardship on private citizens, there has been no pressure to manage for recovery on private lands. The extent to which private lands are involved in recovery will depend on voluntary sacrifice by individual citizens and mutually beneficial agreements between landowners and the government. Given this background, perhaps it is not surprising that recent developments related to the conservation of Red-cockaded Woodpeckers on private lands run the gamut from wildly encouraging to depressingly discouraging.

The dominant trend in conservation efforts on private lands recently has been the development of agreements between landowners and the U.S. Fish and Wildlife Service whereby the landowners agree to a management plan in return for guarantees that they will not be prosecuted for "take" violations or be required to consult about individual management actions so long as they follow the plan. These agreements take the form of a memorandum of agreement (MOA) or a habitat conservation plan (HCP). They vary widely in the extent to which the new management strategy is incorporated into the management plan as a function of the economic incentives involved and other considerations. Again, the objective of the U.S. Fish and Wildlife Service seems to be to eliminate conflict and achieve some level of conservation, but not to demand that conservation be optimized at the expense of other considerations. In some instances, the goal seems to be to placate landowners who are politically troublesome rather than achieve anything of significance with respect to conservation.

Conflicts between conservation and economic incentives on private lands are perhaps greatest on industrial forestlands. Several timber companies have entered into MOAs involving some or all of their pine-producing lands in the Southeast. These vary greatly in their objectives, from Champion's MOA in Texas, which attempts to establish a population in a longleaf area on company land, to Weyerhaeuser's MOA in North Carolina, which protects the company from liability due to the foraging needs of woodpecker groups from the adjacent Croatan National Forest in exchange for providing some foraging habitat for these groups, but includes no provisions for housing woodpecker groups on company land. The most widely publicized agreement, and the most important, is the MOA involving Georgia-Pacific. It is the most important agreement because it applies to all of this company's lands within the woodpecker's range and includes a management plan that is being widely touted as the model upon which management of other industrial forests with Red-cockaded Woodpeckers will be based. Therefore, it is worth examining the extent to which this MOA includes the new management strategy.

In the MOA the U.S. Fish and Wildlife Service states that the Georgia-Pacific management plan is deemed sufficient to ensure the survival of existing Red-cockaded Woodpeckers on the relevant forests. The U.S. Fish and Wildlife Service therefore promises that the company will not be prosecuted for "take" violations so long as it follows the management plan. The MOA allows the U.S. Fish and Wildlife Service access to Georgia-Pacific lands to assess compliance. The objective of the MOA is to sustain the existing population indefinitely, but not to increase the population. Hence one does not expect those elements of the new management strategy that promote population increase to be included in management policy under the MOA, but those elements that reduce cluster abandonment should be.

The MOA is based on a management plan written by Gene Wood and John Kleinhofs entitled "Integration of Timber Management and Red-cockaded Woodpecker Conservation Goals on Georgia-Pacific Corporation Timberlands in the Southern United States." The plan outlines a strategy whereby, according to Wood and Kleinhofs, Georgia-Pacific can comply with the Endangered Species Act while continuing to practice forestry, demonstrating that integration of intensive timber management and woodpecker conservation can be ecologically and economically sound. As Wood and Kleinhofs eloquently state, "Meaningful conservation practices are not cheap. If they were, then the idea of conservation would be trivial, which it is not." The company claims that the management included in the plan represents a sacrifice of profit margin in order to ensure a high probability of sustaining the current woodpecker populations on its lands.

The management plan calls for prescribed burning to control hardwood development and for use of cavity restrictors to protect cavities from Pileated Woodpeckers. It also contains provisions for sustaining modest amounts of foraging habitat, and for growing and protecting old trees in cluster areas and replacement stands near clusters. Thus the company has committed, at its expense, to conducting activities that promote Red-cockaded Woodpeckers and to reducing harvesting in foraging and especially cluster areas. The result will be some economic sacrifice on the part of the company and increased persistence of existing groups, since some causes of loss are countered by the appropriate components of the new management strategy.

However, some causes of population decline are not addressed sufficiently, namely loss of geographically isolated groups and loss of groups due to cavity tree death. Old trees are provided in which the birds can construct replacement cavities, but there are no provisions to provide artificial cavities to groups that lose their cavities suddenly or do not happen to have old trees available to them yet. For isolated clusters, the plan includes several provisions that allow them to quickly lose protected status when they are abandoned, but no provisions to reduce isolation. The plan seems designed to eliminate isolated groups, and groups are considered isolated when only 5–8 km from their nearest neighbors.

Overall, then, the plan has no provisions to expand populations or to stabilize populations with poor spatial distributions. It addresses some causes of group loss but not others. Thus it incorporates some appropriate elements of the new management strategy but omits others. It is not a prescription for increasing populations, but it was not intended to be. It was intended to be a prescription for population stability, but it is not that either. It is instead a prescription for slow population decline, which still is an improvement over the prescription for rapid population decline that previous woodpecker management on many industrial forests represented.

The Georgia-Pacific plan should not be viewed as the salvation of private timber lands, as some have portrayed it, nor as the perfect model for other companies to emulate. It is better than what existed before, but there is much room for improvement, if other companies want to make larger economic sacrifices to achieve greater success in conservation. Although the U.S. Fish and Wildlife Service has agreed that this plan will not result in "take," in reality it will result in loss of population. Just as prosecution for "take" violations is the primary weapon the U.S. Fish and Wildlife Service has to coerce landowners to manage for Red-cockaded Woodpeckers, removing this threat is the primary reward the agency can offer in exchange for voluntary management on private land. Rather than attempt to force optimal management, which would require extensive legal proceedings and could produce political backlash, the agency seems willing to accept suboptimal management. However, there does not seem to be any consistent standard as to what constitutes acceptable suboptimal management. As a result, the management included in MOAs and HCPs is quite variable.

Another example that illustrates this point is the HCP for the North Carolina Sandhills population. This HCP involves the one population among the 15 recovery populations in which a significant number of birds (about 15%) occur on private lands. The special provisions of this HCP have more to do with recovery and population expansion than with maintaining the existing population. The HCP was developed to alleviate the fears of landowners who might be inclined to take action to reduce the future value of their land to woodpeckers, and hence its value to recovery, in order to avoid increased liability under the ESA. The HCP provides a vehicle, known as the Safe Harbor Program, whereby landowners can obtain an incidental take permit from the U.S. Fish and Wildlife Service for any new groups that establish themselves on their land in return for adopting a positive management program that will improve the land as Red-cockaded Woodpecker habitat. The landowner retains existing responsibilities for providing nesting and foraging habitat for the birds that already occur on the land. These are identified as the baseline responsibilities and follow the standard guidelines the U.S. Fish and Wildlife Service has adopted for private lands (see below). In addition to identifying baseline responsibilities, a Safe Harbor agreement also identifies the habitat improvements to be made by the landowner and allows the U.S. Fish and Wildlife Service access to the land to verify compliance with the terms of the agreement. The program is open to managers of nonfederal public land, notably state-owned land, as well as to private landowners.

Safe Harbor agreements provide landowners who want to help the birds with a vehicle to do so without incurring additional liability for them, their successors, or their neighbors. If new groups establish themselves on the land, neither the landowners nor their neighbors, nor those who inherit or buy the property from them, are required to provide nesting or foraging habitat for the birds. Harming the new birds is not allowed, but even cutting their active cavity trees is permitted, provided this is done outside the nesting season and the U.S. Fish and Wildlife Service is notified so that they may translocate birds if the cluster is to be destroyed. Of course, the hope is that few landowners will want to eliminate new groups that they have gone to some effort to establish. The critical concept in Safe Harbors is that good stewardship imposes no economic loss on the landowner. Landowners can always obtain full value for their land, and any sacrifice they wish to make in order to promote conservation is voluntary and is not imposed on their neighbors or successors. On both legal and philosophical grounds, this makes effective conservation more likely.

At this writing, many landowners in the Sandhills area have joined the Safe Harbor Program and signed agreements. Considerable amounts of land are owned by most participants, for example, the owners of many of the golf courses in the golfing resort community of Pinehurst. Collectively, these lands are more than enough to make a significant contribution to recovery. In practice, the habitat improvements in these agreements consist largely of the placement of artificial cavities in abandoned clusters and occasionally in new areas. Even though the net benefits of the program in theory are restricted to improvements that are not subsequently eliminated through incidental "take," when viewed against the conceptual framework embodied in the new management strategy, the program seems likely to produce substantial benefits in practice. The key parameter in the strategy is the number of suitable territories, and the number of new territories to be added on private land in the Sandhills through recruitment cluster construction under completed Safe Harbor agreements is already more than 50% of the currently existing number. Isolation of groups is bound to be reduced by the addition of more groups. Indeed, there is some possibility that Safe Harbor may eventually result in a linking between the two Sandhills subpopulations, which currently are independent demographically due to a gap between them of 8–10 km of private land on which few birds occur. The gap is only 2 km in one stretch, where groups occur on two tracts of private land, one on either side of the gap, and it could be reduced to 2–5 km over a significant span through Safe Harbor.

One weakness of Safe Harbor is that required management does little to maintain the quality of existing territories other than prevent cavity trees from being cut. There is some reason to believe that management will be better than what is required. Although not required to remove hardwoods, most landowners desire to maintain an open aspect to their forests, either for aesthetic reasons (e.g., golf courses) or economic ones (i.e., to enable pine-straw raking). Although not required to replace cavity trees that are killed or to protect cavities from enlargement, many participants, already paying for cavity construction and cavity restriction in recruitment clusters, may be willing to add a little more work to the bill to provide for existing territories.

Overall, Safe Harbor is likely to result in an increase in the number of suitable territories and thus an increased population. The magnitude of this increase will depend on the number of new territories added relative to the number of poorly protected existing territories lost. Provisions for incidental "take" do not appear to compromise this much. Landowners are un-likely to cut all the cavity trees used by new groups, although they may cut one or two. Neighbors may cut most of those established on their land, but presumably many groups will retain a sufficient number. Cutting of forag-ing habitat is much more likely, and it is almost certain that the foraging habitat available to many of the new groups established will fall well below the usual standards, even the reduced standards used for private land (see below).

To some degree, then, the success of Safe Harbor agreements will de-pend on the validity of an assumption of the new short-term management strategy, namely that the amount of foraging habitat does not limit popu-lations currently. If we are to observe the limit at which lack of foraging habitat renders territories unsuitable, it may be with some of the groups created through this program. Also, it is likely that some groups will be below normal in productivity (see Chapter 7). Still, a rather unproductive group is better than no group. Overall, despite the seemingly less restric-tive nature of the Sandhills HCP compared to the Georgia-Pacific MOA, the former likely will result in better population performance than the latter.

Although a variety of additional conservation measures have been adopted through the MOA and HCP process, the basic requirements for management of woodpeckers on private land are those contained in the Fish and Wildlife Service's *Draft Red-cockaded Woodpecker Procedures Manual for Private Lands.* These requirements have been followed for sev-eral years, even though they remain in draft status. They are not available for public review, nor are they official policy, but they clearly are unofficial policy and are cited by name in most HCPs and MOAs. For example, in the Sandhills HCP they are the basis for identifying baseline responsibilities, and in the Georgia-Pacific MOA they are cited as equivalent in their effect to the management provisions of the MOA. Although we cannot comment on the details of this as yet unpublished document, its key components are identifiable and can be discussed in relation to the new management strategy.

The guidelines for private lands protect cavity trees, but they contain no mandatory provisions for replacing lost cavity trees or constructing re-cruitment clusters to expand populations. The predominant philosophy implicit in the guidelines is that the government should not impose eco-nomic hardship on private landowners in order to recover endangered species, but should instead restrict itself to protecting such species to the extent dictated by law. Where the law is not totally clear, the philosophy seems to be to interpret in favor of the landowner in order to avoid negative political repercussions. For example, the foraging habitat requirements contained in the guidelines are less stringent than those used on public land.

Perhaps the most controversial aspect of the guidelines for private land is that there are provisions that allow small numbers of isolated groups to be designated unviable, and thus subject to less protection than usual. The U.S. Fish and Wildlife Service maintains that if the long-term prospects of such groups are poor, the greatest benefit to the species is gained by moving the birds involved to other areas, where they can contribute to the viability of larger populations. Usually this involves translocating the young produced by such groups, but in some circumstances moving the resident adults is permitted. In either case, no attempt is made to increase the likely tenure of the resident groups or to increase the viability of the small population. Instead, the guidelines favor actions that cause groups and populations to be lost more rapidly in order to add birds to other, more viable populations. In many instances, the guidelines provide landowners, at some expense, with a vehicle for ridding themselves of the birds on their land.

Whether the strategy for handling small populations on private lands results in a net benefit for the species is debatable. Certainly it is less beneficial than stabilizing small populations, but the argument is that achieving stability would be too costly to private landowners. For a small population, how large, how aggregated, and how near larger populations it must be to be viable over a reasonable period is not clear. Certainly small populations of Red-cockaded Woodpeckers persist much longer than populations of the typical small bird, due to the buffering effect of helpers on population dynamics. Recent computer simulations by Crowder, Priddy and Walters suggest that completely isolated populations of as few as five groups have a reasonable chance of persisting for 20 years, although not indefinitely. Whether moving birds is a net benefit to the species depends on what is added to other populations, as well as the prospects of the small population. Certainly moving a couple of fledglings cannot compensate for the elimination of a group. These fledglings would merely compete with individuals in the existing population and therefore add little of value to a healthy recipient population. In some cases, however, the U.S. Fish and Wildlife Service requires that recruitment clusters be constructed to provide for new groups, as well as that birds be moved. If only one new group is added, there is no net benefit, but there may be if more than one new group is added. Still, there is bound to be loss of genetic diversity as small populations are eliminated and gene flow into the remaining larger populations is reduced.

It is difficult to say how the loss of a population of five groups and the genetic material contained therein compares to the addition of six or seven groups to a larger population.

Overall, the guidelines, when evaluated against the view of population dynamics we advocate, are expected to lead to population decline—accelerated decline where isolated groups and small populations are involved, and slow decline where the removal of birds is not permitted. It is only through HCPs and MOAs that involve additional voluntary management— Safe Harbor, for example—that management sufficient for population stability or increase might occur on private lands.

CHAPTER TWELVE

An Uncertain Future

No one can predict the future with certainty, and in a nation as politically volatile as the United States, predicting the future direction of conservation is especially difficult. But in the case of the Red-cockaded Woodpecker, at least the potential futures and the forces that will dictate which comes to pass have crystallized in the last few years. As detailed in Chapter 11, recent research has led to the development of a new management strategy that has been astoundingly effective where employed, indicating that we now have the knowledge and the tools to recover the species. On the other hand, it is not at all clear that we have the political will to accomplish recovery, and it is possible that changes in the law and in policy will send the Red-cockaded Woodpecker to extinction. In this chapter, we explore these possibilities.

Overall, the extent to which the new short-term management strategy is incorporated into recently developed management guidelines is encouraging (see Chapter 11). In some cases the new strategy is essentially required, and in others it is at least encouraged or permitted. Even on private land, some elements are included in some management plans. Certainly new guidelines are much more likely to result in positive population dynamics than the failed guidelines that preceded them. The mistakes of the past are not being repeated, and new guidelines reflect an understanding of the implications of the new information that is available. At the level of management agencies at least, the optimistic future appears more likely than the pessimistic one. Time will tell whether the same will to undertake effective conservation exists at the level of local managers and within the highest levels of government.

We will return to this issue, but first we will examine the various conflicts that might undermine conservation of Red-cockaded Woodpeckers.

Society's will to practice the best possible conservation depends to a large degree on the nature and magnitude of conflicts between conservation and other uses of land. Therefore, it is worthwhile to assess current knowledge about the primary conflicts involved in the conservation of Red-cockaded Woodpeckers.

Conflicts with Other Conservation Priorities

One type of potential conflict is with other conservation priorities. It is possible that conservation of one species—in this case the Red-cockaded Woodpecker—could adversely impact the ecosystem as a whole or particular other species, leading to opposition in the conservation community to management designed to promote woodpeckers. The issue is the appropriateness of single-species management as opposed to ecosystem management. Because of the power of the Endangered Species Act, conservation strategies in the United States tend to focus on individual species, rather than entire ecosystems. Often the ecosystem is considered indirectly as habitat for the endangered species, but typically this results in intensive focus on some components of the ecosystem while others go ignored, rather than equitable treatment for all. More insidious than neglect is the problem that management of one species will affect other species with which it has strong interactions within the ecosystem. Some of the effects will be positive, others negative. Finally, the strongest effects of single-species management occur when species that have negative interactions with the target species are managed as enemies, that is, are deliberately and directly eliminated.

Does implementation of the new management strategy raise valid concerns about adverse impacts of single-species management on the ecosystem or other elements within the ecosystem? In this section we argue that fortunately there are no valid major concerns of this sort—that there need be no conflicts between management of Red-cockaded Woodpeckers and other conservation agendas within southern pine ecosystems. In fact, implementation of the new management strategy is an effective way to manage these fire-maintained ecosystems.

EFFECTS ON THE ECOSYSTEM

In the case of the Red-cockaded Woodpecker, many elements of its management benefit the ecosystem as well as the bird. Of particular note in this regard is growing-season burning, which is critical to maintain appropriate plant and animal diversity within southern pine ecosystems, particularly longleaf pine ecosystems. Fire frequency is an important variable to which species in these ecosystems are adapted, and suppressing fire favors only that small subset of species that specialize in occupying sites that have not burned for long intervals. Reintroduction of fire favors the vast majority of species, including the Red-cockaded Woodpecker, which are adapted to historically more common and shorter interfire intervals. Prescribed fire returns those species adapted to the longest intervals to their former, lower levels of abundance. Fire makes young oaks rarer, for example, and lovers of small oaks might object to this, but it does not eliminate them, nor does it destroy the few oaks that manage to survive long enough to become large. Rather than suppressing fire to make a few species superabundant, using fire to maintain diversity is easily defended on biological grounds. Growing-season burning is the cornerstone of ecosystem management in fire-maintained pine ecosystems, as well as of woodpecker management, and this fact alone makes the two highly compatible.

No other element of woodpecker management affects the total ecosystem as much as burning. Most elements impact only a few other species, although some of these might be affected strongly. Fortunately, elements included in the new management strategy generally affect the community as a whole in a positive way. Cavity construction benefits all secondary cavity users that sometimes use Red-cockaded Woodpecker cavities, and there are many such species in southern pine ecosystems (see Chapter 5). The exception is use of cavity restrictors, which exclude larger species from the cavities. Restrictors can be viewed as favoring the Red-cockaded Woodpecker, as well as small secondary cavity users such as flying squirrels, at the expense of larger species such as Red-bellied Woodpeckers, Pileated Woodpeckers, American Kestrels, Wood Ducks, and fox squirrels. Where Red-cockaded Woodpecker populations are small or otherwise imperiled, few would object to this cost to larger cavity users, since these other species can still use or excavate holes in dead trees or hardwoods, as they do in other habitats. Where Red-cockaded Woodpecker populations are sufficiently large and healthy, however, some level of cavity enlargement should be permitted to avoid adverse impacts on larger secondary cavity users. This is particularly important where growing-season fire is frequent, as fire may reduce the availability of snags and hardwoods. Despite this caveat, we conclude that the new management strategy does not result in significant costs to other species within the pine ecosystems and can be viewed as favorable for the ecosystem as a whole.

EFFECTS ON GAME SPECIES

Although we conclude that the new management strategy benefits other species within southern pines ecosystems, there nevertheless have been two objections to Red-cockaded Woodpecker management because of threats posed to other species, and both have involved growing-season fire. First, some game biologists and hunters have objected to the destruction of oaks by fire because of the value of oak mast to wildlife species, notably white-tailed deer (*Odocoileus virginianus*) and turkey. This objection can be easily dismissed, provided woodpecker management is implemented correctly. In southern pine ecosystems such as longleaf sandhills and flatwoods, the oak species that occur—for example, blue jack (*Quercus incana*), post (*Q. stellata*) and turkey oak (*Q. laevis*)—do not produce mast until they have grown large enough to withstand fire fairly well. The multitudes of saplings and young trees that occur in areas of fire suppression and that are so vulnerable to growing-season burning do not produce mast. It is the sparse, scattered old trees that produce mast, and that they do so in abundance is evidenced by legions of progeny that sprout up any place the fires bypass for a few years.

These scattered old trees pose no threat to Red-cockaded Woodpeckers or other fire-dependent species, and are an important component of the ecosystem. Although they generally survive fire in regularly burned, open parklike habitat, overzealous managers may remove them by chemical or mechanical means. This is both unnecessary and grounds for valid objection. However, the dependence of game species on acorns in southern pine ecosystems has been overstated. It is based on extrapolation from hardwood forests in other regions, rather than on direct evidence from southern pine ecosystems. In longleaf ecosystems especially, deer and game birds do quite well under a growing-season fire regime. Effects on game species remain a concern to some, but this concern is not valid if the new management strategy is implemented properly.

EFFECTS ON NEOTROPICAL MIGRATORY BIRDS

The second objection comes from those who are concerned with songbirds, especially Neotropical migrants, and worry that fire during the growing season will destroy nests and thereby reduce populations. Hunters echo this objection because of similar concerns about the nests of quail and turkey. This objection has led to legal action in Texas (*Sierra Club et al. v. Lyng et al.*). For species that thrive in longleaf and other southern pine ecosystems, the objection is invalid, because the benefit of improved habitat condition resulting from fire greatly exceeds the cost of occasional nest loss. Individuals that lose nests can renest, and the availability of suitable habitat and its quality (assessed by the density of individuals it can support) likely are much more important determinants of population size than is the rate of nest loss. Many songbirds, as well as quail and turkey, fall into the category of benefiting from an improved habitat resulting from a growing-season fire regime.

Species that do not normally do well in pine ecosystems are another matter. Not only are their nests destroyed by fire, but fire reduces the ability of the habitat to support them. In the extreme case, fire suppression allows succession from pines to hardwoods and the replacement of species adapted to pines by those adapted to hardwoods. There are places where this has occurred, and many more where the process has progressed to the point that some hardwood-adapted species have appeared and coexist to some degree, typically in a spatial mosaic, with pine-adapted species. For example, the Ovenbird (*Seiurus aurocapillus*) formerly did not occur in the North Carolina Sandhills, but currently it is found in fire-suppressed areas where large hardwoods occur. The issue is, not only the extent to which Ovenbirds lose nests to fire, but also whether management for Red-cockaded Woodpeckers specifically and the longleaf ecosystem generally should be allowed to eliminate the Ovenbird from the North Carolina Sandhills. The objection is to reintroducing a formerly common but recently rare process, fire, because it will favor species adapted to that process over those not adapted. Certainly some of the species that will suffer will be Neotropical migrants, and they will suffer more from resulting habitat change than from direct mortality of nests.

Managing to maximize local diversity is a poor strategy if the goal is regional diversity. It will inevitably result in promoting species that are locally rare but regionally common over species that are locally common but regionally rare. It would be possible to maintain species found in the southern Appalachians and Piedmont in longleaf areas if one were willing to reduce populations of pine-adapted species such as the Red-cockaded Woodpecker. On the other hand, it would not be possible to maintain the Red-cockaded Woodpecker or other pine specialists in the mountains. The best strategy, in our opinion, is to manage pine areas to maximize populations of species adapted to those habitats, and not to worry about whether the reversal of habitat degradation leads to a loss of species using degraded areas as secondary habitat. Management for the latter species should occur in their primary habitat types, wherever they may occur. A regional management strategy should dictate local strategies, rather than vice versa.

Most of the birds that currently occur in Red-cockaded Woodpecker habitat, including Neotropical migrants, likely will persist under the ecosystem management regime advocated for southern pine habitats. Research by Todd Engstrom on the Wade Tract indicates that many species, including quail and Neotropical migrants such as Eastern Wood-Pewees, Blue Grosbeaks (*Guiraca caerulea*), Summer Tanagers, Great Crested Flycatchers, and others, achieve high population densities in well-maintained old-growth longleaf (see Chapter 2). The Bachman's Sparrow, another species of concern found in southern pine habitats, is known to benefit greatly from growing-season burning and should thrive under the sort of ecosystem management we have described.

Research in the North Carolina Sandhills by Sharlene Kreiger, a graduate student working with Jaime Collazo at North Carolina State University, and Jennifer Allen, a graduate student working with Walters at Virginia Tech, indicates that many migrants and other species are strongly associated with streams and hillside drains. Examples include Eastern Towhees, White-eyed Vireos and Hooded Warblers (*Wilsonia citrina*) (see also Chapter 2). Vegetation along streams and drains is characterized by a dense shrub layer, and often by hardwood midstory, beneath a canopy of pond pines, in contrast to the surrounding upland community dominated by longleaf pine and wiregrass. Interestingly, despite their association with midstory within drains, few of these drain specialists extend into the uplands in fire-suppressed areas with dense oak midstory. Densities of many are actually higher in drains in regularly burned areas than in drains in fire-suppressed areas. These results suggest that although a few drain-inhabiting species benefit from fire suppression, most species will continue to do well, perhaps better, under a growing-season burning regime, and that there is little benefit to the bird community by suppressing fire in upland areas.

Objections to growing season fire because of adverse effects on other bird species appear to us misguided, based either on lack of understanding of local population dynamics of the supposedly affected species or lack of a sufficiently broad conservation perspective. The only species that will clearly suffer are species like the Ovenbird and Wood Thrush (*Hylocichla mustelina*), which are associated with hardwood forests and whose presence in southern pine ecosystems depends on succession from pine to hardwood, and a few drain specialists that expand their niche under fire suppression. In longleaf areas, promoting such succession to benefit these species seems to us a poor conservation strategy. However, in some other habitat types, notably loblolly pine, in which mature hardwoods are more prominent, this is a legitimate issue. In these areas it is probably appropriate to retain a mix of hardwoods and pines in order to provide both for species like Ovenbirds and for Red-cockaded Woodpeckers and their associates. Other species that one might think of as associated with hardwoods, like Hooded Warblers and even Summer Tanagers, in fact thrive in regularly burned pine ecosystems. As for loss of nests to fire, there is no evidence that such losses reduce the populations of any species and little reason to suspect that they would. We thus conclude that the impact of fire on other bird species is not a valid issue in most ecosystems. In those few ecosystems where it is (i.e., loblolly-shortleaf), it should be relatively simple to promote conservation of both songbirds and Red-cockaded Woodpeckers by varying the management strategy among management units.

CONTROL OF SNAKES AND FLYING SQUIRRELS

There are two common management practices not included in the new management strategy that clearly have adverse effects on other species. These are (1) placing snake-excluder devices, and in some instances snake nets, on cavity trees and (2) trapping and killing flying squirrels. The first is intended to reduce predation on adults and nests; the second, cavity usurpation and, to some extent, nest loss. These practices clearly are detrimental to the target species, but the effect on snake and squirrel populations has never been measured. The number of squirrels killed in some locations, especially the Savannah River Site, is staggering. Snake excluders and nets are only beginning to be employed, so the amount of mortality they may introduce is as yet unclear. Lowering snake and squirrel populations may also have indirect effects, both negative and positive, on their competitors, their predators, and—in the case of snakes—alternative prey species. Indeed, rat snakes probably prey on flying squirrels more often than on Red-cockaded Woodpeckers, so killing snakes and excluding them from cavities may actually increase squirrel numbers and thus squirrel use of cavities. The group of scientists from the University of Arkansas led by James Withgott, Joseph Neal, and Warren Montague, who invented and first employed snake excluders and nets, recommended using only excluders in order to minimize the impact on snakes. In the absence of nets, the only effect on snakes is some interference with their ability to feed, which probably has a relatively small effect on their populations, since they normally cannot pass the pine resin barrier anyway.

The use of antisnake devices and killing of squirrels are clear cases of favoring one species over another and of interfering with the normal balance between them that exists within the ecosystem. Proponents of ecosystem management can raise valid objections to these practices. That these practices are unlikely to benefit the Red-cockaded Woodpecker significantly, according to the view of population dynamics that we espouse, provides further grounds for objecting to them. It seems to us impossible to justify killing flying squirrels on ethical or biological grounds. This is equally true for the use of snake nets. In most regions, it will be difficult to justify using snake excluders simply because of the small expected benefit relative to the large cost of placing and maintaining the devices. However, there may be some tangible benefit from these devices in areas with extremely high nest predation rates. Areas without longleaf pine where the woodpeckers are dependent on pines with inferior gum flow, such as Arkansas perhaps, may qualify. However, it has yet to be demonstrated that nest predation rates reach levels sufficient to have a significant impact on Red-cockaded Woodpecker population dynamics anywhere. Appropriate use of these devices may be limited to extremely small populations of Red-cockaded Woodpeckers in which every nest and every individual is critical. In such cases, causing negative impacts on other species may be necessary to ensure the persistence of the woodpecker population. The goal, however, should be to quickly reach the point where these devices are no longer necessary. The goal of woodpecker recovery should be to return the Red-cockaded Woodpecker to its natural role in the ecosystem, which includes providing cavities for flying squirrels and prey for snakes. The goal of managers should be to maintain populations that have no need for devices like snake nets or even cavity restrictors, populations in which losses to squirrels and snakes occur regularly but have no significant impact on woodpecker numbers.

Thus there are elements of Red-cockaded Woodpecker management—namely, control of snakes and flying squirrels—that conflict with other conservation priorities. However, these elements are not included in the new management strategy, and we do not advocate their use. At most, their use should be restricted to a temporary basis within critically endangered populations.

Conflicts with Other Land Use Priorities

Implementation of the new management strategy for Red-cockaded Woodpeckers does not conflict with other conservation needs within southern pine ecosystems. Hence conflicts are limited to those between conservation and other uses of the land. We now examine those conflicts and their possible resolution.

In recent years, Red-cockaded Woodpecker habitats, particularly longleaf ecosystems, have been used increasingly for pine straw production. The needles that fall each year from the pines are raked, baled, and shipped to markets where the straw is sold as mulch. Pine straw raking can be quite lucrative, and in fact sometimes is a more profitable use of land than timber harvest.

Pine straw raking affects Red-cockaded Woodpeckers in several ways, some direct and some indirect. The primary direct impact is through behavioral disturbance when raking occurs around cavity trees. Because of this effect, most managers include a provision in contracts with pine straw rakers that straw cannot be raked within a specified radius around cavity trees during the breeding season or that raking cannot occur at all during the breeding season. This causes only minimal conflict, because the straw can always be raked at another time. That is, the impact is on scheduling the activity, not on whether it can occur or how it must be done. The economic effect, if any, is minimal.

More problematic is the impact of straw removal on the ecosystem and thus indirectly on the woodpecker. If straw is raked each year, considerable nutrients are removed from the ecosystem and the ability of the area to carry a fire is reduced. Fire normally returns the nutrients contained within the needles to the soil. The long-term health of the plant community, including the pines, may be affected by continued removal of pine straw. In many areas, a cycle of raking in some years and burning unraked straw in others has been adopted. This represents some reduction of potential profits from raking over the short term in order to protect the habitat and the woodpecker over the long term. By promoting the continued health of pines, it may also represent a reduction of short-term profits in order to increase profits over the long-term.

The pine-straw-raking industry, although it has developed to the point that it includes pine straw thieves, remains a rather small one. Therefore, conflicts between industry and conservation have not had the political repercussions or impacts on management policy that conflicts involving military training or timber harvesting have had. Perhaps as the industry increases its capacity to harvest straw, its willingness to harvest at less than maximum levels will wane and conflict will increase. It will be necessary to increase our knowledge of the impacts of raking on the ecosystem and individual species within it if we are to address this conflict in a rational way. It seems likely that some level of harvest will be compatible with woodpecker conservation but inevitable that this level will be well below the maximum harvest, and hence that there will be at least short-term economic costs to conservation. The conflict may actually be less if the economics of pine straw harvesting are measured over longer time frames so that sustainability becomes an issue. Nevertheless, this conflict may limit conservation efforts on private land as much as conflict with timber harvesting and development over the next few decades. On public lands where economic incentives are less critical, it should be easier to arrive at policies that allow some level of pine straw harvesting without compromising woodpecker conservation.

MILITARY TRAINING

As is discussed in Chapter 11, the conservation of Red-cockaded Woodpeckers appears to be compatible with military training to a large degree. The birds have thrived on some of the military bases most heavily used for training, and many training impacts are not expected to have large effects on population dynamics. The most essential restrictions on training are those that protect cavity trees from the damage caused by tracked vehicles and those that constrain the placement of new ranges. It appears possible to design training grounds that both retain realism and protect cavity tree clusters, and to work with trainers so that ranges are shifted rather than eliminated. Thus, the impact on training likely is manageable and non-threatening to the military mission, even on bases where the forests are heavily used for training. The new army guidelines and the Camp Lejeune management plan should result in the collection of data that will increase our knowledge of the impacts of training on Red-cockaded Woodpeckers, and therefore our ability to minimize conflict between conservation and the military mission.

Obstacles to resolving this conflict appear to lie more with philosophical and political differences than the needs of woodpeckers or military trainers. There is resistance on many military bases, typically originating from their civilian employees rather than the military command, to the notion that conservation needs should be a consideration in planning military training. The heightened controversy at Fort Bragg compared to other bases is not a result of greater conflict with training, but rather of inappropriate scientific advice from key civilian employees and their ability to influence the military command. On the other hand, within some environmental organizations there is resistance to the idea that the military mission is important. Training often impacts the ecosystem more than it does the Red-cockaded Woodpecker specifically. This sets the stage for conflict between the military and environmentalists over the relative merits of using military bases as habitat preserves versus training areas, a conflict that extends far beyond the Red-cockaded Woodpecker. Is it really necessary to national security to allow tracked vehicles to tear up the ground in longleaf habitat? The conflict inherent in such questions will remain even where management plans allow woodpecker populations to grow rapidly in the presence of intense training activity. There appears to be no inherent incompatibility between military training and effective woodpecker conservation, but the larger conflict between the military and environmentalists likely will continue to influence the development of land use policy that promotes effective multiple use of military lands.

TIMBER HARVESTING

We end with what has always been, and still remains, the primary conflict between conservation of Red-cockaded Woodpeckers and other uses of the land, the conflict with timber harvesting. It is the resolution of this conflict that will best reflect the values of our society. The Red-cockaded Woodpecker is arguably the best species to use in developing general conservation policy in which timber harvesting is an issue because changes in the population dynamics of Red-cockaded Woodpeckers as a function of the level of timber harvesting can be measured with some accuracy. This relationship is much less clear in other, more famous conflicts, such as that between timber harvesting and conservation of the Spotted Owl. As is discussed above, solutions that allow multiple use of the land are possible, and increasing woodpecker populations while continuing to harvest a significant amount of timber is realistic. What remains to be decided is the appropriate density of woodpecker populations and level of timber harvest on different kinds of land.

Should industrial forestlands be used for anything other than short-term economic gain? Georgia-Pacific plans to reduce gain by the amount required to maintain the slowly declining, low-density woodpecker populations that currently exist on their lands. Is this sufficient, or do we as a society demand further reductions in profits to promote conservation? Or perhaps Georgia-Pacific sacrifices too much; perhaps as a society we do not care whether anything survives on industrial forestlands except timber. Do we care whether the system may eventually collapse under the high-intensity production practices currently used because of loss of site productivity—that is, do we care that high levels of timber production may not be sustainable, as long as the short-term profit margin is favorable?

And what of individual landowners with a small amount of timber on their land? Should the activities of such individuals be constrained if Red-cockaded Woodpeckers exist on their property, even if this results in economic loss? Should the landowners be compensated for these losses? If so, can we establish a system for evaluating such claims that is not open to regular abuse? The Environmental Defense Fund has proposed an innovative program that involves establishing a habitat mitigation market. In essence, a landowner desiring to conduct an activity that would eliminate a woodpecker group could do so by paying another landowner to increase (by one) the number of groups the other is willing to support on his or her land. The value of a group in this program would be determined by the market. Thus some landowners, who perhaps wanted to maintain their forests anyway, could earn money by practicing effective conservation. Proposed programs like this one and existing programs like Safe Harbor (see Chapter 11), which promote maintenance and even an increase of existing populations on private land, contrast with many recent HCPs, which sanction the elimination of birds on private lands in order to facilitate conservation on public lands. The issue on private lands is not so much compatibility between timber production (or other land uses) and woodpecker conservation as it is whether private lands should be included in conservation efforts at all.

On national forests there is no conflict between conservation and private property rights, but there is potential conflict between conservation and the legally mandated mission of the Forest Service to produce timber. Here the issue is indeed compatibility, since there is no mandate to maximize either conservation or timber production, but only to accomplish both. As discussed above, the current Forest Service guidelines make it possible to increase woodpecker populations at current densities while harvesting a considerable amount of timber. As time passes, the density targets for Red-cockaded Woodpeckers will be a good measure of the relative values placed on conservation and timber production on national forests.

More controversial than the issue of whether timber production precludes woodpecker conservation is the issue of which silvicultural methods are appropriate. The method of choice on national forests in the past has been clear-cutting, a form of even-aged management that results in single-age stands. Other methods that result in stands with a mix of ages, such as single-tree selection, group selection, or irregular shelterwood cuts in which the residual shelterwood trees are retained, mimic natural disturbances much better, and therefore result in a forest much more like those that existed in earlier times. These methods are thus more consistent with the long-term ecosystem management strategy that we advocate. A concern with these methods, however, is the frequency of disturbance due to silvicultural activities within stands. Where woodpecker populations are healthy (recovered), the Forest Service guidelines allow shelterwood cuts without retention of the shelterwood trees, a version of even-aged management similar in impact to clear-cutting, and merely extend rotation ages to provide old trees. Although this system does produce old trees, they are to be harvested just as they become highly attractive as foraging and cavity excavation sites. This seems to be another insidious by-product of the tradition of treating minima as optima, that is, of treating trees of the minimum age at which they become frequently used (100–120 years for longleaf), instead of trees of the most preferred age (apparently >150 years), as what the birds need. The guidelines allow the retention of shelterwood trees, as well as the uneven-aged techniques of single-tree and group selection, but this is at the discretion of the local manager.

We know something about the compatibility of timber production and woodpecker conservation, and of trade-offs between timber volume and woodpecker density, for forests in which the silvicultural methods typical of the last several decades are used. We remain ignorant about compatibility and trade-offs for forests where other silvicultural practices, such as irregular shelterwood or uneven-aged methods, are used. What are critically needed are at least some forests managed in alternative ways, so that imagined costs and benefits can be replaced with real information.

The Forest Service guidelines require either irregular shelterwood with the retention of residual trees or uneven-aged methods where woodpecker populations are less healthy (i.e., small or exhibiting an unfavorable population trend). Thus we stand to learn something about what happens under silvicultural regimes other than clear-cutting and regular shelterwood on some national forests. Other federal lands, such as military bases and wildlife refuges, seem to be following the Forest Service's lead in that irregular shelterwood cuts (with 15–30 residuals per hectare) emerge as the preferred silvicultural method. There is, however, language in the military guidelines that encourages emulation of earlier forest conditions and the use of uneven-aged silvicultural techniques. Other techniques are allowed in the refuge guidelines, but most are discouraged. For example, uneven-aged management is discouraged because of the possibility that regenerating pines will obstruct cavity entrances. At least in longleaf pine, young trees are intolerant of shading, and this prevents saplings from growing close to the adult. Fear of uneven-aged management appears widespread but misguided. According to many arguments against it, the woodpeckers and the forests themselves cannot survive without more structured, intense management. This begs the question of how the ecosystem persisted for thousands of years without the assistance of forest managers. Hopefully, at least some forest managers on military and refuge lands will employ the less practiced, more controversial silvicultural methods. Only then will we begin to truly understand the full extent of the relationship between conservation and timber production in the southern pine ecosystems.

Legal Challenges

In view of the above considerations, it is surprising that a potential disruption of Red-cockaded Woodpecker management and of the restoration of fire-maintained ecosystems in the region comes from conservation and environmental groups. The Texas Committee on Natural Resources and Edward C. (Ned) Fritz, a founder and key member of the organization, object to several key aspects of Red-cockaded Woodpecker management. It is ironic that Fritz and the Texas Committee on Natural Resources pose such an impediment, as they were the primary advocates supporting the establishment of wilderness areas on national forests in Texas. The Texas Committee on Natural Resources and Fritz were also the primary plaintiffs in the court proceedings that resulted in the 1988 court rulings that found the U.S. Forest Service in violation of the Endangered Species Act in relation to Red-cockaded Woodpeckers (see Chapter 10). In fact, these recent legal proceedings technically represent a continuation of that same case! It is ironic that the case that initially raised the profile of Red-cockaded Woodpeckers and ushered in the modern era of intensified management now represents a threat to that management.

The Texas Committee on Natural Resources now vigorously opposes the use of frequent prescribed fire, hardwood midstory reduction, and silvicultural systems other than uneven-aged ones to manage Red-cockaded Woodpeckers and restore fire-maintained pine ecosystems on appropriate portions of the landscape. This opposition is based, in our view, on a failure to understand processes operating within fire-maintained pine ecosystems, especially their dependence on frequent fire, and the biology and evolutionary history of Red-cockaded Woodpeckers. The opposition to prescribed fire is of special concern and most perplexing, in view of the massive alteration of biodiversity and natural communities that suppression and alteration of fire regimes have had in the southeastern United States and around the globe. Should their views prevail in the courts, they would pose insurmountable difficulties in the efforts to recover Red-cockaded Woodpeckers and restore fire-maintained pine ecosystems and the biodiversity they support.

Prospectus

It appears that there exists an effective management strategy for the Red-cockaded Woodpecker that is compatible with ecosystem management and lacking in significant adverse effects on other species within the ecosystem. This makes recovery of the species, and a degree of restoration of the pine ecosystems on which it depends, a real possibility. The barriers to this occurring are political and practical rather than biological. The first step toward recovery is development of an appropriate, effective management strategy that is universally accepted. This seemingly has been accomplished, with the caveat that the ongoing legal challenge in Texas just discussed proves to be only a temporary, local setback. The second step is incorporation of the management strategy into management requirements. This step is proceeding rapidly through development of management guidelines by the various agencies responsible for the lands on which the species occurs. Although the outcome is not optimal, it is at least favorable. Not all elements of the management strategy have always been incorporated, or if incorporated, properly emphasized and prioritized. Also, not everyone understands or subscribes to the view of population dynamics on which the new strategy is based. Still, several agencies follow this strategy fairly closely, and all have responded quickly to the new knowledge that led to the strategy by changing their guidelines to incorporate elements of the strategy to some degree. In the end, it is our view that the guidelines produced are sufficient to promote recovery of the species in the long term, except for those guidelines developed for private land (e.g., the Georgia-Pacific plan).

What remains is the third step, implementation of the guidelines that now exist. We might object to the fact that the decision has been made to rely on public lands for recovery, that we can only hope for, but not demand, contributions from private land through optional programs such as Safe Harbor. And we might wish that the new management strategy were adopted more completely or more clearly in various management guidelines. But still, what exists seems sufficient to accomplish recovery, if implemented properly. Implementation depends on the competence and will of local managers to carry out the guidelines in the manner intended by those who developed them. It also depends on the influence of local, regional, and national politics on the will and ability of local managers to implement the guidelines. Human error, capability, and value judgments are as important as the quality of the management guidelines in this last and critical step toward recovery. That one base (Fort Bragg) has a slightly declining woodpecker population and seeks an exemption from the Endangered Species Act while another (Camp Lejeune) with similar training needs wins environmental awards, based to a large degree on its highly successful Red-cockaded Woodpecker program, illustrates how different outcomes can occur at the implementation stage. It would be naive to expect anything better in the future.

What happens next depends on the conflicts between conservation of Red-cockaded Woodpeckers and other interests that exist, and how those conflicts are resolved. This will affect not only the development of policy but also the will of local managers to implement effective management. Conflict with military training appears to be more a matter of conflicting values than of real incompatibilities, and is perhaps best solved through education and communication. A demonstration of compatibility between military training and woodpecker conservation on a few exemplary bases such as Eglin and Camp Lejeune may effectively resolve this conflict. Measuring the impacts of military training activities on the birds also will contribute to a resolution. Conflicts involving timber harvesting and to a lesser extent pine straw raking are, in contrast, real. Resolving them will require compromise in which there is a trade-off between conservation and economic gain. It is the resolution of these conflicts that will reveal our priorities as a society.

Our message is that recovery is possible, and that the necessary elements to bring it about are available. Whether or not it occurs depends on whether we as a society want it to occur. There was reason for optimism in the early and mid-1990s, an exciting period in woodpecker conservation. There was new research and a new management strategy, and formerly declining populations turned into increasing ones. However, as the new millennium opens, that momentum is dissipating. Some of the increasing populations are now only stable or are declining again, and the conflicts we have discussed are the cause of this loss of momentum. At Camp Lejeune, management to increase the population further was postponed until conflicts with military training were resolved. In Texas, hardwood encroachment again has the potential to threaten populations as the will to implement effective management wanes, partly due to pressure from misguided conservationists. Meanwhile, Eglin and a few others continue to push forward toward recovering their populations.

We cannot predict the eventual outcome, or even discern in which direction we are headed at the moment. Our hope is that at the least, the outcome will depend on rational decisions about our priorities—about the value of recovered Red-cockaded Woodpecker populations compared to other uses of the land. We hope that we will not be distracted from rational decision making by debates over other issues such as habitat preservation versus multiple use, disguised as argument about woodpecker management, or by hidden agendas. We have precious few opportunities to make such clear choices. What happens to the Red-cockaded Woodpecker in the coming decades will be a critical test of our society's philosophy and commitment to conservation, and will determine the heritage that we leave for future generations.

Common and Scientific Names of Species Mentioned in Text

Species Common Name	Scientific Name	Ch./Pg.
Acorn Woodpecker	*Melanerpes formicivorus*	6/127
African elephant	*Loxodonta africana*	2/20
African hunting dog	*Lycaon pictus*	6/116
ambrosia beetle	*Platypus flavicornis*	8/208
American beech	*Fagus grandifolia*	2/14
American bison	*Bison bison*	2/19
American black bear	*Ursus americanus*	6/160
American Crow	*Corvus brachyrhynchos*	6/116
American Kestrel	*Falco sparverius*	2/38
arboreal ant	*Crematogaster ashmeadi*	7/176
Arizona Woodpecker	*Picoides arizonae*	3/49
Bachman's Sparrow	*Aimophila aestivalis*	2/37
Bachman's Warbler	*Vermivora bachmanii*	4/66
baldcypress	*Taxodium distichum*	2/24
Black-backed Woodpecker	*Picoides arcticus*	3/49
black cherry	*Prunus serotina*	7/177
black gum	*Nyssa sylvatica*	7/177
black turpentine beetle	*Dendroctonus terebrans*	5/108
Blue Grosbeak	*Guiraca caerulea*	12/305
blue jack oak	*Quercus incana*	12/303
bluestain fungus	*Ophiostoma minus*	8/203
broad-headed skink	*Eumeces laticeps*	5/95
Brown Creeper	*Certhia americana*	7/194
Brown-headed Nuthatch	*Sitta pusilla*	2/37
Brown Pelican	*Pelecanus occidentalis*	9/220
brown spot fungus	*Scirrhia acicola*	2/42
California Condor	*Gymnogyps californianus*	2/19
camel	*Tanupolama mirifica*	3/48
camelid	*Hemiauchenia macrocephala*	2/19

Species Common Name	Scientific Name	Ch./Pg.
camelid	*Palaeolama mirifica*	2/19
Carolina Chickadee	*Poecile carolinensis*	7/194
Carolina Parakeet	*Conuropsis carolinensis*	4/66
Carolina Wren	*Thryothorus ludovicianus*	9/236
checkered clerid beetle	*Thanasimus dubius*	8/207
Chuck-will's-widow	*Caprimulgus carolinensis*	2/38
Common Nighthawk	*Chordeiles minor*	2/38
Common Yellowthroat	*Geothlypis trichas*	2/38
Cooper's Hawk	*Accipiter cooperii*	6/140
corn earworm	*Heliothis zea*	4/62
corn snake	*Elaphe guttata guttata*	5/91
dire wolf	*Canis dirus*	2/19
Dodo	*Raphus cucullatus*	10/240
Downy Woodpecker	*Picoides pubescens*	3/49
Eastern Bluebird	*Sialia sialis*	2/38
eastern diamondback rattlesnake	*Crotalus adamanteus*	2/35
eastern indigo snake	*Drymarchon couperi*	2/35
Eastern Screech-Owl	*Otus asio*	5/95
Eastern Towhee	*Pipilo erythrophthalmus*	2/38
Eastern Wood-Pewee	*Contopus virens*	2/38
engraver beetle (four-spined)	*Ips avulsus*	8/201
engraver beetle (five-spined)	*Ips grandicollis*	8/201
engraver beetle (six-spined)	*Ips calligraphus*	8/201
European Starling	*Sturnus vulgaris*	5/98
flatwoods salamander	*Ambystoma cingulatum*	2/35
Florida Scrub-Jay	*Aphelocoma coerulescens*	6/119
fox squirrel	*Sciurus niger*	2/39
giant ground sloth	*Eremotherium rusconii*	2/19
giant ground sloth	*Megalonyx jeffersonii*	2/19
Giant Moa	*Megalapteryx didinus*	10/240
glyptodonts	*Glyptotherium floridanum*	2/19
Golden-crowned Kinglet	*Regulus satrapa*	7/194
gopher tortoise	*Gopherus polyphemus*	2/35
gray squirrel	*Sciurus carolinenis*	2/39
gray treefrog	*Hyla versicolor, H. chrysoscelis*	5/95
Great Auk	*Pinguinus impennis*	10/240
Great Crested Flycatcher	*Myiarchus crinitus*	2/38
Hairy Woodpecker	*Picoides villosus*	3/49
heartwood-decaying fungus	*Spongipellis pachyodon*	5/79
Henslow's Sparrow	*Ammodramus henslowii*	2/37
honeybee	*Apis mellifera*	5/95
Hooded Warbler	*Wilsonia citrina*	12/305
House Sparrow	*Passer domesticus*	5/98
human	*Homo sapiens*	1/4

Species Common Name	Scientific Name	Ch./Pg.
Ivory-billed Woodpecker	Campephilus principalis	4/66
jack pine	Pinus banksiana	2/15
Ladder-backed Woodpecker	Picoides scalaris	3/49
Lewis's Woodpecker	Melanerpes lewis	7/173
little bluestem	Schizachyrium scoparium	2/33
loblolly pine	Pinus taeda	1/4
long-horned sawyer beetle	Neacanthocinus obsoletus	8/208
longleaf pine	Pinus palustris	1/3
Louisiana pine snake	Pituophis ruthveni	Plate 6
mastodon	Mammut americanum	2/19
Merriam's Teratorn	Teratornis merriami	2/19
mycangial fungus	Ceratocystiopsis ranaculosus	8/203
mycangial fungus	Entomocorticium sp. A	8/203
naked mole-rat	Heterocephalus glaber	6/116
Northern Bobwhite	Colinus virginianus	2/37
Northern Cardinal	Cardinalis cardinalis	6/160
Northern Flicker	Colaptes auratus	5/93
Nuttall's Woodpecker	Picoides nuttallii	3/49
Osprey	Pandion haliaetus	9/220
Ostrich	Struthio camelus	6/116
Ovenbird	Seiurus aurocapillus	12/304
Palila	Loxioides bailleui	10/254
Passenger Pigeon	Ectopistes migratorius	4/66
Pileated Woodpecker	Dryocopus pileatus	2/38
pine barrens treefrog	Hyla andersonii	2/35
pine snake	Pituophis ruthveni, P. melanoleucus	2/35
Pine Warbler	Dendroica pinus	2/37
pine woods treefrog	Hyla femoralis	2/35
pitcher plant	Sarracenia flava	Plate 3
pitch pine	Pinus rigida	2/35
poison ivy	Rhus toxicodendron	7/177
pokeberry	Phytolacca americana	7/177
pond pine	Pinus serotina	2/34
post oak	Quercus stellata	12/303
Puerto Rican Parrot	Amazona vittata	10/240
raccoon	Procyon lotor	6/160
rat snake	Elaphe obsoleta	5/91
Red-bellied Woodpecker	Melanerpes carolinus	5/93
Red-breasted Nuthatch	Sitta canadensis	5/92
Red-cockaded Woodpecker	Picoides borealis	1/1
Red-headed Woodpecker	Melanerpes erythrocephalus	5/93
red heart fungus	Phellinus pini	1/2
Reunion Solitaire	Raphus solitarius	10/240
Rodrigues Solitaire	Pezophaps solitaria	10/240

Species Common Name	Scientific Name	Ch./Pg.
root-decaying fungus	*Heterobasidium annosum*	5/109
Roptrocerus	*Roptrocerus xylophagorum*	8/209
Ruby-crowned Kinglet	*Regulus calendula*	7/194
sabertooth cat	*Smilodon fatalis*	2/19
Sanderling	*Calidris alba*	5/98
sand pine	*Pinus clausa*	2/43
Sharp-shinned Hawk	*Accipiter striatus*	6/140
shortleaf pine	*Pinus echinata*	2/29
slash pine	*Pinus elliottii*	2/33
southern flying squirrel	*Glaucomys volans*	5/95
southern magnolia	*Magnolia grandiflora*	2/14
southern pine beetle	*Dendroctonus frontalis*	2/24
southern pine sawyer beetle	*Monochamus titillator*	8/208
Spotted Owl	*Strix occidentalis*	1/1
Strickland's Woodpecker	*Picoides stricklandi*	3/49
Summer Tanager	*Piranga rubra*	2/38
tapir	*Tapirus viroensis*	3/48
Three-toed Woodpecker	*Picoides tridactylus*	3/49
Tufted Titmouse	*Baeolophus bicolor*	5/93
turkey oak	*Quercus laevis*	12/303
Virginia pine	*Pinus virginiana*	2/34
wax myrtle	*Myrica cerifera, M. inodora*	7/177
western redcedar	*Thuja plicata*	10/261
White-breasted Nuthatch	*Sitta carolinensis*	5/93
White-eyed Vireo	*Vireo griseus*	2/38
White-headed Woodpecker	*Picoides albolarvatus*	3/49
white oak	*Quercus alba*	8/213
white-tailed deer	*Odocoileus virginianus*	12/303
Wild Turkey	*Meleagris gallopavo*	4/66
wiregrass	*Aristida stricta, A. beyrichiana*	2/27
wood-decaying fungus	*Lentinus lepidius*	5/85
wood-decaying fungus	*Lenzites saepiaria*	5/85
wood-decaying fungus	*Phaeolus schweinitzii*	5/85
wood-decaying fungus	*Phlebia radiata*	5/85
Wood Duck	*Aix sponsa*	2/44
Wood Thrush	*Hylocichla mustelina*	12/305
Yellow-rumped Warbler	*Dendroica coronata*	7/194

Selected References and Additional Readings

We have tried to choose a style of writing in this book to enhance its readability. References are not cited as is normally done in scientific literature but are referred to in the text by the author(s) and their affiliation at the time the work was completed. References and recommended readings for each chapter are presented below.

1. An Introduction

Allen, D. H. 1991. An insert technique for constructing artificial Red-cockaded Woodpecker cavities. U.S. Dept. Agric., For. Serv. Gen. Tech. Rep. SE-73.

Conner, R. N., and K. A. O'Halloran. 1987. Cavity-tree selection by Red-cockaded Woodpeckers as related to growth dynamics of southern pines. *Wilson Bulletin* 99:398–412.

Conner, R. N., and D. C. Rudolph. 1989. Red-cockaded Woodpecker colony status and trends on the Angelina, Davy Crockett and Sabine National Forests. U.S. Dept. Agric., For. Serv. Res. Pap. SO-250.

Copeyon, C. K. 1990. A technique for constructing cavities for the Red-cockaded Woodpecker. *Wildlife Society Bulletin* 18:303–311.

Copeyon, C. K., J. R. Walters, and J. H. Carter III. 1991. Induction of Red-cockaded Woodpecker group formation by artificial cavity construction. *Journal of Wildlife Management* 55:549–556.

Jackson, J. A. 1971. The evolution, taxonomy, distribution, past populations and current status of the Red-cockaded Woodpecker. Pages 4–26 *in* The ecology and management of the Red-cockaded Woodpecker (R. L. Thompson, ed.). Tallahassee, Florida: Bureau of Sport Fisheries and Wildlife, U.S. Dept. Interior, and Tall Timbers Research Station.

James, F. C. 1991. Signs of trouble in the largest remaining population of Red-cockaded Woodpeckers. *Auk* 108:419–423.

―――. 1995. The status of the Red-cockaded Woodpecker in 1990 and the prospect for recovery. Pages 439–451 *in* Red-cockaded Woodpecker: recovery, ecology and management (D. L. Kulhavy, R. G. Hooper, and R. Costa, eds.). Nacogdoches, Texas: Center for Applied Studies, College of Forestry, Stephen F. Austin State Univ.

Kulhavy, D. L., R. G. Hooper, and R. Costa, eds. 1995. Red-cockaded Woodpecker: recovery, ecology and management. Nacogdoches, Texas: Center for Applied Studies, College of Forestry, Stephen F. Austin State Univ.

Lay, D. W., and D. N. Russell. 1970. Notes on the Red-cockaded Woodpecker in Texas. *Auk* 87:781–786.

Ligon, J. D. 1968. Sexual differences in foraging behavior in two species of *Dendrocopos* woodpeckers. *Auk* 85:203–215.

———. 1970. Behavior and breeding biology of the Red-cockaded Woodpecker. *Auk* 87:255–278.

Maxwell, R. S., and R. D. Baker. 1983. Sawdust empire: the Texas lumber industry, 1830–1940. College Station: Texas A&M Univ. Press.

McFarlane, R. W. 1992. A stillness in the pines. New York: W. W. Norton.

Steirly, C. C. 1949. A note on the Red-cockaded Woodpecker. *Raven* 20:6–7.

———. 1950. Nest cavities of the Red-cockaded Woodpecker. *Raven* 21:2–3.

———. 1957. Nesting ecology of the Red-cockaded Woodpecker in Virginia. *Raven* 28:24–36. Also published in *Atlantic Naturalist* 12(1957):280–292.

Stouffer, M. 1982. At the crossroads: the story of America's endangered species. Wild America Film Series, Film No. 106. Aspen, Colorado: Marty Stouffer Productions.

Thompson, R. L., ed. 1971. The ecology and management of the Red-cockaded Woodpecker. Tallahassee, Florida: Bureau of Sport Fisheries and Wildlife, U.S. Dept. Interior, and Tall Timbers Research Station.

U.S. Department of Defense. Army. 1996. 1996 management guidelines for the Red-cockaded Woodpecker on army installations. DOD, Department of the Army. (unpublished)

U.S. Fish and Wildlife Service. 1979. Red-cockaded Woodpecker recovery plan. Atlanta, Georgia: U.S. Fish and Wildl. Serv.

———. 1985. Red-cockaded Woodpecker recovery plan. Atlanta, Georgia: U.S. Fish and Wildl. Serv.

U.S. Forest Service. 1995. Final environmental impact statement for the management of the Red-cockaded Woodpecker and its habitat on national forests in the southeast region. U.S. Dept. Agric., For. Serv. Manage. Bull. R8-MB 73. Atlanta, Georgia: Southern Region.

Walters, J. R. 1990. Red-cockaded Woodpeckers: a 'primitive' cooperative breeder. Pages 67–101 *in* Cooperative breeding in birds: long-term studies of ecology and behavior (P. B. Stacey and W. D. Koenig, eds.). Cambridge, Massachusetts: Cambridge Univ. Press.

Walters, J. R., P. D. Doerr, and J. H. Carter III. 1988. The cooperative breeding system of the Red-cockaded Woodpecker. *Ethology* 78:275–305.

———. 1992. A test of the ecological basis of cooperative breeding in Red-cockaded Woodpeckers. *Auk* 109:90–97.

Webb, T., III. 1986. Vegetational change in eastern North America from 18,000 to 500 Yr B.P. Pages 63–69 *in* Climate-vegetation interactions: proceedings of a workshop (C. Rosenzweig and R. Dickinson, eds.). Rep. OIES-2. Boulder, Colorado: Office for Interdisciplinary Earth Studies and Univ. Corporation for Atmospheric Research.

Webb, T., and P. J. Bartlein. 1992. Global changes during the last three million years: climatic controls and biotic responses. *Annual Review of Ecology and Systematics* 23:141–173.

Wood, D. A., ed. 1983. Red-cockaded Woodpecker symposium II proceedings. Tallahassee: State of Florida Game and Fresh Water Fish Comm.

2. Fire-Maintained Pine Ecosystems

Bartram, W. [1791] 1988. Travels through North and South Carolina, Georgia, east and west Florida, the Cherokee country, the extensive territories of the Muscogulies, or Creek Confederacy, and the country of the Chactaws. New York: Penguin Books.

Bridges, E. L., and S. L. Orzell. 1989. Longleaf pine communities of the west Gulf Coastal Plain. *Natural Areas Journal* 9:246–253.

Burns, R. M., and B. H. Honkala, eds. 1990. Silvics of North America. Vol. 1, Conifers. Washington, D.C.: U.S. Dept. Agric., Forest Service.

Catesby, M. 1743. The natural history of Carolina, Florida and the Bahama Islands. Vol. 2. London: M. Catesby.

Chapman, H. H. 1909. A method of studying growth of longleaf pine applied in Tyler Co., Texas. *Society of American Foresters Proceedings* 4:207–220.

Christensen, N. L. 1993. The effects of fire on nutrient cycles in longleaf pine ecosystems. *Proceedings of Tall Timbers Fire Ecology Conference* 18:205–225.

Clewell, A. F. 1989. Natural history of wiregrass (*Aristida stricta* Michx., Gramineae). *Natural Areas Journal* 9:223–233.

Conner, R. N., and D. C. Rudolph. 1991. Forest habitat loss, fragmentation and Red-cockaded Woodpecker populations. *Wilson Bulletin* 103:446–457.

Delcourt, P. A. 1980. Goshen Springs: late Quaternary vegetation record for southern Alabama. *Ecology* 61:371–386.

Engstrom, R. T. 1993. Characteristic mammals and birds of longleaf pine forests. *Proceedings of Tall Timbers Fire Ecology Conference* 18:127–138.

Folkerts, G. W. 1982. The Gulf Coast pitcher plant bogs. *American Scientist* 70:260–267.

Frost, C. C. 1993. Four centuries of changing landscape patterns in the longleaf pine ecosystem. *Proceedings of Tall Timbers Fire Ecology Conference* 18:17–43.

Glitzenstein, J. S., W. J. Platt, and D. R. Streng. 1995. Effects of fire regime and habitat on tree dynamics in north Florida longleaf pine savannas. *Ecological Monographs* 65:441–476.

Hays, J. D., J. Imbrie, and N. Shackleton. 1976. Variations in the earth's orbit: pacemaker of the ice volume cycle. *Science* 194:1121–1132.

Hermann, S. M., ed. 1993. The longleaf pine ecosystem: ecology, restoration and management. *Proceedings of Tall Timbers Fire Ecology Conference 18*. Tallahassee, Florida: Tall Timbers Research Station.

Hooper, R. G., and C. J. McAdie. 1995. Hurricanes as a factor in the long-term management of Red-cockaded Woodpeckers. Pages 148–166 in Red-cockaded Woodpecker: recovery, ecology and management (D. L. Kulhavy, R. G. Hooper, and R. Costa, eds.). Nacogdoches, Texas: Center for Applied Studies, College of Forestry, Stephen F. Austin State Univ.

Imbrie, J. 1985. A theoretical framework for the Pleistocene ice ages. *Journal of the Geological Society of London* 142:417–432.

Komarek, E. V., Jr. 1964. The natural history of lightning. *Proceedings of Tall Timbers Fire Ecology Conference* 3:139–183.

Little, E. L., Jr. 1971. Atlas of United States trees. Vol. 1, Conifers and important hardwoods. Washington, D.C.: U.S. Dept. Agric., For. Serv. Misc. Publ. No. 46.

Lyell, C. 1845. Travels in North America in the years 1841–42: with geological observations on the United States, Canada, and Nova Scotia. New York: Wiley and Putnam.

Martin, P. S., and R. G. Klein, eds. 1984. Quaternary extinctions: a prehistoric revolution. Tucson: Univ. Arizona Press.

Peet, R. K. 1993. A taxonomic study of *Aristida stricta* and *A. beyrichiana*. *Rhodora* 95:25–37.

Platt, W. J., G. W. Evans, and S. L. Rathbun. 1988. The population dynamics of a long-lived conifer (*Pinus palustris*). *American Naturalist* 131:491–525.

Rostlund, E. 1960. The geographic range of the historic bison in the Southeast. *Annals of the Association of American Geographers* 50:395–407.

Schmidtling, R. C., and V. Hipkins. 1998. Genetic diversity in longleaf pine (*Pinus palustris*): influence of historical and prehistorical events. *Canadian Journal of Forest Research* 28:1135–1145.

Smith, D. M. 1986. The practice of silviculture. 8th ed. New York: John Wiley.

Streng, D. R., J. S. Glitzenstein, and W. J. Platt. 1993. Evaluating effects of season of burn in longleaf pine forests: a critical literature review and some results from an ongoing long-term study. *Proceedings of Tall Timbers Fire Ecology Conference* 18:227–263.

Wahlenberg, W. G. 1946. Longleaf pine: its use, ecology, regeneration, protection, growth and management. Washington, D.C.: Charles Lathrop Pack Forestry Foundation.

———. 1960. Loblolly pine: its use, ecology, regeneration, protection, growth and management. Durham, North Carolina: School of Forestry, Duke Univ.

Walker, J. L. 1993. Rare vascular plant taxa associated with the longleaf pine ecosystem. *Proceedings of Tall Timbers Fire Ecology Conference* 18:105–125.

Watts, W. A. 1980. The late Quaternary vegetation of the eastern United States. *Annual Review of Ecology and Systematics* 11:387–409.

Webb, T., III. 1986. Vegetational change in eastern North America from 18,000 to 500 Yr B.P. Pages 63–69 in Climate-vegetation interactions: proceedings of a workshop (C. Rosenzweig and R. Dickinson, eds.). Rep. OIES-2. Boulder, Colorado: Office for Interdisciplinary Earth Studies and Univ. Corporation for Atmospheric Research.

Webb, T., and P. J. Bartlein. 1992. Global changes during the last three million years: climatic controls and biotic responses. *Annual Review of Ecology and Systematics* 23:141–173.

Weigl, P. D., M. A. Steele, L. J. Sherman, J. C. Ha, and T. L. Sharpe. 1989. The ecology of the fox squirrel (*Sciurus niger*) in North Carolina: implications for survival in the Southeast. *Bulletin of Tall Timbers Research Station* No. 24.

Wells, O. O., G. L. Switzer, and R. C. Schmidtling. 1991. Geographic variation in Mississippi loblolly pine and sweetgum. *Silvae Genetica* 40:105–118.

Whitehead, D. R. 1965. Palynology and Pleistocene phytogeography of unglaciated eastern North America. Pages 417–432 in The Quaternary of the United States (H. E. Wright, Jr. and D. G. Frey, eds.). Princeton, New Jersey: Princeton Univ. Press.

3. Evolution, Taxonomy, and Morphology of the Red-cockaded Woodpecker

American Ornithologists' Union. 1957. Check-list of North American birds. 5th ed. Baltimore, Maryland: Port City Press.

———. 1976. Thirty-third supplement to the American Ornithologists' Union check-list of North American birds. *Auk* 93:875–879.

———. 1983. Check-list of North American birds. 6th ed. Lawrence, Kansas: American Ornithologists' Union, Allen Press.

Baird, S. F., T. M. Brewer, and R. Ridgway. 1874. A history of North American birds. Vol. 2. Boston: Little, Brown.

Bock, W. J., and W. D. Miller. 1959. The scansorial foot of the woodpeckers, with comments on the evolution of perching and climbing feet in birds. *American Museum Novitates* No. 1931.

Bond, J. 1971. Birds of the West Indies. 3rd ed. Boston: Houghton Mifflin.

Conner, R. N., D. Saenz, D. C. Rudolph, W. G. Ross, and D. L. Kulhavy. 1998. Red-cockaded Woodpecker nest-cavity selection: relationships with cavity age and resin production. *Auk* 115:447–454.

Delacour, J. 1951. The significance of the number of toes in some woodpeckers and kingfishers. *Auk* 68:49–51.

Dott, R. H., Jr., and R. L. Batten. 1971. Evolution of the earth. New York: McGraw-Hill.

Emerson, K. C., and J. C. Johnson. 1961. The genus *Penenirmus* (Mallophaga) found on North American woodpeckers. *Journal of the Kansas Entomological Society* 34:34–43.

Emlen, J. T. 1986. Responses of breeding Cliff Swallows to nidicolous parasite infestations. *Condor* 88:110–111.

Garrod, A. H. 1872. Note on some of the cranial peculiarities of the woodpecker. *Ibis*, 3rd Ser. 2:357–360.

Gold, C. S., and D. L. Dahlsten. 1983. Effects of parasitic flies (*Protocalliphora* spp.) on nestlings of Mountain and Chestnut-backed chickadees. *Wilson Bulletin* 95:560–572.

Goodwin, D. 1968. Notes on woodpeckers (Picidae). *Bulletin of the British Museum of Natural History, Zoology* 17:1–44.

Jackson, J. A. 1971. The evolution, taxonomy, distribution, past populations and current status of the Red-cockaded Woodpecker. Pages 4–26 *in* The ecology and management of the Red-cockaded Woodpecker (R. L. Thompson, ed.). Tallahassee, Florida: Bureau of Sport Fisheries and Wildlife, U.S. Dept. Interior, and Tall Timbers Research Station.

———. 1979. Age characteristics of Red-cockaded Woodpeckers. *Bird-Banding* 50:23–29.

———. 1993. Born of fire, under fire. *Birder's World* 7(6):12–16.

———. 1994. Red-cockaded Woodpecker (*Picoides borealis*). The birds of North America, No. 85 (A. Poole and F. Gill, eds.). Philadelphia: Academy of Natural Sciences; Washington, D.C.: American Ornithologists' Union.

LaBranche, M. S., and J. R. Walters. 1994. Patterns of mortality in nests of Red-cockaded Woodpeckers in the Sandhills of southcentral North Carolina. *Wilson Bulletin* 106:258–271.

Ligon, J. D. 1968. Sexual differences in foraging behavior in two species of *Dendrocopos* woodpeckers. *Auk* 85:203–215.

———. 1970. Behavior and breeding biology of the Red-cockaded Woodpecker. *Auk* 87:255–278.

———. 1971. Some factors influencing numbers of the Red-cockaded Woodpecker. Pages 30–43 *in* The ecology and management of the Red-cockaded Woodpecker (R. L. Thompson, ed.). Tallahassee, Florida: Bureau of Sport Fisheries and Wildlife, U.S. Dept. Interior, and Tall Timbers Research Station.

Lucas, F. A. 1895. The tongues of woodpeckers. Washington, D.C.: U.S. Dept. Agric., Ornithology and Mammalogy Div., Bull. No. 7.

Mengel, R. M., and J. A. Jackson. 1977. Geographic variation of the Red-cockaded Woodpecker. *Condor* 79:349–355.

Miller, A. H. 1955. A hybrid woodpecker and its significance in speciation in the genus *Dendrocopos*. *Evolution* 9:317–321.

Møller, A. P. 1990. Effects of parasitism by a haematophagous mite on reproduction in the Barn Swallow. *Ecology* 71:2345–2357.

Olson, S. L. 1985. The fossil record of birds. Pages 79–238 *in* Avian biology. Vol. 3 (D. S. Farner, J. R. King, and K. C. Parkes, eds.). New York: Academic Press.

Parker, W. K. 1879. On the morphology of the skull in the woodpeckers (Picidae) and wrynecks (Yungidae). *Transactions of the Linnean Society* 1:1–22.

Pence, D. B. 1972. *Picicnemidocoptes dryocopae* gen. et sp. n. (Acarina: Knemidokoptidae) from the Pileated Woodpecker, *Dryocopus pileatus* L., with a new record for *Knemidokoptes jamaicensis* Turk. *Journal of Parasitology* 58:339–342.

Peters, H. S. 1936. A list of external parasites from birds of the eastern part of the United States. *Bird-Banding* 7:9–27.

Pizzoni-Ardemani, A. 1990. Sexual dimorphism and geographic variation in the Red-cockaded Woodpecker (*Picoides borealis*). M.S. thesis, North Carolina State Univ., Raleigh.

Price, R. D., and K. C. Emerson. 1975. The *Menacanthus* (Mallophaga: Menoponidae) on the Piciformes (Aves). *Annals of the Entomological Society of America* 68:779–785.

Rendell, W. B., and N. A. M. Verbeek. 1996. Old nest material in nestboxes of Tree Swallows: effects on reproductive success. *Condor* 98:142–152.

Richardson, F. 1942. Adaptive modifications for tree-trunk foraging in birds. *University of California Publications in Zoology* 46:317–368.

Ridgway, R. 1914. The birds of North and Middle America. Part 4. *U.S. National Museum Bulletin* No. 50.

Short, L. L. 1970. Reversed sexual dimorphism in tail length and foraging differences in woodpeckers. *Bird-Banding* 41:85–92.

———. 1971. The systematics and behavior of some North American woodpeckers, genus *Picoides* (Aves). *Bulletin of the American Museum of Natural History* 145:1–118.

———. 1982. Woodpeckers of the world. Delaware Museum of Natural History Monogr. Ser. No. 4. Greenville: Delaware Museum of Natural History.

Sibley, C. G., and J. E. Ahlquist. 1990. Phylogeny and classification of birds. New Haven, Connecticut: Yale Univ. Press.

Sibley, C. G., and B. L. Monroe, Jr. 1990. Distribution and taxonomy of birds of the world. New Haven, Connecticut: Yale Univ. Press.

Spring, L. W. 1965. Climbing and pecking adaptations in some North American woodpeckers. *Condor* 67:457–488.

Stresemann, E., and V. Stresemann. 1966. Die Mauser der Vögel. *Journal für Ornithologie* No. 107. *Sonderheft.*

Todd, W. E. C. 1946. Critical notes on the woodpeckers. *Annals of the Carnegie Museum* 30:297–317.

Vieillot, L. J. P. 1807. Histoire naturelle des oiseaux de l'Amérique septentrionale, vol. 2. Paris: Chez Desray.

Voous, K. H., Jr. 1947. On the history of the distribution of the genus *Dendrocopos*. *Limosa* 20:1–142.

Wagler, J. G. 1827. Systema avium. Pars prima. Stuttgart: J. G. Cottae.

Wetmore, A. 1941. Notes on the birds of North Carolina. *Proceedings of the U.S. National Museum* 90:483–530.

Wilson, A. 1810. American ornithology, vol. 2. Philadelphia: Bradford and Inskeep.

Wilson, N., and E. L. Bull. 1977. Ectoparasites found in the nest cavities of Pileated Woodpeckers in Oregon. *Bird-Banding* 48:171–173.

Woolfenden, G. E. 1959. A Pleistocene avifauna from Rock Spring, Florida. *Wilson Bulletin* 71:183–187.

4. Red-cockaded Woodpecker Distribution

Audubon, J. J. [1839] 1949. Ornithological biography. Vol. 5. Edinburgh: Adam and Charles Black.

Baird, S. F., T. M. Brewer, and R. Ridgway. 1874. A history of North American birds. Vol. 2. Boston: Little, Brown.

Baker, W. W. 1971. Observations of the food habits of the Red-cockaded Woodpecker. Pages 100–107 *in* The ecology and management of the Red-cockaded Woodpecker (R. L. Thompson, ed.). Tallahassee, Florida: Bureau of Sport Fisheries and Wildlife, U.S. Dept. Interior, and Tall Timbers Research Station.

Bartram, W. [1791] 1988. Travels through North and South Carolina, Georgia, east and west Florida, the Cherokee country, the extensive territories of the Muscogulies, or Creek Confederacy, and the country of the Chactaws. New York: Penguin Books.

Beal, F. E. L. 1911. Food of the woodpeckers of the United States. U.S. Dept. Agric., Biol. Survey Bull. No. 37.

Beal, F. E. L., W. L. McAtee, and E. R. Kalmbach. 1916. Common birds of the southeastern United States in relation to agriculture. U.S. Dept. Agric., Farmer's Bull. 755.

Bent, A. C. 1939. Life histories of North American woodpeckers. *U.S. National Museum Bulletin* No. 174.

Conner, R. N., and D. C. Rudolph. 1989. Red-cockaded Woodpecker colony status and trends on the Angelina, Davy Crockett and Sabine national forests. U.S. Dept. Agric., For. Serv. Res. Pap. SO-250.

———. 1991. Effects of midstory reduction and thinning in Red-cockaded Woodpecker cavity tree clusters. *Wildlife Society Bulletin* 19:63–66.

Costa, R., and R. E. F. Escaño. 1989. Red-cockaded Woodpecker: status and management in the southern region in 1986. Atlanta, Georgia: U.S. Dept. Agric., For. Serv. Tech. Publ. R8-TP 12.

Engstrom, R. T., and F. J. Sanders. 1997. Red-cockaded Woodpecker foraging ecology in an old-growth longleaf pine forest. *Wilson Bulletin* 109:203–217.

Hooper, R. G., and M. R. Lennartz. 1995. Short-term response of a high density Red-cockaded Woodpecker population to loss of foraging habitat. Pages 283–289 *in* Red-cockaded Woodpecker: recovery, ecology and management (D. L. Kulhavy, R. G. Hooper, and R. Costa, eds.). Nacogdoches, Texas: Center for Applied Studies, College of Forestry, Stephen F. Austin State Univ.

Jackson, J. A. 1971. The evolution, taxonomy, distribution, past populations and current status of the Red-cockaded Woodpecker. Pages 4–29 *in* The ecology and management of the Red-cockaded Woodpecker (R. L. Thompson, ed.). Tallahassee, Florida: Bureau of Sport Fisheries and Wildlife, U.S. Dept. Interior, and Tall Timbers Research Station.

———. 1978. Analysis of the distribution and population status of the Red-cockaded Woodpecker. Pages 101–110 *in* Proceedings of rare and endangered wildlife symposium (R. R. Odum and L. Landers, eds.). Georgia Dept. Natural Resources, Game and Fish Div. Tech. Bull. WL 4.

James, F. C. 1995. The status of the Red-cockaded Woodpecker in 1990 and the prospect for recovery. Pages 439–451 *in* Red-cockaded Woodpecker: recovery, ecology and management (D. L. Kulhavy, R. G. Hooper, and R. Costa, eds.). Nacogdoches, Texas: Center for Applied Studies, College of Forestry, Stephen F. Austin State Univ.

Lennartz, M. R., P. H. Geissler, R. F. Harlow, R. C. Long, K. M. Chitwood, and J. A. Jackson. 1983. Status of the Red-cockaded Woodpecker populations on federal lands in the South. Pages 7–12 *in* Red-cockaded Woodpecker symposium II proceedings (D. A. Wood, ed.). Tallahassee: State of Florida Game and Fresh Water Fish Comm.

Lennartz, M. R., H. A. Knight, J. P. McClure, and V. A. Rudis. 1983. Status of Red-cockaded Woodpecker nesting habitat in the South. Pages 13–19 *in* Red-cockaded Woodpecker symposium II proceedings (D. A. Wood, ed.). Tallahassee: State of Florida Game and Fresh Water Fish Comm.

Maxwell, R. S., and R. D. Baker. 1983. Sawdust empire: the Texas lumber industry, 1830–1940. College Station: Texas A&M Univ. Press.

Oberholser, H. C. 1938. The bird life of Louisiana. New Orleans: State of Louisiana Dept. Conservation.

Platt, W. J., G. W. Evans, and S. L. Rathbun. 1988. The population dynamics of a long-lived conifer (*Pinus palustris*). *American Naturalist* 131:491–525.

Platt, W. J., J. S. Glitzenstein, and K. R. Streng. 1989. Evaluating pyrogenicity and its effects on vegetation in longleaf pine savannas. *Proceedings of Tall Timbers Fire Ecology Conference* 17:143–191.

Schaefer, R. R. 1996. Red-cockaded Woodpecker reproduction and provisioning of nestlings in relation to habitat. M.S. thesis, Stephen F. Austin State Univ., Nacogdoches, Texas.

Schorger, A. W. 1955. The Passenger Pigeon, its natural history and extinction. Madison: Univ. Wisconsin Press.

Short, L. L. 1982. Woodpeckers of the world. Delaware Museum of Natural History Monogr. Ser. No. 4. Greenville: Delaware Museum of Natural History.

Tanner, J. T. [1942] 1966. The Ivory-billed Woodpecker. National Audubon Society Research Rep. No. 1. New York: Dover Publications.

Vieillot, L. J. P. 1807. Histoire naturelle des oiseaux de l'Amérique septentrionale. 2:66. Paris: Chez Desray.

Wahlenberg, W. G. 1946. Longleaf pine: its use, ecology, regeneration, protection, growth, and management. Washington, D.C.: Charles Lathrop Pack Forestry Foundation.

Webb, T., III. 1986. Vegetational change in eastern North America from 18,000 to 500 Yr B.P. Pages 63–69 *in* Climate-vegetation interactions: proceedings of a workshop (C. Rosenzweig and R. Dickinson, eds.). Rep. OIES-2. Boulder, Colorado: Office for Interdisciplinary Earth Studies and Univ. Corporation for Atmospheric Research.

Wilson, A., and C. L. Bonaparte. 1831. American ornithology, or, The natural history of the birds of the United States. Vol. 1. Edinburgh: Constable and Company.

Wood, D. A. 1983. Foraging and colony habitat characteristics of the Red-cockaded Woodpecker in Oklahoma. Pages 51–58 *in* Red-cockaded Woodpecker symposium II proceedings (D. A. Wood, ed.). Tallahassee: State of Florida Game and Fresh Water Fish Comm.

———. 1983. Observations on the behavior and breeding biology of the Red-cockaded Woodpecker in Oklahoma. Pages 92–94 *in* Red-cockaded Woodpecker symposium II proceedings (D. A. Wood, ed.). Tallahassee: State of Florida Game and Fresh Water Fish Comm.

Woodruff, E. S. 1908. A preliminary list of the birds of Shannon and Carter Counties, Missouri. *Auk* 25:191–214.

5. Cavity Trees in Fire-Maintained Southern Pine Ecosystems

Bendel, P. R., and J. E. Gates. 1987. Home range and microhabitat partitioning of the southern flying squirrel *Glaucomys volans. Journal of Mammalogy* 68:243–255.

Chapman, F. M. 1909. Nesting of the Red-cockaded Woodpecker. Pages 265–266 *in* Bird lore (M. O. Wright and W. Dutcher, eds.). Philadelphia: Macmillan.

Conner, R. N. 1989. Injection of 2,4-D to remove hardwood midstory within Red-cockaded Woodpecker colony areas. U.S. Dept. Agric., For. Serv. Res. Pap. SO-251.

Conner, R. N., and B. A. Locke. 1979. Effects of a prescribed burn on cavity trees of Red-cockaded Woodpeckers. *Wildlife Society Bulletin* 7:291–293.

Conner, R. N., and K. A. O'Halloran. 1987. Cavity-tree selection by Red-cockaded Woodpeckers as related to growth dynamics of southern pines. *Wilson Bulletin* 99:398–412.

Conner, R. N., and D. C. Rudolph. 1989. Red-cockaded Woodpecker colony status and trends on the Angelina, Davy Crockett and Sabine national forests. U.S. Dept. Agric., For. Serv. Res. Pap. SO-250.

———. 1991. Effects of midstory reduction and thinning in Red-cockaded Woodpecker cavity tree clusters. *Wildlife Society Bulletin* 19:63–66.

———. 1995. Excavation dynamics and use patterns of Red-cockaded Woodpecker cavities: relationships with cooperative breeding. Pages 343–352 *in* Red-cockaded Woodpecker: recovery, ecology and management (D. L. Kulhavy, R. G. Hooper, and R. Costa, eds.). Nacogdoches, Texas: Center for Applied Studies, College of Forestry, Stephen F. Austin State Univ.

———. 1995. Wind damage to Red-cockaded Woodpecker cavity trees on eastern Texas national forests. Pages 183–190 *in* Red-cockaded Woodpecker: recovery, ecology and management (D. L. Kulhavy, R. G. Hooper, and R. Costa, eds.). Nacogdoches, Texas: Center for Applied Studies, College of Forestry, Stephen F. Austin State Univ.

Conner, R. N., D. C. Rudolph, D. L. Kulhavy, and A. E. Snow. 1991. Causes of mortality of Red-cockaded Woodpecker cavity trees. *Journal of Wildlife Management* 55:531–537.

Conner, R. N., D. C. Rudolph, D. Saenz, and R. R. Schaefer. 1994. Heartwood, sapwood, and fungal decay associated with Red-cockaded Woodpecker cavity trees. *Journal of Wildlife Management* 58:728–734.

Conner, R. N., D. Saenz, D. C. Rudolph, W. G. Ross, and D. L. Kulhavy. 1998. Red-cockaded Woodpecker nest-cavity selection: relationships with cavity age and resin production. *Auk* 115:447–454.

Conner, R. N., A. E. Snow, and K. A. O'Halloran. 1991. Red-cockaded Woodpecker use of seed-tree/shelterwood cuts in eastern Texas. *Wildlife Society Bulletin* 19:67–73.

DeLotelle, R. S., and R. J. Epting. 1988. Selection of old trees for cavity excavation by Red-cockaded Woodpeckers. *Wildlife Society Bulletin* 16:48–52.

Dennis, J. V. 1971. Species using Red-cockaded Woodpecker holes in northeastern South Carolina. *Bird-Banding* 42:79–87.

Engstrom, R. T., and G. W. Evans. 1990. Hurricane damage to Red-cockaded Woodpecker *Picoides borealis* cavity trees. *Auk* 107:608–610.

Field, R., and B. K. Williams. 1985. Age of cavity trees and colony stands selected by Red-cockaded Woodpeckers. *Wildlife Society Bulletin* 13:92–96.

Harding, S. R. 1997. The dynamics of cavity excavation and use by the Red-cockaded Woodpecker (*Picoides borealis*). M.S. thesis, Virginia Polytechnic Institute and State Univ., Blacksburg.

Harlow, R. F., and A. T. Doyle. 1990. Food habits of southern flying squirrels *Glaucomys volans* collected from Red-cockaded Woodpecker *Picoides borealis* colonies in South Carolina. *American Midland Naturalist* 124:187–191.

Harlow, R. F., and M. R. Lennartz. 1983. Interspecific competition for Red-cockaded Woodpecker cavities during the nesting season in South Carolina. Pages 41–43 *in* Red-cockaded Woodpecker symposium II proceedings (D. A. Wood, ed.). Tallahassee: State of Florida Game and Fresh Water Fish Comm.

Hooper, R. G. 1982. Use of dead cavity trees by Red-cockaded Woodpeckers. *Wildlife Society Bulletin* 10:163–164.

———. 1988. Longleaf pines used for cavities by Red-cockaded Woodpeckers. *Journal of Wildlife Management* 52:392–398.

———. 1995. Short-term response of a high density population of Red-cockaded Woodpeckers to loss of foraging habitat. Pages 283–289 *in* Red-cockaded Woodpecker: recovery, ecology and management (D. L. Kulhavy, R. G. Hooper, and R. Costa, eds.). Nacogdoches, Texas: Center for Applied Studies, College of Forestry, Stephen F. Austin State Univ.

Hooper, R. G., D. L. Krusac, and D. L. Carlson. 1991. An increase in a population of Red-cockaded Woodpeckers. *Wildlife Society Bulletin* 19:277–286.

Hooper, R. G., and M. R. Lennartz. 1983. Roosting behavior of Red-cockaded Woodpecker clans with insufficient cavities. *Journal of Field Ornithology* 54:72–76.

Hooper, R. G., M. R. Lennartz, and H. D. Muse. 1991. Heart rot and cavity tree selection by Red-cockaded Woodpeckers. *Journal of Wildlife Management* 55:323–327.

Hooper, R. G., and C. J. McAdie. 1995. Hurricanes as a factor in the long-term management of Red-cockaded Woodpeckers. Pages 148–166 *in* Red-cockaded Woodpecker: recovery, ecology and management (D. L. Kulhavy, R. G. Hooper, and R. Costa, eds.). Nacogdoches, Texas: Center for Applied Studies, College of Forestry, Stephen F. Austin State Univ.

Hooper, R. G., A. F. Robinson, Jr., and J. A. Jackson. 1980. The Red-cockaded Woodpecker: notes on life history and management. U.S. Dept. Agric., For. Serv. Gen. Rep. SA-GR9.

Jackson, J. A. 1974. Gray rat snakes versus Red-cockaded Woodpecker: predator-prey adaptations. *Auk* 91:342–347.

———. 1976. Relative climbing tendencies of gray *Elaphe obsoleta spiloides* and black rat snakes, *E. o. obsoleta. Herpetologica* 32:359–361.

———. 1977. Red-cockaded Woodpeckers and pine red heart disease. *Auk* 94:160–163.

———. 1978. Competition for cavities and Red-cockaded Woodpecker management. Pages 103–112 *in* Endangered birds, management techniques for preserving threatened species (S. A. Temple, ed.). Madison: Univ. of Wisconsin Press.

———. 1978. Predation by a gray rat snake on Red-cockaded Woodpecker nestlings. *Bird-Banding* 49:187–188.

Jackson, J. A., M. R. Lennartz, and R. G. Hooper. 1979. Tree age and cavity initiation by Red-cockaded Woodpeckers. *Journal of Forestry* 77:102–103.

Jones, H. K., Jr., and F. T. Ott. 1973. Some characteristics of Red-cockaded Woodpecker cavity trees in Georgia. *Oriole* 38:33–39.

Kalisz, P. J., and S. E. Boettcher. 1991. Active and abandoned Red-cockaded Woodpecker habitat in Kentucky. *Journal of Wildlife Management* 55:146–154.

Kappes, J. J., Jr. 1997. Defining cavity-associated interactions between Red-cockaded Woodpeckers and other cavity-dependent species: interspecific competition or cavity kleptoparasitism? *Auk* 114:778–780.

Laves, K. S. 1996. Effects of southern flying squirrels, *Glaucomys volans,* on Red-cockaded Woodpecker, *Picoides borealis,* reproductive success. M.S. thesis, Clemson Univ., Clemson, South Carolina.

Lay, D. W., and D. N. Russell. 1970. Notes on the Red-cockaded Woodpecker in Texas. *Auk* 87:781–786.

Ligon, J. D. 1970. Behavior and breeding biology of the Red-cockaded Woodpecker. *Auk* 87:255–278.

Locke, B. A., R. N. Conner, and J. C. Kroll. 1983. Factors influencing colony site selection by Red-cockaded Woodpecker. Pages 46–50 *in* Red-cockaded Woodpecker symposium II proceedings (D. A. Wood, ed.). Tallahassee: State of Florida Game and Fresh Water Fish Comm.

Loeb, S. C., and R. G. Hooper. 1997. An experimental test of interspecific competition for Red-cockaded Woodpecker cavities. *Journal of Wildlife Management* 61:1268–1280.

Mitchell, L. R., D. C. Carlile, and C. R. Chandler. 1999. Effects of southern flying squirrels on nest success of Red-cockaded Woodpeckers. *Journal of Wildlife Management* 63:538–545.

Neal, J. C., W. G. Montague, and D. A. James. 1993. Climbing by black rat snakes on cavity trees of Red-cockaded Woodpeckers. *Wildlife Society Bulletin* 21:160–165.

Ross, W. G., D. L. Kulhavy, and R. N. Conner. 1995. Vulnerability and resistance of Red-cockaded Woodpecker cavity trees to southern pine beetles in Texas. Pages 410–414 *in* Red-cockaded Woodpecker: recovery, ecology and management (D. L. Kulhavy, R. G. Hooper, and R. Costa, eds.). Nacogdoches, Texas: Center for Applied Studies, College of Forestry, Stephen F. Austin State Univ.

———. 1997. Stand conditions and tree characteristics affect quality of longleaf pine for Red-cockaded Woodpecker cavity trees. *Forest Ecology and Management* 91:145–154.

Rudolph, D. C., and R. N. Conner. 1991. Cavity tree selection by Red-cockaded Woodpeckers in relation to tree age. *Wilson Bulletin* 103:458–467.

Rudolph, D. C., R. N. Conner, and R. R. Schaefer. 1995. Red-cockaded Woodpecker detection of red heart infection. Pages 338–342 *in* Red-cockaded Woodpecker: recovery, ecology and management (D. L. Kulhavy, R. G. Hooper, and R. Costa, eds.). Nacogdoches, Texas: Center for Applied Studies, College of Forestry, Stephen F. Austin State Univ.

Rudolph, D. C., R. N. Conner, and J. Turner. 1990. Competition for Red-cockaded Woodpecker roost and nest cavities: effects of resin age and entrance diameter. *Wilson Bulletin* 102:23–26.

Rudolph, D. C., H. Kyle, and R. N. Conner. 1990. Red-cockaded Woodpeckers vs. rat snakes: the effectiveness of the resin barrier. *Wilson Bulletin* 102:14–22.

Saenz, D., R. N. Conner, C. E. Shackelford, and D. C. Rudolph. 1998. Pileated Woodpecker damage to Red-cockaded Woodpecker cavity trees in eastern Texas. *Wilson Bulletin* 110:362–367.

Stabb, M. A., M. E. Gartshore, and P. L. Aird. 1989. Interactions of southern flying squirrels, *Glaucomys volans*, and cavity-nesting birds. *Canadian Field Naturalist* 103:401–402.

Steirly, C. C. 1949. A note on the Red-cockaded Woodpecker. *Raven* 20:6–7.

———. 1950. Nest cavities of the Red-cockaded Woodpecker. *Raven* 21:2–3.

———. 1957. Nesting ecology of the Red-cockaded Woodpecker in Virginia. *Raven* 28:24–36. Also published in *Atlantic Naturalist* 12(1957):280–292.

———. 1966. Measurements of the nest cavity of the Red-cockaded Woodpecker. *Raven* 37:64.

Teitelbaum, R. D., and W. P. Smith. 1985. Cavity-site characteristics of the Red-cockaded Woodpecker in Fontainebleau State Park, Louisiana. *Proceedings of the Louisiana Academy of Science* 48:116–122.

Thompson, R. L., ed. 1971. The ecology and management of the Red-cockaded Woodpecker. Tallahassee, Florida: Bureau of Sport Fisheries and Wildlife, U.S. Dept. Interior, and Tall Timbers Research Station.

U.S. Fish and Wildlife Service. 1979. Red-cockaded Woodpecker recovery plan. Atlanta, Georgia: U.S. Fish and Wildl. Serv.

———. 1985. Red-cockaded Woodpecker recovery plan. Atlanta, Georgia: U.S. Fish and Wildl. Serv.

Williams, B. K., and R. Field. 1986. Response: evidence for selection of old trees by Red-cockaded Woodpeckers. *Wildlife Society Bulletin* 14:322–325.

Wood, D. A., ed. 1983. Red-cockaded Woodpecker symposium II proceedings. Tallahassee: State of Florida Game and Fresh Water Fish Comm.

Zwicker, S. M. 1995. Selection of pines for foraging and cavity excavation by Red-cockaded Woodpeckers. M.S. thesis, North Carolina State Univ., Raleigh.

6. Social Behavior and Population Biology

Brown, J. L. 1987. Helping and communal breeding in birds: ecology and evolution. Princeton, New Jersey: Princeton Univ. Press.

Choudhury, S. 1995. Divorce in birds: a review of the hypotheses. *Animal Behavior* 50:413–429.

Copeyon, C. K. 1990. A technique for constructing cavities for the Red-cockaded Woodpecker. *Wildlife Society Bulletin* 18:303–311.

Daniels, S. J. 1997. Female dispersal and inbreeding in the Red-cockaded Woodpecker. M.S. thesis, Virginia Polytechnic Institute and State Univ., Blacksburg.

Daniels, S. J., J. A. Priddy, and J. R. Walters. 2000. Inbreeding in small populations of Red-cockaded Woodpeckers: insights from a spatially-explicit simulation model. Pages 129–147 *in* Genetics, demography and viability of fragmented populations (A. G. Young and G. M. Clarke, eds.).

Daniels, S. J., and J. R. Walters. 2000. Inbreeding depression and its effects on natal dispersal in Red-cockaded Woodpeckers. *Condor* 102:482–491.

———. 2000. Between-year breeding dispersal in Red-cockaded Woodpeckers: multiple causes and estimated cost. *Ecology* 81:2473–2484.

DeLotelle, R. S., and R. J. Epting. 1992. Reproduction of the Red-cockaded Woodpecker in central Florida. *Wilson Bulletin* 104:285–294.

DeLotelle, R. S., R. J. Epting, and G. Demuth. 1995. A 12-year study of Red-cockaded Woodpeckers in central Florida. Pages 259–269 *in* Red-cockaded Woodpecker: recovery, ecology and management (D. L. Kulhavy, R. G. Hooper, and R. Costa, eds.). Nacogdoches, Texas: Center for Applied Studies, College of Forestry, Stephen F. Austin State Univ.

Emlen, S. T. 1982. The evolution of helping II: the role of behavioral conflict. *American Naturalist* 119:40–53.

———. 1991. Evolution of cooperative breeding in birds and mammals. Pages 301–337 *in* Behavioral ecology: an evolutionary approach. 3rd ed. (J. R. Krebs and N. B. Davies, eds.). Oxford: Blackwell Scientific Publications.

Emlen, S. T., J. M. Emlen, and S. A. Levin. 1986. Sex-ratio selection in species with helpers-at-the-nest. *American Naturalist* 127:1–8.

Epting, R. J., and R. S. DeLotelle. Adaptive sex ratios of fledgling Red-cockaded Woodpeckers in central Florida. *Wilson Bulletin.* (unpublished manuscript)

Fisher, R. A. 1930. The genetical theory of natural selection. Oxford: Oxford University Press.

Fitzpatrick, J. W., and G. E. Woolfenden. 1989. Florida Scrub Jay. Pages 201–218 *in* Lifetime reproduction in birds (I. Newton, ed.). New York: Academic Press.

Gowaty, P. A., and M. R. Lennartz. 1985. Sex ratios of nestling and fledgling Red-cockaded Woodpeckers (*Picoides borealis*) favor males. *American Naturalist* 126:347–353.

Haig, S. M., J. R. Belthoff, and D. H. Allen. 1993. Examination of population structure in Red-cockaded Woodpeckers using DNA profiles. *Evolution* 47:185–194.

Haig, S. M., R. Bowman, and T. D. Mullins. 1996. Population structure of Red-cockaded Woodpeckers in South Florida: RAPDs revisited. *Molecular Ecology* 5:725–734.

Haig, S. M., J. M. Rhymer, and D. G. Heckel. 1994. Population differentiation in randomly amplified polymorphic DNA of Red-cockaded Woodpeckers, *Picoides borealis*. *Molecular Ecology* 3:581–595.

Haig, S. M., J. R. Walters, and J. H. Plissner. 1994. Genetic evidence for monogamy in the cooperatively breeding Red-cockaded Woodpecker. *Behavioral Ecology and Sociobiology* 34:295–303.

Hannon, S. J., R. L. Mumme, W. D. Koenig, and F. A. Pitelka. 1985. Replacement of breeders and within-group conflict in the cooperatively breeding Acorn Woodpecker. *Behavioral Ecology and Sociobiology* 17:303–312.

Hardesty, J. L., K. E. Gault, and H. F. Percival. 1997. Trends, status and aspects of demography of the Red-cockaded Woodpecker (*Picoides borealis*) in the Sandhills of Florida's panhandle. Part II. Final report, Research Work Order 146. Gainesville: Florida Cooperative Fish and Wildlife Research Unit, University of Florida.

Heppell, S. S., J. R. Walters, and L. B. Crowder. 1994. Evaluating management alternatives for Red-cockaded Woodpeckers: a modeling approach. *Journal of Wildlife Management* 58:479–487.

Hooper, R. G. 1983. Colony formation by Red-cockaded Woodpeckers: hypotheses and management implications. Pages 72–77 in Red-cockaded Woodpecker symposium II proceedings (D. A. Wood, ed.). Tallahassee: State of Florida Game and Fresh Water Fish Comm.

Hooper, R. G., D. L. Krusac, and D. L. Carlson. 1991. An increase in a population of Red-cockaded Woodpeckers. *Wildlife Society Bulletin* 19:277–286.

Hooper, R. G., and C. J. McAdie. 1995. Hurricanes and the long-term management of the Red-cockaded Woodpecker. Pages 148–166 in Red-cockaded Woodpecker: recovery, ecology and management (D. L. Kulhavy, R. G. Hooper, and R. Costa, eds.). Nacogdoches, Texas: Center for Applied Studies, College of Forestry, Stephen F. Austin State Univ.

Hooper, R. G., J. C. Watson, and R. E. F. Escaño. 1990. Hurricane Hugo's initial effects on Red-cockaded Woodpeckers in the Francis Marion National Forest. *North American Wildlife and Natural Resources Conference* 55:220–224.

Jackson, J. A. 1994. Red-cockaded Woodpecker (*Picoides borealis*). The birds of North America, No. 85 (A. Poole and F. Gill, eds.). Philadelphia: Academy of Natural Sciences; Washington, D.C.: American Ornithologists' Union.

James, F. C. 1991. Signs of trouble in the largest remaining population of Red-cockaded Woodpeckers. *Auk* 108:419–423.

Jamieson, I. G., and J. L. Craig. 1987. Critique of helping behavior in birds: a departure from functional explanations. Pages 79–98 in Perspectives in ethology. Vol. 7 (P. Bateson and P. Klopfer, eds.). New York: Plenum Press.

Khan, M. Z., and J. R. Walters. 1997. Is helping a beneficial learning experience for Red-cockaded Woodpecker (*Picoides borealis*) helpers? *Behavioral Ecology and Sociobiology* 41:69–73.

———. 2000. An analysis of reciprocal exchange of helping behavior in the Red-cockaded Woodpecker. *Behavioral Ecology and Sociobiology* 47:376–381.

———. Effects of helpers on breeder survival in the Red-cockaded Woodpecker (*Picoides borealis*). *Behavioral Ecology and Sociobiology*. (in press)

Koenig, W. D., W. J. Carmen, R. L. Mumme, and M. T. Stanback. 1992. The evolution of delayed dispersal in cooperative breeders. *Quarterly Review of Biology* 67:111–150.

Koenig, W. D., and R. L. Mumme. 1987. Population ecology of the cooperatively breeding Acorn Woodpecker. Princeton, New Jersey: Princeton Univ. Press.

———. 1990. Levels of analysis and the functional significance of helping behavior. Pages 269–303 in Interpretation and explanation in the study of animal behavior (M. Bekoff and D. Jamieson, eds.). Boulder, Colorado: Westview Press.

Koenig, W. D., and J. R. Walters. 1999. Sex-ratio selection in species with helpers at the nest: the repayment model revisited. American Naturalist 153:124–130.

LaBranche, M. S. 1988. Reproductive ecology of the Red-cockaded Woodpecker in the Sandhills of North Carolina. M.S. thesis, North Carolina State Univ., Raleigh.

———. 1992. Asynchronous hatching, brood reduction and sex ratio biases in Red-cockaded Woodpeckers. Ph.D. dissertation, North Carolina State Univ., Raleigh.

LaBranche, M. S., and J. R. Walters. 1994. Patterns of mortality in nests of Red-cockaded Woodpeckers in the Sandhills of southcentral North Carolina. Wilson Bulletin 106:258–271.

LaBranche, M. S., J. R. Walters, and K. S. Laves. 1994. Double brooding in Red-cockaded Woodpeckers. Wilson Bulletin 106:403–408.

Lande, R. 1988. Genetics and demography in biological conservation. Science 241: 1455–1460.

———. 1995. Mutation and conservation. Conservation Biology 9:782–791.

Lennartz, M., and R. F. Harlow. 1979. The role of parent and helper Red-cockaded Woodpeckers at the nest. Wilson Bulletin 91:331–335.

Lennartz, M. R., and D. G. Heckel. 1987. Population dynamics of a Red-cockaded Woodpecker population in Georgia Piedmont loblolly pine habitat. Pages 48–55 in Proceedings of the third southeastern nongame and endangered wildlife symposium (R. R. Odom, K. A. Riddleberger, and J. C. Ozier, eds.). Columbia: South Carolina Wildlife and Marine Resources Department.

Lennartz, M., R. G. Hooper, and R. F. Harlow. 1987. Sociality and cooperative breeding of Red-cockaded Woodpeckers, Picoides borealis. Behavioral Ecology and Sociobiology 20:77–88.

Letcher, B. H., J. A. Priddy, J. R. Walters, and L. B. Crowder. 1998. An individual-based, spatially-explicit simulation model of the population dynamics of the endangered Red-cockaded Woodpecker, Picoides borealis. Biological Conservation 86:1–14.

Ligon, J. D. 1970. Behavior and breeding biology of the Red-cockaded Woodpecker. Auk 87:255–278.

Maguire, L. A., G. F. Wilhere, and Q. Dong. 1995. Population viability analysis for Red-cockaded Woodpeckers in the Georgia Piedmont. Journal of Wildlife Management 59:533–542.

Martin, T. E., and P. Li. 1992. Life history traits of open versus cavity-nesting birds. Ecology 73:579–592.

McGowan, K. J., and G. E. Woolfenden. 1989. A sentinel system in the Florida Scrub Jay. Animal Behavior 37:1000–1006.

Neal, J. C., D. A. James, W. G. Montague, and J. E. Johnson. 1993. Effects of weather and helpers on survival of nestling Red-cockaded Woodpeckers. Wilson Bulletin 105:666–673.

Phillips, L. F., Jr., J. Tomcho, Jr., and J. R. Walters. 1998. Red-cockaded Woodpeckers in northwestern Florida produce a second brood. Florida Field Naturalist 26:109–113.

Reed, J. M., P. D. Doerr, and J. R. Walters. 1988. Minimum viable population size of the Red-cockaded Woodpecker. Journal of Wildlife Management 52:385–391.

Reed, J. M., J. R. Walters, T. E. Emigh, and D. E. Seaman. 1993. Effective population size in Red-cockaded Woodpeckers: population and model differences. *Conservation Biology* 7:302–308.

Rossell, C. R., Jr., and J. J. Britcher. 1994. Evidence of plural breeding by Red-cockaded Woodpeckers. *Wilson Bulletin* 106:557–559.

Shaffer, M. L. 1981. Minimum population sizes for species conservation. *BioScience* 31:131–134.

Short, L. L. 1982. Woodpeckers of the world. Delaware Museum Natural History Monograph Ser. No. 4. Greenville: Delaware Museum of Natural History.

Stacey, P. B., and J. D. Ligon. 1991. The benefits of philopatry hypothesis for the evolution of cooperative breeding: variation in territory quality and group size effects. *American Naturalist* 137:831–846.

Stangel, P. W., and P. M. Dixon. 1995. Associations between fluctuating asymmetry and heterozygosity in the Red-cockaded Woodpecker. Pages 239–247 *in* Red-cockaded Woodpecker: recovery, ecology and management (D. L. Kulhavy, R. G. Hooper, and R. Costa, eds.). Nacogdoches, Texas: Center for Applied Studies, College of Forestry, Stephen F. Austin State Univ.

Stangel, P. W., M. R. Lennartz, and M. H. Smith. 1992. Genetic variation and population structure of Red-cockaded Woodpeckers. *Conservation Biology* 6:283–292.

Stoleson, S. H., and S. R. Beissinger. 1995. Hatching asynchrony and the onset of incubation in birds, revisited? When is the critical period? *Current Ornithology* 12:191–270.

Walters, J. R. 1990. Red-cockaded Woodpeckers: a 'primitive' cooperative breeder. Pages 67–101 *in* Cooperative breeding in birds: long-term studies of ecology and behavior (P. B. Stacey and W. D. Koenig, eds.). Cambridge, England: Cambridge Univ. Press.

———. 1991. Application of ecological principles to the management of endangered species: the case of the Red-cockaded Woodpecker. *Annual Review of Ecology and Systematics* 22:505–523.

Walters, J. R., J. H. Carter III, P. D. Doerr, and C. K. Copeyon. 1995. Response to drilled artificial cavities by Red-cockaded Woodpeckers in the North Carolina Sandhills: 4-year assessment. Pages 380–384 *in* Red-cockaded Woodpecker: recovery, ecology and management (D. L. Kulhavy, R. G. Hooper, and R. Costa, eds.). Nacogdoches, Texas: Center for Applied Studies, College of Forestry, Stephen F. Austin State Univ.

Walters, J. R., C. K. Copeyon, and J. H. Carter III. 1992. Test of the ecological basis of cooperative breeding in Red-cockaded Woodpeckers. *Auk* 109:90–97.

Walters, J. R., L. B. Crowder, and J. A. Priddy. Population viability analysis for Red-cockaded Woodpeckers using an individual-based model. *Ecological Applications*. (in press)

Walters, J. R., P. D. Doerr, and J. H. Carter III. 1988. The cooperative breeding system of the Red-cockaded Woodpecker. *Ethology* 78:275–305.

———. 1992. Delayed dispersal and reproduction as a life history tactic in cooperative breeders: fitness calculations from Red-cockaded Woodpeckers. *American Naturalist* 139:623–643.

Watson, J. C., R. G. Hooper, D. L. Carlson, W. E. Taylor, and T. E. Milling. 1995. Restoration of the Red-cockaded Woodpecker population on the Francis Marion National Forest: three years post Hugo. Pages 172–182 *in* Red-cockaded Woodpecker: recovery, ecology and management (D. L. Kulhavy, R. G. Hooper, and R. Costa, eds.). Nacogdoches, Texas: Center for Applied Studies, College of Forestry, Stephen F. Austin State Univ.

Winkler, H., and L. L. Short. 1978. A comparative analysis of acoustical signals in pied woodpeckers (Aves, *Picoides*). *Bulletin of the American Museum of Natural History* 160:1–110.

Woolfenden, G. E., and J. W. Fitzpatrick. 1984. The Florida Scrub Jay: demography of a cooperative-breeding bird. Princeton, New Jersey: Princeton Univ. Press.

———. 1990. Florida Scrub Jays: a synopsis after 18 years of study. Pages 240–266 *in* Cooperative breeding in birds: long-term studies of ecology and behavior (P. B. Stacey and W. D. Koenig, eds.). Cambridge, England: Cambridge Univ. Press.

7. Foraging Ecology of Red-cockaded Woodpeckers

Baker, W. W. 1971. Observations of the food habits of the Red-cockaded Woodpecker. Pages 100–107 *in* The ecology and management of the Red-cockaded Woodpecker (R. L. Thompson, ed.). Tallahassee, Florida: Bureau of Sport Fisheries and Wildlife, U.S. Dept. Interior, and Tall Timbers Research Station.

Beal, F. E. L., W. L. McAtee, and E. R. Kalmbach. 1916. Common birds of the southeastern United States in relation to agriculture. U.S. Dept. Agric., Farmer's Bull. 755.

Beckett, T. 1971. A summary of Red-cockaded Woodpecker observations in South Carolina. Pages 87–95 *in* The ecology and management of the Red-cockaded Woodpecker (R. L. Thompson, ed.). Tallahassee, Florida: Bureau of Sport Fisheries and Wildlife, U.S. Dept. Interior, and Tall Timbers Research Station.

Collins, C. S. 1998. The influence of hardwood midstory and pine species on pine bole arthropod communities in eastern Texas. M.S. thesis, Stephen F. Austin State Univ., Nacogdoches, Texas.

Conner, R. N., and D. C. Rudolph. 1989. Red-cockaded Woodpecker colony status and trends on the Angelina, Davy Crockett and Sabine national forests. U.S. Dept. Agric., For. Serv. Res. Pap. SO-250.

Conner, R. N., D. C. Rudolph, R. R. Schaefer, D. Saenz, and C. E. Shackelford. 1999. Relationships among Red-cockaded Woodpecker group density, nestling provisioning rates, and habitat. *Wilson Bulletin* 111:494–498.

DeLotelle, R. S., and R. J. Epting. 1992. Reproduction of the Red-cockaded Woodpecker in central Florida. *Wilson Bulletin* 104:285–294.

DeLotelle, R. S., R. J. Epting, and J. R. Newman. 1987. Habitat use and territory characteristics of Red-cockaded Woodpeckers in central Florida. *Wilson Bulletin* 99:202–217.

Engstrom, T. R., and F. J. Sanders. 1997. Red-cockaded Woodpecker foraging ecology in an old-growth longleaf pine forest. *Wilson Bulletin* 109:203–217.

Grubb, T. C., Jr. 1989. Ptilochronology: feather growth bars as indicators of nutritional status. *Auk* 106:314–320.

Hanula, J. L., and K. E. Franzreb. 1995. Arthropod prey of nestling Red-cockaded Woodpeckers in the upper coastal plain of South Carolina. *Wilson Bulletin* 107:485–495.

Hardesty, J. L., K. E. Gault, and F. P. Percival. 1997. Ecological correlates of Red-cockaded Woodpecker (*Picoides borealis*) foraging preference, habitat use, and home range size in northwest Florida (Eglin Air Force Base). Final Report Research Work Order 99. Gainesville: Florida Cooperative Fish and Wildlife Research Unit, University of Florida.

Harlow, R. F., and M. R. Lennartz. 1977. Foods of nestling Red-cockaded Woodpeckers in coastal South Carolina. *Auk* 94:376–377.

Hess, C. A., and F. C. James. 1998. Diet of the Red-cockaded Woodpecker in the Apalachicola National Forest. *Journal of Wildlife Management* 62:509–517.

Hogstad, O. 1976. Sexual dimorphism and divergence in winter foraging behavior of Three-toed Woodpeckers, *Picoides tridactylus. Ibis* 118:41–50.

Hooper, R. G. 1996. Arthropod biomass in winter and the age of longleaf pines. *Forest Ecology and Management* 82:115–131.

Hooper, R. G., and R. F. Harlow. 1986. Forest stands selected by foraging Red-cockaded Woodpeckers. U.S. Dept. Agric., For. Serv. Res. Pap. SE-259.

Hooper, R. G., and M. R. Lennartz. 1981. Foraging behavior of the Red-cockaded Woodpecker in South Carolina. *Auk* 98:321–334.

Hooper, R. G., L. J. Niles, R. F. Harlow, and G. W. Wood. 1982. Home ranges of Red-cockaded Woodpeckers in coastal South Carolina. *Auk* 99:675–682.

Hovis, J. A., and R. F. Labisky. 1985. Vegetative associations of Red-cockaded Woodpecker colonies in Florida. *Wildlife Society Bulletin* 13:307–314.

Jackson, J. A., and S. D. Parris. 1995. The ecology of Red-cockaded Woodpeckers associated with construction and use of a multi-purpose range complex at Fort Polk, Louisiana. Pages 277–282 *in* Red-cockaded Woodpecker: recovery, ecology and management (D. L. Kulhavy, R. G. Hooper, and R. Costa, eds.). Nacogdoches, Texas: Center for Applied Studies, College of Forestry, Stephen F. Austin State Univ.

James, F. C., C. A. Hess, B. C. Kicklighter, and R. A. Thum. 2001. Ecosystem management and the niche gestalt of the Red-cockaded Woodpecker in longleaf pine forests. *Ecological Applications.* (in press)

James, F. C., C. A. Hess, and D. Kufrin. 1997. Species-centered environmental analysis: indirect effects of fire history on Red-cockaded Woodpeckers. *Ecological Applications* 7:118–129.

Jenkins, J. M. 1979. Foraging behavior of male and female Nuttall Woodpeckers. *Auk* 96:418–420.

Kroll, J. C., and R. R. Fleet. 1979. Impact of woodpecker predation on over-wintering within-tree populations of the southern pine beetle (*Dendroctonus frontalis*). Pages 269–281 *in* The role of insectivorous birds in forest ecosystems (J. G. Dickson, R. N. Conner, R. R. Fleet, J. C. Kroll, and J. A. Jackson, eds.). New York: Academic Press.

Ligon, J. D. 1970. Behavior and breeding biology of the Red-cockaded Woodpecker. *Auk* 87:255–278.

McFarlane, R. W. 1995. The relationship between body size, trophic position and foraging territory among woodpeckers. Pages 303–308 *in* Red-cockaded Woodpecker: recovery, ecology and management (D. L. Kulhavy, R. G. Hooper, and R. Costa, eds.). Nacogdoches, Texas: Center for Applied Studies, College of Forestry, Stephen F. Austin State Univ.

Morse, D. H. 1970. Ecological aspects of some mixed-species foraging flocks of birds. *Ecological Monographs* 40:119–168.

———. 1972. Habitat utilization of the Red-cockaded Woodpecker during the winter. *Auk* 89:429–435.

New, K. C., and J. L. Hanula. 1998. Effect of time elapsed after prescribed burning in longleaf pine stands on potential prey of the Red-cockaded Woodpecker. *Southern Journal of Applied Forestry* 22:175–183.

Porter, M. L., and R. L. Labisky. 1986. Home range and foraging habitat of Red-cockaded Woodpeckers in northern Florida. *Journal of Wildlife Management* 50:239–247.

Ramey, P. 1980. Seasonal, sexual, and geographical variation in the foraging ecology of Red-cockaded Woodpeckers (*Picoides borealis*). M.S. thesis, Mississippi State Univ., Starkville.

Repasky, R. R., R. J. Blue, and P. D. Doerr. 1991. Laying Red-cockaded Woodpeckers cache bone fragments. *Condor* 93:458–461.

Schaefer, R. R. 1996. Red-cockaded Woodpecker reproduction and provisioning in relation to habitat. M.S. thesis, Stephen F. Austin State Univ., Nacogdoches, Texas.

Short, L. L. 1982. Woodpeckers of the world. Delaware Museum of Natural History Monogr. Ser. No. 4. Greenville: Delaware Museum of Natural History.

Sibley, C. G., and J. E. Ahlquist. 1990. Phylogeny and classification of birds. New Haven, Connecticut: Yale Univ. Press.

Skorupa, J. P. 1979. Foraging ecology of Red-cockaded Woodpeckers in South Carolina. M.S. thesis, Univ. of California, Davis.

Skorupa, J. P., and R. W. McFarlane. 1976. Seasonal variation in foraging territory of Red-cockaded Woodpeckers. *Wilson Bulletin* 88:662–665.

Van Balen, J. B., and P. D. Doerr. 1978. The relationship of understory vegetation to Red-cockaded Woodpecker activity. *Proceedings of the Annual Conference, Southeastern Association of Fish and Wildlife Agencies* 32:82–92.

Walters, J. R., S. J. Daniels, J. H. Carter III, P. D. Doerr, K. Brust, and J. M. Mitchell. 2000. Foraging habitat resources, preferences and fitness of Red-cockaded Woodpeckers in the North Carolina Sandhills. (unpublished report, submitted to DOD, Department of the Army, Fort Bragg).

Ward, B. 1930. Red-cockaded Woodpeckers on corn. *Bird-Lore* 32:127–128.

Wood, D. A. 1983. Foraging and colony habitat characteristics of the Red-cockaded Woodpecker in Oklahoma. Pages 51–58 *in* Red-cockaded Woodpecker symposium II proceedings (D. A. Wood, ed.). Tallahassee: State of Florida Game and Fresh Water Fish Comm.

Zwicker, S. M. 1995. Selection of pines for foraging and cavity excavation by Red-cockaded Woodpeckers. M.S. thesis, North Carolina State Univ., Raleigh.

Zwicker, S., and J. R. Walters. 1999. Selection of pines for foraging by Red-cockaded Woodpeckers. *Journal of Wildlife Management* 63:843–852.

8. Red-cockaded Woodpeckers and Bark Beetles

Ayres, M. P., R. T. Wilkens, J. J. Ruel, M. J. Lombardero, and E. Vallery. 2000. Nitrogen budgets of phloem-feeding bark beetles with and without symbiotic fungi. *Ecology* 81:2198–2210.

Barras, S. J. 1970. Antagonism between *Dendroctonus frontalis* and the fungi *Ceratocystis minor. Annals of the Entomological Society of America* 63:1187–1190.

———. 1973. Reduction of progeny and development in the southern pine beetle following removal of symbiotic fungi. *Canadian Entomologist* 105:1295–1299.

Beal, J. A. 1927. Weather as a factor in southern pine beetle control. *Journal of Forestry* 25:741–742.

———. 1933. Temperature extremes as a factor in the ecology of the southern pine beetle. *Journal of Forestry* 31:328–336.

Belanger, R. P., R. L. Hedden, and M. R. Lennartz. 1988. Potential impact of the southern pine beetle on Red-cockaded Woodpecker colonies in the Georgia Piedmont. *Southern Journal of Applied Forestry* 12:194–199.

Belanger, R. P., and B. F. Malac. 1980. Silviculture can reduce losses from the southern pine beetle. U.S. Dept. Agric., Comb. For. Pest Res. Dev. Prog. Tech. Bull. 1612.

Berisford, C. W., and R. T. Franklin. 1971. Attack patterns of *Ips avulsus* and *Ips grandicollis* (Coleoptera: Scolytidae) on four species of southern pines. *Annals of the Entomological Society of America* 64:894–897.

Billings, R. J. 1979. Detecting and aerially evaluating southern pine beetle outbreaks. *Southern Journal of Applied Forestry* 3:50–54.

———. 1980. Direct control. Pages 179–192 *in* The southern pine beetle (R. C. Thatcher, J. L. Searcy, J. E. Coster, and G. D. Hertel, eds.). U.S. Dept. Agric., For. Serv. Sci. Educ. Adm. Tech. Bull. 1631.

Billings, R. F., C. M. Bryant V, and K. H. Wilson. 1985. Development, implementation and validation of a large area hazard- and risk-rating system for the southern pine beetle. Pages 226–232 *in* Integrated pest management research symposium: the proceedings (S. J. Branham and R. C. Thatcher, eds.). U.S. Dept. Agric., For. Serv. Gen. Tech. Rep. SO-56.

Billings, R. F., and C. Doggett. 1980. An aerial observer's guide for recognizing and reporting southern pine beetle spots. U.S. Dept. Agric., Comb. For. Pest Res. and Dev. Prog. Agric. Handb. No. 560.

Billings, R. F., and H. A. Pase III. 1979. A field guide for ground checking southern pine beetle spots. U.S. Dept. Agric., Comb. For. Pest Res. and Dev. Prog. Agric. Handb. No. 558.

———. 1979. Spot proliferation patterns as a measure of the area-wide effectiveness of southern pine beetle control tactics. Pages 86–97 *in* Evaluating control tactics for the southern pine beetle: symposium proceedings (J. E. Coster and J. L. Searcy, eds.). U.S. Dept. Agric., Comb. For. Pest Res. Dev. Prog. Tech. Bull. 1613.

Billings, R. F., and F. E. Varner. 1986. Why control southern pine beetle infestations in wilderness areas? The Four Notch and Huntsville State Park experience. Pages 129–134 *in* Wilderness and natural areas in the eastern United States: a management challenge (D. L. Kulhavy and R. N. Conner, eds.). Nacogdoches, Texas: School of Forestry, Stephen F. Austin State Univ.

Blackford, J. L. 1955. Woodpecker concentration in burned forest. *Condor* 57:28–30.

Bridges, J. R. 1983. Mycangial fungi of *Dendroctonus frontalis* (Coleoptera: Scolytidae) and their relationship to beetle population trends. *Environmental Entomology* 12:858–861.

Bridges, J. R., W. A. Nettleton, and M. D. Connor. 1985. Southern pine beetle (Coleoptera: Scolytidae) infestations without the bluestain fungus, *Ceratocystis minor*. *Journal of Economic Entomology* 78:325–327.

Brown, M. W., T. E. Nebeker, and C. R. Honea. 1987. Thinning increases loblolly pine vigor and resistance to bark beetles. *Southern Journal of Applied Forestry* 11:28–31.

Bryant, C. M., V. 1983. Pine engraver and black turpentine beetles associated with endemic populations of southern pine beetles in east Texas. M.S. thesis, Stephen F. Austin State Univ., Nacogdoches, Texas.

Conner, R. N., and D. C. Rudolph. 1995. Losses of Red-cockaded Woodpecker cavity trees to southern pine beetles. *Wilson Bulletin* 107:81–92.

Conner, R. N., D. C. Rudolph, D. L. Kulhavy, and A. E. Snow. 1991. Causes of mortality of Red-cockaded Woodpecker cavity trees. *Journal of Wildlife Management* 55:531–537.

Cooper, M. E., and F. M. Stephen. 1978. Parent adult re-emergence in southern pine beetle populations. *Environmental Entomology* 7:574–577.

Coster, J. E., R. R. Hicks, Jr., and K. G. Watterston. 1978. Directional spread of southern pine beetle (Coleoptera: Scolytidae) infestations in East Texas. *Journal of the Georgia Entomological Society* 13:315–321.

Coster, J. E., and P. C. Johnson. 1979. Characterizing flight aggregation of the southern pine beetle. *Environmental Entomology* 8:381–387.

Coster, J. E., T. L. Payne, E. R. Hart, and L. J. Edson. 1977. Seasonal variations in mass attack behavior of the southern pine beetle. *Journal of the Georgia Entomological Society* 12:204–211.

Coster, J. E., and J. P. Vité. 1972. Effects of feeding and mating on pheromone release in the southern pine beetle. *Annals of the Entomological Society of America* 65:263–266.

Coulson, R. N. 1979. Population dynamics of bark beetles. *Annual Review of Entomology* 24:417–447.

Coulson, R. N., J. W. Fitzgerald, F. L. Oliveria, R. N. Conner, and D. C. Rudolph. 1995. Red-cockaded Woodpecker habitat management and southern pine beetle infestations. Pages 191–195 *in* Red-cockaded Woodpecker: recovery, ecology and management (D. L. Kulhavy, R. G. Hooper, and R. Costa, eds.). Nacogdoches, Texas: Center for Applied Studies, College of Forestry, Stephen F. Austin State Univ.

Coulson, R. N., M. D. Guzman, K. Skordinski, J. W. Fitzgerald, R. N. Conner, D. C. Rudolph, F. L. Oliveria, D. F. Wunneburger, and P. E. Pulley. 1999. Forest landscapes: their effect on the interaction of the southern pine beetle and Red-cockaded Woodpecker. *Journal of Forestry* 97(10):4–11.

Coulson, R. N., P. B. Hennier, R. O. Flamm, E. J. Rykiel, L. C. Hu, and T. L. Payne. 1983. The role of lightning in the epidemiology of the southern pine beetle. *Zeitschrift für angewandte Entomologie* 96:182–193.

Dixon, J. C., and E. A. Osgood. 1961. Southern pine beetle: review of present knowledge. U.S. Dept. Agric., For. Serv. Res. Pap. SE-128.

Fargo, W. S., R. N. Coulson, P. E. Pulley, D. N. Pope, and C. L. Kelley. 1978. Spatial and temporal patterns of within-tree colonization by *Dendroctonus frontalis. Canadian Entomologist* 110:1213–1232.

Gara, R. I., and J. E. Coster. 1968. Studies on the attack behavior of the southern pine beetle. III. Sequence of tree infestation within stands. *Contributions from Boyce Thompson Institute* 24:77–86.

Goyer, R. A., G. J. Lenhard, T. E. Nebeker, and L. D. Jarrard. 1981. How to identify common insect associates of the southern pine beetle. U.S. Dept. Agric., Comb. For. Pest Res. and Dev. Prog. Agric. Handb. No. 563.

Hayes, J. L., and B. L. Strom. 1994. 4-Allylanisole as an inhibitor of bark beetle (Coleoptera: Scolytidae) aggregation. *Journal of Economic Entomology* 87:1586–1594.

Hayes, J. L., B. L. Strom, L. Roton, and L. L. Ingram, Jr. 1994. Repellent properties of a novel host compound to southern pine beetle. *Journal of Chemical Ecology* 20:1595–1615.

Hicks, R. R., Jr., J. E. Coster, and G. N. Mason. 1987. Forest insect hazard rating. *Journal of Forestry* 85(10):20–26.

Hodges, J. D., W. N. Elam, and W. F. Watson. 1977. Physical properties of the oleoresin system of the four major southern pines. *Canadian Journal of Forestry Research* 7:520–525.

Hodges, J. D., W. W. Elam, W. F. Watson, and T. E. Nebeker. 1979. Oleoresin characteristics and susceptibility of four southern pines to southern pine beetle (Coleoptera: Scolytidae) attacks. *Canadian Entomologist* 111:889–896.

Hodges, J. D., and P. L. Lorio, Jr. 1973. Comparison of oleoresin composition in declining and healthy loblolly pines. U.S. Dept. Agric., For. Serv. Res. Note SO-158.

Hodges, J. D., and L. S. Pickard. 1971. Lightning in the ecology of the southern pine beetle, *Dendroctonus frontalis* (Coleoptera: Scolytidae). *Canadian Entomologist* 103:44–51.

Hodges, J. D., and R. C. Thatcher. 1976. Southern pine beetle survival in trees felled by the cut and top-cut and leave method. U.S. Dept. Agric., For. Serv. Res. Note SO-219.

Klepzig, K. D. 1998. Competition between a biological control fungus, *Ophiostoma piliferum*, and symbionts of the southern pine beetle. *Mycologia* 90:69–75.

Klepzig, K. D., and R. T. Wilkens. 1997. Competitive interactions among symbiont fungi of the southern pine beetle. *Applied and Environmental Microbiology* 63: 621–627.

Koplin, J. R. 1969. The numerical response of woodpeckers to insect prey in a sub-alpine forest in Colorado. *Condor* 71:436–438.

Kroll, J. C., R. N. Conner, and R. R. Fleet. 1980. Woodpeckers and the southern pine beetle. U.S. Dept. Agric., Comb. For. Pest Res. and Dev. Prog. Agric. Handb. No. 564.

Kroll, J. C., and R. R. Fleet. 1979. Impact of woodpecker predation on over-wintering within-tree populations of southern pine beetle (*Dendroctonus frontalis*). Pages 269–281 *in* The role of insectivorous birds in forest ecosystems (J. G. Dickson, R. N. Conner, R. R. Fleet, J. A. Jackson, and J. C. Kroll, eds.). New York: Academic Press.

Kulhavy, D. L., J. H. Mitchell, and R. N. Conner. 1988. The southern pine beetle and the Red-cockaded Woodpecker: potential for interaction. Pages 337–343 *in* Integrated control of scolytid bark beetles (T. L. Payne and H. Saarenmaa, eds.). Blacksburg: Virginia Polytechnic Institute and State Univ.

Lindgren, B. S. 1990. Ambrosia beetles. *Journal of Forestry* 88(2):8–11.

Lorio, P. L., Jr. 1984. Should small infestations of southern pine beetle receive control priority? *Southern Journal of Applied Forestry* 8:201–204.

Lorio, P. L., Jr., and J. D. Hodges. 1968. Oleoresin exudation pressure and relative water content of inner bark as indicators of moisture stress in loblolly pines. *Forest Science* 14:392–398.

Lorio, P. L., Jr., G. N. Mason, and G. L. Autry. 1982. Stand risk rating for the southern pine beetle: integrating pest management with resource management. *Journal of Forestry* 80:202–214.

Lorio, P. L., Jr., and R. A. Sommers. 1981. Use of available resource data to rate stands for southern pine beetle risk. Pages 75–78 *in* Hazard-rating systems in forest insect pest management (R. L. Hedden, S. J. Barras, and J. E. Coster, eds.). U.S. Dept. Agric. Tech. Rep. WO-27.

Lorio, P. L., Jr., R. A. Sommers, C. A. Blanche, J. D. Hodges, and T. E. Nebeker. 1990. Modeling pine resistance to bark beetles based on growth and differentiation balance principles. Pages 402–409 *in* Process modeling of forest growth responses to environmental stress (R. K. Dixon, R. S. Meldahl, G. A. Ruark, and W. G. Warren, eds.). Portland, Oregon: Timber Press.

Lorio, P. L., Jr., and S. J. Zarnoch. 1984. Calculating tree bole surface area for estimating populations of the southern pine beetle (Coleoptera: Scolytidae). *Environmental Entomology* 13:1069–1073.

Moore, G. E. 1972. Southern pine beetle mortality in North Carolina caused by parasites and predators. *Environmental Entomology* 1:58–65.

Moser, J. C., R. A. Sommers, P. L. Lorio, Jr., J. R. Bridges, and J. J. Witcosky. 1987. Southern pine beetles attack felled green timber. U.S. Dept. Agric., For. Serv. Res. Note SOH-342.

Nebeker, T. E., and J. D. Hodges. 1983. Influence of forestry practices on host susceptibility to bark beetles. *Zeitschrift für angewandte Entomologie* 96:194–208.

———. 1985. Thinning and harvesting practices to minimize site and stand disturbance and susceptibility to bark beetle and disease attacks. Pages 263–271 *in* Proc. integrated pest management research symposium (S. J. Branham and R. C. Thatcher, eds.). U.S. Dept. Agric., For. Serv. Gen. Tech. Rep. SO-56.

Nebeker, T. E., and G. C. Purser. 1980. Relationship of temperature and prey type to developmental time of the bark beetle predator *Thanasimus dubius* (F.) (Coleoptera: Cleridae). *Canadian Entomologist* 112:179–184.

Overgaard, N. A. 1970. Control of the southern pine beetle by woodpeckers in central Louisiana. *Journal of Economic Entomology* 63:1016–1017.

Paine, T. D., F. M. Stephen, and H. A. Taha. 1984. Conceptual model of infestation probability based on bark beetle abundance and host tree susceptibility. *Environmental Entomology* 13:619–624.

Payne, T. L., and R. F. Billings. 1989. Evaluation of (S)-verbenone applications for suppressing southern pine beetle (Coleoptera: Scolytidae) infestations. *Journal of Economic Entomology* 82:1702–1708.

Payne, T. L., J. E. Coster, J. V. Richerson, L. J. Edson, and E. R. Hart. 1978. Field response of the southern pine beetle to behavioral chemicals. *Environmental Entomology* 7:578–582.

Payne, T. L., L. H. Kudon, K. D. Walsh, and C. W. Berisford. 1985. Influence of infestation density on suppression of *D. frontalis* infestations with attractant. *Zeitschrift für angewandte Entomologie* 99:39–43.

Payne, T. L., and J. V. Richerson. 1985. Pheromone-mediated competitive replacement between two bark beetle populations: influence on infestation suppression. *Zeitschrift für angewandte Entomologie* 99:131–138.

Prentice, S. A. 1977. Frequency of lightning discharges. Pages 465–496 *in* Lightning. Vol. 1 (R. H. Golde, ed.). New York: Academic Press.

Richerson, J. V., F. A. McCarty, and T. L. Payne. 1980. Disruption of southern pine beetle infestations with Frontalure. *Environmental Entomology* 9:90–93.

Richerson, J. V., and T. L. Payne. 1979. Effects of bark beetle inhibitors on landing and attack behavior of the southern pine beetle and beetle associates. *Environmental Entomology* 8:360–364.

Rudolph, D. C., and R. N. Conner. 1995. The impact of southern pine beetle induced mortality on Red-cockaded Woodpecker cavity trees. Pages 208–213 *in* Red-cockaded Woodpecker: recovery, ecology and management (D. L. Kulhavy, R. G. Hooper, and R. Costa, eds.). Nacogdoches, Texas: Center for Applied Studies, College of Forestry, Stephen F. Austin State Univ.

Schowalter, T. D., R. N. Coulson, and D. A. Crossley, Jr. 1981. Role of southern pine beetle and fire in maintenance of structure and function of the southeastern coniferous forest. *Environmental Entomology* 10:821–825.

Shook, R. S., and P. H. Baldwin. 1970. Woodpecker predation on bark beetles in Engelmann spruce logs as related to stand density. *Canadian Entomologist* 102:1345–1354.

Sierra Club, The Wilderness Society, and Texas Committee on Natural Resources v. Froehlke. 816 F.2d 205, New Orleans, Louisiana (Fifth Circuit Appellate Court 1987).

Sierra Club, The Wilderness Society, and Texas Committee on Natural Resources v. Richard E. Lyng, F. Dale Robertson, John E. Alcock, and William M. Lannan. Civil Action No. L-85-69-CA, Tyler, Texas (District of East Texas 1985–2001).

Steirly, C. C. 1965. Role of woodpeckers in the control of southern pine beetles in Virginia. *Raven* 36:55–59.

Swain, K. M., Sr., and M. C. Remion. 1981. Direct control methods for the southern pine beetle. U.S. Dept. Agric., Comb. For. Pest Res. and Dev. Prog., Agric. Handb. No. 575.

Taylor, A. R. 1977. Lightning and trees. Pages 831–849 *in* Lightning. Vol. 2 (R. H. Golde, ed.). New York: Academic Press.

Thatcher, R. C. 1960. Bark beetles affecting southern pines: a review of current knowledge. U.S. Dept. Agric., For. Serv. So. For. Exp. Sta. Occas. Pap. 180.

Thatcher, R. C., and M. D. Connor. 1985. Identification and biology of southern pine bark beetles. U.S. Dept. Agric., Coop. State Res. Serv. Agric. Handb. No. 634.

Thatcher, R. C., G. N. Mason, and G. D. Hertel. 1986. Integrated pest management in southern pine forests. U.S. Dept. Agric., Coop. State Res. Serv. Agric. Handb. 650.

Thatcher, R. C., and L. S. Pickard. 1964. Seasonal variations in activity of the southern pine beetle in East Texas. *Journal of Economic Entomology* 57:840–842.

———. 1967. Seasonal development of the southern pine beetle in East Texas. *Journal of Economic Entomology* 60:656–658.

Thatcher, R. C., J. L. Searcy, J. E. Coster, and G. D. Hertel, eds. 1980. The southern pine beetle. U.S. Dept. Agric., For. Serv. Sci. Ed. Adm. Tech. Bull. 1631.

Thompson, W. A., and J. C. Moser. 1986. Temperature thresholds related to flight of *Dendroctonus frontalis* Zimm. (Coleoptera: Scolytidae). *Agronomie* 6:905–910.

U.S. Forest Service. 1987. Final environmental impact statement for the suppression of the southern pine beetle. U.S. Dept. Agric., For. Serv. Manage. Bull. R-8-MB-2.

———. 1987. Final environmental impact statement—land and resource management plan, National Forests and Grasslands in Texas. U.S. Dept. Agric., For. Serv. Manage. Bull. R-8-MB-9.

Vité, J. P. 1970. Pest management systems using synthetic pheromones, symposium on population attractant. *Contributions from Boyce Thompson Institute* 24:343–350.

Vité, J. P., and R. G. Crozier. 1968. Studies on the attack behavior of the southern pine beetle. IV. Influence of host condition on aggregation pattern. *Contributions from Boyce Thompson Institute* 24:87–94.

Vité, J. P., and W. Francke. 1976. The aggregation pheromones of bark beetles: progress and problems. *Naturwissenschaften* 63:550–555.

Vité, J. P., R. I. Gara, and H. D. von Scheller. 1964. Field observations on the response to attractants of bark beetles infesting southern pines. *Contributions from Boyce Thompson Institute* 22:261–270.

Vité, J. P., and J. A. Renwick. 1971. Inhibition of *Dendroctonus frontalis* response to frontalin by isomers of brevicomin. *Naturwissenschaften* 8:418–419.

Vité, J. P., and D. L. Williamson. 1970. *Thanasimus dubius:* prey perception. *Journal of Insect Physiology* 16:233–239.

Wahlenberg, W. G. 1960. Loblolly pine. Durham, North Carolina: School of Forestry, Duke Univ.

Yeager, L. E. 1955. Two woodpecker populations in relation to environmental change. *Condor* 57:148–153.

9. The Causes of Population Declines

Baker, W. W. 1983. Decline and extirpation of a population of Red-cockaded Woodpeckers in northwest Florida. Pages 44–45 *in* Red-cockaded Woodpecker symposium II proceedings (D. A. Wood, ed.). Tallahassee: State of Florida Game and Fresh Water Fish Comm.

Carter, J. H., III, J. R. Walters, S. H. Everhart, and P. D. Doerr. 1989. Restrictors for Red-cockaded Woodpecker cavities. *Wildlife Society Bulletin* 17:68–72.

Conner, R. N., and D. C. Rudolph. 1989. Red-cockaded Woodpecker colony status and trends on the Angelina, Davy Crockett, and Sabine national forests. U.S. Dept. Agric., For. Serv. Res. Paper SO-250.

———. 1991. Forest habitat loss, fragmentation, and Red-cockaded Woodpecker populations. *Wilson Bulletin* 103:446–457.

Doerr, P. D., J. R. Walters, and J. H. Carter III. 1989. Reoccupation of abandoned clusters of cavity trees (colonies) by Red-cockaded Woodpeckers. *Proceedings of* the *Annual Conference, Southeastern Association of Fish and Wildlife Agencies* 43:326–336.

Engstrom, R. T., and F. J. Sanders. 1997. Red-cockaded Woodpecker foraging ecology in an old-growth longleaf pine forest. *Wilson Bulletin* 109:203–217.

Franzreb, K. E. 1997. Success of intensive management of a critically imperiled population of Red-cockaded Woodpeckers in South Carolina. *Journal of Field Ornithology* 68:458–470.

Hagan, J. M., and J. M. Reed. 1988. Red color bands reduce fledging success in Red-cockaded Woodpeckers. *Auk* 105:498–503.

Haig, S. M., J. R. Belthoff, and D. H. Allen. 1993. Population viability analysis for a small population of Red-cockaded Woodpeckers and an evaluation of enhancement strategies. *Conservation Biology* 7:289–301.

Hooper, R. G., D. L. Krusac, and D. L. Carlson. 1991. An increase in a population of Red-cockaded Woodpeckers. *Wildlife Society Bulletin* 19:277–286.

Hooper, R. G., and M. R. Lennartz. 1995. Short-term response of a high density Red-cockaded Woodpecker population to loss of foraging habitat. Pages 283–289 *in* Red-cockaded Woodpecker: recovery, ecology and management (D. L. Kulhavy, R. G. Hooper, and R. Costa, eds.). Nacogdoches, Texas: Center for Applied Studies, College of Forestry, Stephen F. Austin State Univ.

Jackson, J. A. 1994. Red-cockaded Woodpecker (*Picoides borealis*). The birds of North America, No. 85 (A. Poole and F. Gill, eds.). Philadelphia: Academy of Natural Sciences; Washington, D.C.: American Ornithologists' Union.

Jackson, J. A., and S. D. Parris. 1995. The ecology of Red-cockaded Woodpeckers associated with construction and use of a multi-purpose range complex at Fort Polk, Louisiana. Pages 277–282 *in* Red-cockaded Woodpecker: recovery, ecology and management (D. L. Kulhavy, R. G. Hooper, and R. Costa, eds.). Nacogdoches, Texas: Center for Applied Studies, College of Forestry, Stephen F. Austin State Univ.

James, F. C. 1991. Signs of trouble in the largest remaining population of Red-cockaded Woodpeckers. *Auk* 108:419–423.

James, F. C., C. A. Hess, B. C. Kicklighter, and R. A. Thum. 2001. Ecosystem management and the niche gestalt of the Red-cockaded Woodpecker in longleaf pine forests. *Ecological Applications.* (in press)

James, F. C., C. A. Hess, and D. Kufrin. 1997. Species-centered environmental analysis: indirect effects of fire history on Red-cockaded Woodpeckers. *Ecological Applications* 7:118–129.

Kappes, J. J., Jr. 1997. Defining cavity-associated interactions between Red-cockaded Woodpeckers and other cavity-dependent species: interspecific competition or cavity kleptoparasitism? *Auk* 114:778–780.

Kelly, J. F. 1991. The influence of habitat quality on the population decline of the Red-cockaded Woodpecker in the McCurtain County Wilderness Area, Oklahoma. M.S. thesis, Oklahoma State Univ., Stillwater.

Kelly, J. F., S. M. Pletschet, and D. M. Leslie, Jr. 1993. Habitat associations of Red-cockaded Woodpecker cavity trees in an old-growth forest of Oklahoma. *Journal of Wildlife Management* 57:122–128.

LaBranche, M. S. 1988. Reproductive ecology of the Red-cockaded Woodpecker in the Sandhills of North Carolina. M.S. thesis, North Carolina State Univ., Raleigh.

Lennartz, M., R. G. Hooper, and R. F. Harlow. 1987. Sociality and cooperative breeding of Red-cockaded Woodpeckers, *Picoides borealis. Behavioral Ecology and Sociobiology* 20:77–88.

Loeb, S. C. 1993. Use and selection of Red-cockaded Woodpecker cavities by southern flying squirrels. *Journal of Wildlife Management* 57:329–335.

Masters, R. E., J. E. Skeen, and J. Whitehead. 1995. Preliminary fire history of McCurtain County Wilderness Area and implications for Red-cockaded Woodpecker management. Pages 290–302 *in* Red-cockaded Woodpecker: recovery, ecology and management (D. L. Kulhavy, R. G. Hooper, and R. Costa, eds.). Nacogdoches, Texas: Center for Applied Studies, College of Forestry, Stephen F. Austin State Univ.

Richardson, D. M., J. W. Bradford, B. J. Gentry, and J. L. Hull. 1998. Evaluation of a pick-up tool for removing Red-cockaded Woodpecker nestlings from cavities. *Wildlife Society Bulletin* 26:855–858.

Rudolph, D. C., and R. N. Conner. 1994. Forest fragmentation and Red-cockaded Woodpecker populations: an analysis at intermediate scale. *Journal of Field Ornithology* 65:365–375.

Schaefer, R. R. 1996. Red-cockaded Woodpecker reproduction and provisioning of nestlings in relation to habitat. M.S. thesis, Stephen F. Austin State Univ., Nacogdoches, Texas.

Van Balen, J. B., and P. D. Doerr. 1978. The relationship of understory vegetation to Red-cockaded Woodpecker activity. *Proceedings of the Annual Conference, Southeastern Association of Fish and Wildlife Agencies* 32:82–92.

Walters, J. R. 1991. Application of ecological principles to the management of endangered species: the case of the Red-cockaded Woodpecker. *Annual Review of Ecology and Systematics* 22:505–523.

Walters, J. R., S. J. Daniels, J. H. Carter III, P. D. Doerr, K. Brust, and J. M. Mitchell. 2000. Foraging habitat resources, preferences and fitness of Red-cockaded Woodpeckers in the North Carolina Sandhills. (unpublished report, submitted to DOD, Department of the Army, Fort Bragg).

Wood, G. W., L. J. Niles, R. M. Hendricks, J. R. Davis, and T. L. Grimes. 1985. Compatibility of even-aged timber management and Red-cockaded Woodpecker conservation. *Wildlife Society Bulletin* 13:5–17.

10. Extinction, Legal Status, and History of Management

Allen, D. H. 1991. An insert technique for constructing artificial Red-cockaded Woodpecker cavities. U.S. Dept. Agric., For. Serv. Gen. Tech. Rep. SE-73.

Babbitt v. Sweet Home Chapter of Communities for a Greater Oregon. Case No. 94-859, Washington, D.C. (U.S. Supreme Court 1994–1995).

Baker, J. B., M. D. Cain, J. M. Guldin, P. A. Murphy, and M. G. Shelton. 1996. Uneven-aged silviculture for the loblolly and shortleaf pine forest cover types. U.S. Dept. Agric., For. Serv. Gen. Tech. Rep. SO-118.

Bigony, M. L. 1991. Controversy in the pines—timber management may determine the Red-cockaded Woodpecker's survival. *Texas Parks and Wildlife* 49(5):13–17.

Briggs, J. C. 1991. A Cretaceous-Tertiary mass extinction? *BioScience* 41:619–624.

Carter, J. H., III, J. R. Walters, S. H. Everhart, and P. D. Doerr. 1989. Restrictors for Red-cockaded Woodpecker cavities. *Wildlife Society Bulletin* 17:68–72.

Cely, J. E. 1982. Comments on relocating Red-cockaded Woodpeckers. *Wildlife Society Bulletin* 11:89.

Conner, R. N. 1989. Injection of 2,4-D to remove hardwood midstory within Red-cockaded Woodpecker colony areas. U.S. Dept. Agric., For. Serv. Res. Pap. SO-251.

Conner, R. N., and B. A. Locke. 1979. Effects of a prescribed burn on cavity trees of Red-cockaded Woodpeckers. *Wildlife Society Bulletin* 7:291–293.

Conner, R. N., and K. A. O'Halloran. 1987. Cavity-tree selection by Red-cockaded Woodpeckers as related to growth dynamics of southern pines. *Wilson Bulletin* 99:398–412.

Conner, R. N., and D. C. Rudolph. 1989. Red-cockaded Woodpecker colony status and trends on the Angelina, Davy Crockett and Sabine national forests. U.S. Dept. Agric., For. Serv. Res. Pap. SO-250.

———. 1991. Effects of midstory reduction and thinning in Red-cockaded Woodpecker cavity tree clusters. *Wildlife Society Bulletin* 19:63–66.

———. 1991. Forest habitat loss, fragmentation, and Red-cockaded Woodpecker populations. *Wilson Bulletin* 103:446–457.

Conner, R. N., A. E. Snow, and K. A. O'Halloran. 1991. Red-cockaded Woodpecker use of seed-tree/shelterwood cuts in eastern Texas. *Wildlife Society Bulletin* 19:67–73.

Copeyon, C. K. 1990. A technique for constructing cavities for the Red-cockaded Woodpecker. *Wildlife Society Bulletin* 18:303–311.

Costa, R., and R. E. F. Escaño. 1989. Red-cockaded Woodpecker: status and management in the Southern Region in 1986. U.S. Dept. Agric., For. Serv. Southern Region Tech. Publ. R8-TP 12.

DeFazio, J. T., Jr., M. A. Hunnicutt, M. R. Lennartz, G. L. Chapman, and J. A. Jackson. 1987. Red-cockaded Woodpecker translocation experiments in South Carolina. *Proceedings of the Annual Conference, Southeastern Association of Fish and Wildlife Agencies* 41:311–317.

Doerr, P. D., J. R. Walters, and J. H. Carter III. 1989. Reoccupation of abandoned clusters of cavity trees (colonies) by Red-cockaded Woodpeckers. *Proceedings of the Annual Conference, Southeastern Association of Fish and Wildlife Agencies* 43:326–336.

Farrar, R. M., Jr., and W. D. Boyer. 1991. Managing longleaf pine under the selection system—promises and problems. Pages 357–368 *in* Proceedings of the sixth biennial southern silvicultural research conference (S. S. Coleman and D. G. Neary, eds.). U.S. Dept. Agric., For. Serv. Gen. Tech. Rep. SO-75.

Fenton, C. L., and M. A. Fenton. 1958. The fossil book: a record of prehistoric life. Garden City, New York: Doubleday.

Heinrichs, J., and D. B. Heinrichs. 1984. The woodpecker and the pines. *American Forests* 90(3):24–26, 46–49.

Henry, V. G. 1989. Guideline for preparation of biological assessments and evaluations for the Red-cockaded Woodpecker. Atlanta, Georgia: U.S. Fish and Wildl. Serv., Southeast Region.

Hooper, R. G. 1983. Colony formation by Red-cockaded Woodpeckers: hypotheses and management implications. Pages 72–77 *in* Red-cockaded Woodpecker symposium II proceedings (D. A. Wood, ed.). Tallahassee: State of Florida Game and Fresh Water Fish Comm.

Hooper, R. G., A. F. Robinson, Jr., and J. A. Jackson. 1980. The Red-cockaded Woodpecker: notes on life history and management. U.S. Dept. Agric., For. Serv., State and Private Forestry Gen. Rep. SA-GR9.

Hooper, R. G., J. C. Watson, and R. E. F. Escaño. 1990. Hurricane Hugo's initial effects on Red-cockaded Woodpeckers in the Francis Marion National Forest. *North American Wildlife and Natural Resources Conference* 55:220–224.

Jackson, J. A. 1971. The evolution, taxonomy, distribution, past populations and current status of the Red-cockaded Woodpecker. Pages 4–26 *in* The ecology and management of the Red-cockaded Woodpecker (R. L. Thompson, ed.). Tallahassee, Florida: Bureau of Sport Fisheries and Wildlife, U.S. Dept. Interior, and Tall Timbers Research Station.

———. 1976. Rights-of-way management for an endangered species—the Red-cockaded Woodpecker. Pages 247–252 *in* Proceedings of the first national symposium on environmental concerns in rights-of-way management (R. Tillman, ed.). Mississippi State: Mississippi State Univ.

————. 1979. Highways and wildlife—some challenges and opportunities for management. Pages 566–571 in The mitigation symposium: a national workshop on mitigating losses of fish and wildlife habitats. U.S. Dept. Agric., For. Serv. Gen. Tech. Rep. RM-65.

————. 1990. Intercolony movements of Red-cockaded Woodpecker in South Carolina. Journal of Field Ornithology 61:149–272.

Jackson, J. A., C. D. Cooley, and M. B. Hays. 1979. A new trap for capturing cavity roosting birds. Inland Bird Banding 51:42–44.

Jackson, J. A., M. R. Lennartz, and R. G. Hooper. 1979. Tree age and cavity initiation by Red-cockaded Woodpeckers. Journal of Forestry 77(2):102–103.

Jackson, J. A., and S. D. Parris. 1986. Biopolitics, management of federal lands, and the conservation of the Red-cockaded Woodpecker. American Birds 40:1162–1168.

————. 1991. A simple, effective net for capturing cavity roosting birds. North American Bird Bander 16:30–31.

Jackson, J. A., B. J. Schardien, and P. R. Miller. 1983. Moving Red-cockaded Woodpecker colonies: relocation or phased destruction? Wildlife Society Bulletin 11: 59–62.

Jackson, J. A., B. J. Schardien, and G. W. Robinson. 1977. A problem associated with the use of radio transmitters on tree surface foraging birds. Inland Bird Banding 49:50–53.

James, F. C. 1991. Signs of trouble in the largest remaining population of Red-cockaded Woodpeckers. Auk 108:419–423.

Lay, D. W. 1969. Destined for oblivion. Texas Parks and Wildlife 27(2):12–15.

————. 1973. Red-cockaded Woodpecker study. Job Completion Rep., Fed. Aid Proj. No. W-80-R-16, Austin, Texas: East Texas Deer Study, Job No. 10.

Lay, D. W., and D. N. Russell. 1970. Notes on the Red-cockaded Woodpecker in Texas. Auk 87:781–786.

Lennartz, M. R., P. H. Geissler, R. F. Harlow, R. C. Long, K. M. Chitwood, and J. A. Jackson. 1983. Status of Red-cockaded Woodpecker populations on federal lands in the South. Pages 7–12 in Red-cockaded Woodpecker symposium II proceedings (D. A. Wood, ed.). Tallahassee: State of Florida Game and Fresh Water Fish Comm.

Lennartz, M. R., H. A. Knight, J. P. McClure, and V. A. Rudis. 1983. Status of Red-cockaded Woodpecker nesting habitat in the South. Pages 13–19 in Red-cockaded Woodpecker symposium II proceedings (D. A. Wood, ed.). Tallahassee: State of Florida Game and Fresh Water Fish Comm.

Leopold, A. 1949. A Sand County almanac and sketches here and there. New York: Oxford Univ. Press.

Ligon, J. D., P. B. Stacey, R. N. Conner, C. E. Bock, and C. S. Anderson. 1986. Report on the American Ornithologists' Union committee for the conservation of Red-cockaded Woodpecker. Auk 103:848–855.

Line, L. 1995. Gone the way of the dinosaurs. International Wildlife 25(4):16–21.

Marsh, G. P. 1864. Man and nature; or, physical geography as modified by human action. New York: C. Scribner.

Mize, R. 1992. The Red-cockaded Woodpecker and the national forests in Texas. Lufkin, Texas: U.S. Forest Service. (draft copy, internal report)

Myers, N. 1989. Extinction rates past and present. BioScience 39(1):39–41.

Nesbitt, S. A., B. A. Harris, R. W. Repenning, and C. B. Brownsmith. 1982. Notes on

Red-cockaded Woodpecker study techniques. *Wildlife Society Bulletin* 10:160–163.

Odum, R. R., J. Rappole, J. Evans, D. Charbonneau, and D. Palmer. 1982. Red-cockaded Woodpecker relocation experiment in coastal Georgia. *Wildlife Society Bulletin* 10:197–203.

Ortego, B., R. N. Conner, and C. Rudolph. 1988. Status of the Red-cockaded Woodpecker in Texas, 1985–87. *Bulletin of the Texas Ornithological Society* 21:22–25.

Palila v. Hawaii Department of Land and Natural Resources. 471 File Supplement 985 and 649 File Supplement 1070, Honolulu, Hawaii (District of Hawaii 1979–1986).

Pinchot, G. 1947. Breaking new ground. Washington, D.C.: Island Press.

Rudolph, D. C., and R. N. Conner. 1991. Cavity tree selection by Red-cockaded Woodpeckers in relation to tree age. *Wilson Bulletin* 103:458–467.

Rudolph, D. C., R. N. Conner, D. K. Carrie, and R. R. Schaefer. 1992. Experimental reintroduction of Red-cockaded Woodpeckers. *Auk* 109:914–916.

Seagle, S. W., R. A. Lancia, D. A. Adams, M. R. Lennartz, and H. A. Devine. 1987. Integrating timber and Red-cockaded Woodpecker habitat management. *Transactions of the North American Wildlife and Natural Resources Conference* 52:41–52.

Shideler, J. C., and R. L. Hendricks. 1991. The legacy of early ideas of conservation. *Journal of Forestry* 89(10):20–23.

Sierra Club, The Wilderness Society, and Texas Committee on Natural Resources v. Richard E. Lyng, F. Dale Robertson, John E. Alcock, and William M. Lannan. Civil Action No. L-85-69-CA, Tyler, Texas (District of East Texas 1985–2001).

Signor, P. W. 1994. Biodiversity in geological time. *American Zoologist* 34:23–32.

Southeast Negotiation Network. 1990. Scientific summit on the Red-cockaded Woodpecker. Atlanta: Georgia Institute of Tech.

Steadman, D. W. 1995. Prehistoric extinctions of Pacific island birds: biodiversity meets zooarchaeology. *Science* 267:1123–1131.

Taylor, W. E., and R. G. Hooper. 1991. A modification of Copeyon's drill technique for making artificial Red-cockaded Woodpecker cavities. U.S. Dept. Agric., For. Serv. Gen. Tech. Rep. SE-72.

Thompson, R. L. 1983. Red-cockaded Woodpecker symposium II—an overview. Pages 3–4 in Red-cockaded Woodpecker symposium II proceedings (D. A. Wood, ed.). Tallahassee: State of Florida Game and Fresh Water Fish Comm.

———, ed. 1971. The ecology and management of the Red-cockaded Woodpecker. Tallahassee, Florida: Bureau of Sport Fisheries and Wildlife, U.S. Dept. Interior, and Tall Timbers Research Station.

U.S. Bureau of Sport Fisheries and Wildlife. 1968. Rare and endangered fish and wildlife of the United States. Res. Publ. 34.

U.S. Fish and Wildlife Service. 1979. Red-cockaded Woodpecker recovery plan. Atlanta, Georgia: U.S. Fish and Wildl. Serv.

———. 1985. Red-cockaded Woodpecker recovery plan. Atlanta, Georgia: U.S. Fish and Wildl. Serv.

U.S. Forest Service. 1969. Forest Service manual—NFT Supplement No. 3. Lufkin, Texas: National Forests and Grasslands in Texas.

———. 1975. Wildlife habitat management handbook, Ch. 400, section 420—Red-cockaded Woodpecker. FSH 2609.23 R. Atlanta, Georgia: Southern Region.

———. 1979. Wildlife habitat management handbook, Ch. 400, section 420—Red-cockaded Woodpecker. FSH 2609.23 R, Amendment 3. Atlanta, Georgia: Southern Region.

———. 1980. Wildlife habitat management handbook, Ch. 400, section 420—Red-cockaded Woodpecker. FSH 2609.23 R, Amendment 6. Atlanta, Georgia: Southern Region.

———. 1981. Wildlife habitat management handbook, Ch. 400, section 420—Redcockaded Woodpecker. FSH 2609.23 R, Amendment 9. Atlanta, Georgia: Southern Region.

———. 1985. Wildlife habitat management handbook, Ch. 400, section 420—Redcockaded Woodpecker. FSH 2609.23 R, Amendment 13. Atlanta, Georgia: Southern Region.

———. 1987. Final environmental impact statement for the suppression of the southern pine beetle. Manage. Bull. R8-MB2. Atlanta, Georgia: Southern Region.

———. 1990. Decision notice of finding of no significant impact and environmental assessment—interim standards and guidelines for the protection and management of Red-cockaded Woodpecker habitat within ¾ mile of colony sites. Atlanta, Georgia: Southern Region.

———. 1991. Decision notice of finding of no significant impact and supplement to the environmental assessment interim standards and guidelines for protection and management of Red-cockaded Woodpecker habitat within ¾ mile of colony sites (as it pertains to the Apalachicola and Kisatchie national forests). Atlanta, Georgia: Southern Region.

———. 1995. Final environmental impact statement for the management of the Redcockaded Woodpecker and its habitat on national forests in the Southeast region. U.S. Dept. Agric., For. Serv. Manage. Bull. R8-MB 73. Atlanta, Georgia: Southern Region.

Wahlenberg, W. G. 1946. Longleaf pine: its use, ecology, regeneration, protection, growth, and management. Washington, D.C.: Charles Lathrop Pack Forestry Foundation.

———. 1960. Loblolly pine: its use, ecology, regeneration, protection, growth, and management. Durham, North Carolina: Duke Univ. School of Forestry.

Walker, J. S. 1995. Forest management, sustained yield, and silviculture for Redcockaded Woodpecker habitat. Pages 112–130 *in* Red-cockaded Woodpecker: recovery, ecology and management (D. L. Kulhavy, R. G. Hooper, and R. Costa, eds.). Nacogdoches, Texas: Center for Applied Studies, College of Forestry, Stephen F. Austin State Univ.

Walters, J. R. 1990. Red-cockaded Woodpeckers: a 'primitive' cooperative breeder. Pages 68–101 *in* Cooperative breeding in birds (P. B. Stacey and W. D. Koenig, eds.). Cambridge, England: Cambridge Univ. Press.

Walters, J. R., P. D. Doerr, and J. H. Carter III. 1988. The cooperative breeding system of the Red-cockaded Woodpecker. *Ethology* 78:275–305.

Watson, J. C., R. G. Hooper, D. L. Carlson, W. E. Taylor, and T. E. Milling. 1995. Restoration of the Red-cockaded Woodpecker population on the Francis Marion National Forest: three years post Hugo. Pages 172–182 *in* Red-cockaded Woodpecker: recovery, ecology and management (D. L. Kulhavy, R. G. Hooper, and R. Costa, eds.). Nacogdoches, Texas: Center for Applied Studies, College of Forestry, Stephen F. Austin State Univ.

Wood, D. A., ed. 1983. Red-cockaded Woodpecker symposium II proceedings. Tallahassee: State of Florida Game and Fresh Water Fish Comm.

11. State-of-the-Art Management

Carrie, N. R., K. R. Moore, S. A. Stephens, and E. L. Keith. 1998. Influence of cavity availability on Red-cockaded Woodpecker group size. *Wilson Bulletin* 110:93–99.

Carter, J. H., III, R. T. Engstrom, and P. M. Purcell. 1995. Use of artificial cavities for Red-cockaded Woodpecker mitigation: two studies. Pages 372–379 *in* Red-cockaded Woodpecker: recovery, ecology and management (D. L. Kulhavy, R. G. Hooper, and R. Costa, eds.). Nacogdoches, Texas: Center for Applied Studies, College of Forestry, Stephen F. Austin State Univ.

Carter, J. H., III, J. R. Walters, S. H. Everhart, and P. D. Doerr. 1989. Restrictors for Red-cockaded Woodpecker cavities. *Wildlife Society Bulletin* 17:68–72.

Conner, R. N. 1988. Wildlife populations: minimally viable or ecologically functional? *Wildlife Society Bulletin* 16:80–84.

Conner, R. N., and D. C. Rudolph. 1989. Red-cockaded Woodpecker colony status and trends on the Angelina, Davy Crockett, and Sabine national forests. U.S. Dept. Agric., For. Serv. Res. Paper SO-250.

Conner, R. N., D. C. Rudolph, and L. H. Bonner. 1995. Red-cockaded Woodpecker population trends and management on Texas national forests. *Journal of Field Ornithology* 66:140–151.

Copeyon, C. K., J. R. Walters, and J. H. Carter III. 1991. Induction of Red-cockaded Woodpecker group formation by artificial cavity construction. *Journal of Wildlife Management* 55:549–556.

Crowder, L. B., J. A. Priddy, and J. R. Walters. 1998. Demographic isolation of Red-cockaded Woodpecker groups: a model analysis. (unpublished report, submitted to U.S. Fish and Wildlife Service)

DeFazio, J. T., Jr., M. A. Hunnicutt, M. R. Lennartz, G. L. Chapman, and J. A. Jackson. 1987. Red-cockaded Woodpecker translocation experiments in South Carolina. *Proceedings of the Annual Conference, Southeastern Association of Fish and Wildlife Agencies* 41:311–317.

Franzreb, K. E. 1997. Success of intensive management of a critically imperiled population of Red-cockaded Woodpeckers in South Carolina. *Journal of Field Ornithology* 68:458–470.

Gaines, G. D., K. E. Franzreb, D. H. Allen, K. S. Laves, and W. L. Jarvis. 1995. Red-cockaded Woodpecker management on the Savannah River site: a management/research success story. Pages 81–88 *in* Red-cockaded Woodpecker: recovery, ecology and management (D. L. Kulhavy, R. G. Hooper, and R. Costa, eds.). Nacogdoches, Texas: Center for Applied Studies, College of Forestry, Stephen F. Austin State Univ.

Haig, S. M., J. R. Belthoff, and D. H. Allen. 1993. Population viability analysis for a small population of Red-cockaded Woodpeckers and an evaluation of enhancement strategies. *Conservation Biology* 7:289–301.

Hardesty, J. L., K. E. Gault, and H. F. Percival. 1997. Trends, status and aspects of demography of the Red-cockaded Woodpecker (*Picoides borealis*) in the Sandhills of Florida's panhandle. Part II. Final report, Research Work Order 146. Gainesville: Florida Cooperative Fish and Wildlife Research Unit, University of Florida.

Mobley, J., J. H. Carter III, A. L. Clarke, and P. D. Doerr. 1995. Evaluation of potential military training activity impacts on the Red-cockaded Woodpecker (*Picoides borealis*) on Fort Bragg. Fort Bragg, North Carolina: U.S. Army. (unpublished report, submitted to U.S. Army–CERL)

Richardson, D. M., and J. Stockie. 1995. Response of a small Red-cockaded Woodpecker population to intensive management at Noxubee National Wildlife Refuge. Pages 98–105 *in* Red-cockaded Woodpecker: recovery, ecology and management (D. L. Kulhavy, R. G. Hooper, and R. Costa, eds.). Nacogdoches, Texas: Center for Applied Studies, College of Forestry, Stephen F. Austin State Univ.

Rudolph, D. C., R. N. Conner, D. K. Carrie, and R. R. Schaefer. 1992. Experimental reintroduction of Red-cockaded Woodpeckers. *Auk* 109:914–916.

Samano, S., D. R. Wood, J. Cole, F. J. Vilella, and L. W. Burger, Jr. 1998. Red-cockaded Woodpeckers ensnared in mesh snake traps. *Wilson Bulletin* 110:564–566.

U.S. Department of Defense. Air Force. 1993. Natural Resources Management Plan, Eglin Air Force Base, Florida. Niceville, Florida: Natural Resources Management Branch, Eglin Air Force Base.

U.S. Department of Defense. Army. 1994. Management guidelines for the Red-cockaded Woodpecker on Army installations. Washington: Department of the Army. (unpublished)

———. 1996. 1996 Management guidelines for the Red-cockaded Woodpecker on Army installations. Washington: Department of the Army. (unpublished)

U.S. Department of Defense. Marine Corps. 1999. Marine Corps Base Camp Lejeune's mission compatible plan for the comprehensive long range management of the Red-cockaded Woodpecker. Jacksonville, North Carolina: Marine Corps Base Camp Lejeune.

U.S. Fish and Wildlife Service. 1985. Red-cockaded Woodpecker recovery plan. Atlanta, Georgia: U.S. Fish and Wildl. Serv.

———. 1992. Draft Red-cockaded Woodpecker procedures manual for private lands. Atlanta, Georgia: U.S. Fish and Wildl. Serv.

———. 1995. Final habitat conservation plan to encourage the voluntary restoration and enhancement of habitat for the Red-cockaded Woodpecker on private and certain other land in the Sandhills Region of North Carolina by providing "Safe Harbor" to participating landowners. (unpublished report)

———. 1996. Draft strategy and guidelines for the recovery and protection of the Red-cockaded Woodpecker on national wildlife refuges. Atlanta, Georgia: U.S. Fish and Wildl. Serv.

U.S. Forest Service. 1995. Final environmental impact statement for the management of the Red-cockaded Woodpecker and its habitat on national forests in the Southeast region. U.S. Dept. Agric., For. Serv. Manage. Bull. R8-MB 73. Atlanta, Georgia: Southern Region.

Walters, J. R. 1991. Application of ecological principles to the management of endangered species: the case of the Red-cockaded Woodpecker. *Annual Review of Ecology and Systematics* 22:505–523.

———. 1992. Evaluation of possible disturbance effects on Red-cockaded Woodpeckers in the mechanized infantry training area, Camp Lejeune. (unpublished report, submitted to DOD, Marine Corps Base Camp Lejeune)

Walters, J. R., P. P. Robinson, W. Starnes, and J. Goodson. 1995. The relative effectiveness of artificial cavity starts and artificial cavities in inducing the formation of new groups of Red-cockaded Woodpeckers. Pages 367–371 in Red-cockaded Woodpecker: recovery, ecology and management (D. L. Kulhavy, R. G. Hooper, and R. Costa, eds.). Nacogdoches, Texas: Center for Applied Studies, College of Forestry, Stephen F. Austin State Univ.

Wood, G. W., and J. Kleinhofs. 1992. Integration of timber management and Red-cockaded Woodpecker conservation goals on Georgia-Pacific Corporation timberlands in the southern United States. (unpublished report)

———. 1995. Integrating timber management and Red-cockaded Woodpecker conservation: the Georgia-Pacific plan. Pages 75–80 in Red-cockaded Woodpecker: recovery, ecology and management (D. L. Kulhavy, R. G. Hooper, and R. Costa, eds.). Nacogdoches, Texas: Center for Applied Studies, College of Forestry, Stephen F. Austin State Univ.

12. An Uncertain Future

Allen, J. C. 2001. Species-habitat relationships of breeding bird species at Fort Bragg, North Carolina. M.S. thesis, Virginia Polytechnic Institute and State Univ., Blacksburg.

Carter, J. H., III, J. R. Walters, S. H. Everhart, and P. D. Doerr. 1989. Restrictors for Red-cockaded Woodpecker cavities. *Wildlife Society Bulletin* 17:68–72.

Conner, R. N., D. C. Rudolph, D. Saenz, and R. R. Schaefer. 1996. Red-cockaded Woodpecker nesting success, forest structure, and southern flying squirrels in Texas. *Wilson Bulletin* 108:697–711.

———. 1997. Species using Red-cockaded Woodpecker cavities in eastern Texas. *Bulletin of the Texas Ornithological Society* 30:11–16.

Dunning, J. B. 1993. Bachman's Sparrow. The birds of North America, No. 38 (A. Poole, P. Stettenheim, and F. Gill, eds.). Philadelphia: Academy of Natural Sciences; Washington, D.C.: American Ornithologists' Union.

Dunning, J. B., and B. D. Watts. 1990. Regional differences in habitat occupancy by Bachman's Sparrow. *Auk* 107:463–472.

Engstrom, R. T. The avian community of an old-growth longleaf pine forest. *Natural Areas Journal.* (in review)

Engstrom, R. T., L. A. Brennan, W. L. Neel, R. M. Farrar, S. T. Lindeman, W. K. Moser, and S. M. Hermann. 1996. Silvicultural practices and Red-cockaded Woodpecker management: a reply to Rudolph and Conner. *Wildlife Society Bulletin* 24:334–338.

Environmental Defense Fund. 1995. Incentives for endangered species management: opportunities in the Sandhills of North Carolina. Raleigh, North Carolina: Environmental Defense Fund.

James, F. C., C. A. Hess, and D. Kufrin. 1997. Species-centered environmental analysis: indirect effects of fire history on Red-cockaded Woodpeckers. *Ecological Applications* 7:118–129.

Kennedy, E. T., R. Costa, and W. M. Smathers, Jr. 1996. Economic incentives: new directions for Red-cockaded Woodpecker habitat conservation. *Journal of Forestry* 94(4):22–26.

Kreiger, S. M. 1997. Abundance, habitat associations and nest success of passerines within a fire maintained longleaf pine (*Pinus palustris*) ecosystem. M.S. thesis, North Carolina State Univ., Raleigh.

Laves, K. 1996. Effects of southern flying squirrels, *Glaucomys volans*, on Red-cockaded Woodpecker, *Picoides borealis*, reproductive success. M.S. thesis, Clemson Univ., Clemson, South Carolina.

Rudolph, D. C., and R. N. Conner. 1996. Red-cockaded Woodpeckers and silvicultural practice: is uneven-aged silviculture preferable to even-aged? *Wildlife Society Bulletin* 24:330–333.

Sierra Club, The Wilderness Society, and Texas Committee on Natural Resources v. Richard E. Lyng, F. Dale Robertson, John E. Alcock, and William M. Lannan. Civil Action No. L-85-69-CA, Tyler, Texas (District of East Texas 1985–2001).

Withgott, J. H., J. C. Neal, and W. G. Montague. 1995. A technique to deter rat snakes from climbing Red-cockaded Woodpecker cavity trees. Pages 394–400 *in* Red-cockaded Woodpecker: recovery, ecology and management (D. L. Kulhavy, R. G. Hooper, and R. Costa, eds.). Nacogdoches, Texas: Center for Applied Studies, College of Forestry, Stephen F. Austin State Univ.

Index